EPIGENETICS AND PATHOLOGY

Exploring Connections Between Genetic Mechanisms and Disease Expression

EPIGENETICS AND PATHOLOGY

Exploring Connections Between Genetic Mechanisms and Disease Expression

Edited by

Kasirajan Ayyanathan, PhD

Apple Academic Press

TORONTO NEW JERSEY

Apple Academic Press Inc. 3333 Mistwell Crescent Oakville, ON L6L 0A2 Canada	Apple Academic Press Inc. 9 Spinnaker Way Waretown, NJ 08758 USA

Exclusive worldwide distribution by CRC Press, a member of Taylor & Francis Group

No claim to original U.S. Government works
Printed in the United States of America on acid-free paper

International Standard Book Number-13: 978-1-926895-78-9 (Hardcover)

Library of Congress Control Number: 2013944970

Library and Archives Canada Cataloguing in Publication

Epigenetics & pathology: exploring connections between genetic mechanisms and disease expression/edited by Kasirajan Ayyanathan, PhD.

Includes bibliographical references and index.
ISBN 978-1-926895-78-9
1. Epigenetics. 2. Pathology. 3. Medical genetics.
I. Ayyanathan, Kasirajan, editor of compilation II. Title: Epigenetics and pathology.

QH450.E63 2013 571.9 C2013-904652-6

Apple Academic Press also publishes its books in a variety of electronic formats. Some content that appears in print may not be available in electronic format. For information about Apple Academic Press products, visit our website at **www.appleacademicpress.com** and the CRC Press website at **www.crcpress.com**

ABOUT THE EDITOR

KASIRAJAN AYYANATHAN, PhD

Kasirajan Ayyanathan received his PhD degree from the Department of Biochemistry, Indian Institute of Science, one of the premier research institutions in India. Subsequently, at Temple University School of Medicine, USA, he conducted post-doctoral research on the signal transduction by purinergic receptors, a class of G-Protein Coupled Receptors (GPCR), in erythroleukemia cancer cells. Next, he was trained as a staff scientist at the Wistar Institute, USA, for almost ten years and studied transcription regulation, chromatin, and epigenetic regulatory mechanisms in cancer before becoming an Associate Professor at Florida Atlantic University (FAU). Currently he is at the Center for Molecular Biology and Biotechnology as a Research Associate Professor at FAU. He is the recipient of Chern memorial award and Howard Temin career research award.

Dr. Ayyanathan is well trained in molecular biology, cell biology, biochemistry with main focus on studying transcription factors, and gene regulation. He has contributed to several projects such as generation of conditional transcriptional repressors that are directed against the endogenous oncogenes to inhibit malignant growth, establishment of stable cell lines that express chromatin integrated transcriptional repressors and reporter genes in order to study the epigenetic mechanisms of KRAB repression, and identification of novel SNAG repression domain interacting proteins in order to understand their roles in transcriptional repression and oncogenesis. Dr. Ayyanathan has published several research articles in peer-reviewed articles in these subject areas.

CONTENTS

ACKNOWLEDGMENT AND HOW TO CITE

The chapters in this book were previously published in various places and in various formats. By bringing them together here in one place, we offer the reader a comprehensive perspective on recent investigations into the connections between epigenetics and pathology. Each chapter adds to a more complete image of this research topic.

We wish to thank the authors who made their research available for this book, whether by granting their permission individually or by releasing their research as Open Source articles. When citing information contained within this book, please do the authors the courtesy of attributing them by name, referring back to their original articles, using the credits provided at the end of each chapter. Copyright for the individual articles remains with the original copyright holders.

LIST OF CONTRIBUTORS

Mahmood Ameen Abdulla
Department of Molecular Medicine, Faculty of Medicine, University of Malaya, Kuala Lumpur 50603, Malaysia.

Felix R. Althaus
Institute of Pharmacology and Toxicology, University of Zurich-Vetsuisse, Winterthurerstrasse 260, 8057 Zurich, Switzerland

Alice Bertaina
Department of Pediatric Hematology-Oncology, Ospedale Pediatrico Bambino Gesù, IRCCS, Piazza S. Onofrio 4, Roma 00165, Italy

Christian Blenn
Institute of Pharmacology and Toxicology, University of Zurich-Vetsuisse, Winterthurerstrasse 260, 8057 Zurich, Switzerland

Giorgia Bracaglia
Department of Pediatric Hematology-Oncology, Ospedale Pediatrico Bambino Gesù, IRCCS, Piazza S. Onofrio 4, Roma 00165, Italy

Sukhmani Brar
Center for Research on Occupational and Environmental Toxicology (CROET), Oregon Health & Science University, Portland, Oregon, USA

David C. Christiani
Environmental and Occupational Medicine and Epidemiology Program, Department of Environmental Health, Harvard School of Public Health, Boston, MA, USA; Department of Medicine, Massachusetts General Hospital/Harvard Medical School, Boston, MA, USA

Paul J. Christine
Department of Pediatrics, University of Iowa, 500 Newton Road, 2182 ML, Iowa City, IA 52242, USA

Adrian Chu
Center for Research on Occupational and Environmental Toxicology (CROET), Oregon Health & Science University, Portland, Oregon, United States of America

Belén Comas
Dirección de Investigación, Centro de Educación Médica e Investigaciones Clínicas (CEMIC), 1431 Buenos Aires, Argentina

Viviana Cosentino
Dirección de Investigación, Centro de Educación Médica e Investigaciones Clínicas (CEMIC), 1431 Buenos Aires, Argentina

James R. Davie
Manitoba Institute of Cell Biology, University of Manitoba, 675 McDermot Avenue, Winnipeg, MB, R3E 0V9, Canada

Geneviève P. Delcuve
Manitoba Institute of Cell Biology, University of Manitoba, 675 McDermot Avenue, Winnipeg, MB, R3E 0V9, Canada

Doriana Fruci
Department of Pediatric Hematology-Oncology, Ospedale Pediatrico Bambino Gesù, IRCCS, Piazza S. Onofrio 4, Roma 00165, Italy

Valentina Gallo
Department of Epidemiology and Public Health, Imperial College, London, UK; Department of Social and Environmental Research, London School of Hygiene and Tropical Medicine, UK

Sara Geneletti
Department of Statistics, London School of Economics, Houghton Street, London, UK

Shahram Golbabapour
Department of Molecular Medicine, Faculty of Medicine, University of Malaya, Kuala Lumpur 50603, Malaysia

Elena Gómez-Díaz
Institute of Biologia Evolutiva, Barcelona, Spain

Maryam Hairezaei
Department of Molecular Medicine, Faculty of Medicine, University of Malaya, Kuala Lumpur 50603, Malaysia

Martin Hirst
Genome Sciences Centre, BC Cancer Agency, 675 West 10th Avenue, Vancouver, BC, V5Z 1L3 Canada.

Mireia Jordâ
Institute of Biologia Evolutiva, Barcelona, Spain

Dilshad H. Khan
Manitoba Institute of Cell Biology, University of Manitoba, 675 McDermot Avenue, Winnipeg, MB, R3E 0V9, Canada

Muin J. Khoury
National Office of Public Health Genomics, Centers for Disease Control and Prevention, Atlanta, USA

Jinsil Kim
Department of Anatomy and Cell Biology, University of Iowa, 500 Newton Road, 2182 ML, Iowa City, IA 52242, USA

Xihong Lin
Department of Biostatistics, Harvard School of Public Health, Boston, MA, USA

Chen-yu Liu
Environmental and Occupational Medicine and Epidemiology Program, Department of Environmental Health, Harvard School of Public Health, Boston, MA, USA; Institute of Environmental Health, College of Public Health, National Taiwan University, Taipei, Taiwan

Franco Locatelli
Department of Pediatric Hematology-Oncology, Ospedale Pediatrico Bambino Gesù, IRCCS, Piazza S. Onofrio 4, Roma 00165, Italy; Department of Pediatric Sciences, University of Pavia, Corso Strada Nuova 65, Pavia I-27100, Italy

Arnab Maity
Department of Biostatistics, Harvard School of Public Health, Boston, MA, USA, Department of Statistics, North Carolina State University, Raleigh, NC, 27695, USA

Marco A. Marra
Dept. of Medical Genetics, Faculty of Medicine, University of British Columbia, Vancouver, British Colombia, Canada

Richard B. Meagher
Genetics Department, Davison Life Sciences Building, University of Georgia, Athens, GA, 30605, USA

Shiraz Mujtaba

Department of Structural and Chemical Biology, Mount Sinai School of Medicine New York, NY 10029 USA

Jeffrey C. Murray
Department of Pediatrics, University of Iowa, 500 Newton Road, 2182 ML, Iowa City, IA 52242, USA

Krostofer J. Müssar
Genetics Department, Davison Life Sciences Building, University of Georgia, Athens, GA, 30605, USA

Jon A. Oyer
Department of Molecular and Medical Genetics, Oregon Health & Science University, Portland, Oregon, United States of America

Jigneshkumar Patel
Department of Structural and Chemical Biology, Mount Sinai School of Medicine New York, NY 10029 USA

Ravi R. Pathak
Department of Structural and Chemical Biology, Mount Sinai School of Medicine New York, NY 10029 USA

Miguel Angel Peinada
Institute of Biologia Evolutiva, Barcelona, Spain

Mitchell M. Pitlik
Department of Pediatrics, University of Iowa, 500 Newton Road, 2182 ML, Iowa City, IA 52242, USA

Miquel Porta
Institut Municipal d'Investigació Mèdica, and School of Medicine, Universitat Autònoma de Barcelona, Catalonia, Spain

Ana Rivero
Maladies Infectieuses et Vecteurs: Écologie, Génétique, Évolution et Contrôle (MIVEGEC, UMR CNRS-UM2-UM1 5290, IRD 224), Centre IRD, Montpellier, France

Paolo Romania
Department of Pediatric Hematology-Oncology, Ospedale Pediatrico Bambino Gesù, IRCCS, Piazza S. Onofrio 4, Roma 00165, Italy

Rosella Rota
Department of Pediatric Hematology-Oncology, Ospedale Pediatrico Bambino Gesù, IRCCS, Piazza S. Onofrio 4, Roma 00165, Italy

Cesar Saleme
Departamento de Neonatología, Instituto de Maternidad y Ginecología Nuestra Señora de las Mercedes, 4000 San Miguel de Tucumán, Argentina

Amanda R. Schaefer
Department of Pediatrics, University of Iowa, 500 Newton Road, 2182 ML, Iowa City, IA 52242, USA

Steven Shoei-Lung Li
Graduate Institute of Clinical Medicine, Kaohsiung Medical University, Kaohsiung 807, Taiwan, Center of Excellence for Environmental Medicine, Kaohsiung Medical University, Kaohsiung 807, Taiwan

Sher Singh
Department of Life Science, College of Science, National Taiwan Normal University, Taipei 116, Taiwan

Mitchell S. Turker
Department of Molecular and Medical Genetics, Oregon Health & Science University, Portland, Oregon; Center for Research on Occupational and Environmental Toxicology (CROET), Oregon Health & Science University, Portland, Oregon, USA

Paulo Vineis
Department of Epidemiology and Public Health, Imperial College, London, UK

Robert O. Wright
Environmental and Occupational Medicine and Epidemiology Program, Department of Environmental Health, Harvard School of Public Health, Boston, MA, USA; Department of Preventive Medicine, Mount Sinai School of Medicine, New York, NY, USA

Philippe Wyrsch
Institute of Pharmacology and Toxicology, University of Zurich-Vetsuisse, Winterthurerstrasse 260, 8057 Zurich, Switzerland

INTRODUCTION

The term "epigenetics" describes cellular modifications caused by mechanisms other than DNA sequence variations that can be heritable and modified by environmental stimuli. Since the beginning of Mendel's era, the influence of genetic changes (alterations in the underlying DNA sequence) on gene expression remains the main thrust of geneticists. For this reason, it is also widely understood. Mendelian inheritance, named after its discoverer Gregory Mendel, refers to the process of transmission of genetic traits from the parents to the offspring. Since then, several researchers have shown that genes serve as blueprints in almost all living organisms from simple prokaryotes to multicellular eukaryotes, evidenced by the dearth of research articles and textbooks on the subject.

However, two other distinct factors can bring out similar changes in gene expression. One such influencing factor is conferred from the inside of a cellular milieu, namely epigenetic modification. Selective epigenetic changes can specifically alter the chromatin environment and thus result in altered gene expression. This also underlies the basis of Lamarckism. Similarly, from the outside of a cellular milieu, several environmental factors can influence gene expression and result in rapid changes in gene expression. This is mostly prevalent in gene expression observed in prokaryotes, wherein quick adjustment to its immediate environment is mandatory for them to thrive effectively.

A growing body of evidence points towards epigenetic mechanisms being responsible for a wide range of biological phenomena, from the plasticity of plant growth and development to the nutritional control of caste determination in honeybees and the etiology of human disease (e.g., cancer). With the (partial) elucidation of the molecular basis of epigenetic variation and the heritability of certain of these changes, the field of evolutionary epigenetics is flourishing. Despite this, the role of epigenetics in shaping host–pathogen interactions has received comparatively little attention. Yet there is plenty of evidence supporting the implication of epigenetic mechanisms in the modulation of the biological interaction between hosts and pathogens. The phenotypic plasticity of many key parasite life-history traits appears to be under epigenetic control. Moreover, pathogen-induced effects in host phenotype may have transgenerational consequences, and the bases of these changes and their heritability probably have an epigenetic component. The significance of epigenetic modifications may, however, go beyond providing a mechanistic basis for host and pathogen plasticity. Epigenetic epidemiology has recently emerged as a promising area for future research on infectious diseases. In addition, the incorporation of epigenetic inheritance and epigenetic plasticity mechanisms to evolutionary models and empirical studies of host–pathogen interactions will provide new insights into the evolution and coevolution of these associations. The goal of this volume is to bring together into a coherent whole some of the most current and relevant research being done in this field.

In chapter 1, "The Epigenetics of Host-Pathogen Interactions," authors Goméz-Díaz et al review the evidence available for the role epigenetics on host–pathogen interactions, and the utility and versatility of the epigenetic technologies available that can be cross-applied to host–pathogen studies. They conclude with recommendations and directions for future research on the burgeoning field of epigenetics as applied to host–pathogen interactions.

In chapter 2, "The Molecular Mechanisms of Epigenetic Regulation," Golbabapour and his colleagues explore the epigenetic regulatory events that occur during the gametogenesis, embryogenesis and placental development. The epigenetic modifications that modulate expression of genes and subsequent reprogramming of the somatic nucleus to pluripotent state are also briefly discussed. The authors' purpose in this chapter is to summarize effective epigenetic events that could increase efficiency of SCNT and to emphasize recent epigenetic findings. They briefly look into transition techniques and highlight epigenetic modifications that happen during the nucleus reprogramming.

Kim and his colleagues focus on the amnion in chapter 3, "Mutation Rate and DNA Methylation in the Human Genome." To investigate the importance of epigenetic events in this tissue in the physiology and pathophysiology of pregnancy, they performed genome-wide DNA methylation profiling of human amnion from term (with and without labor) and preterm deliveries. Using the Illumina Infinium HumanMethylation27 BeadChip, they identified genes exhibiting differential methylation associated with normal labor and preterm birth. Their work provides preliminary evidence that DNA methylation changes in the amnion may be at least partially involved in the physiological process of labor and the etiology of preterm birth, suggesting that DNA methylation profiles, in combination with other biological data, may provide valuable insight into the mechanisms underlying normal and pathological pregnancies.

Chapter 4, "Causal Relationships in Gene-Environment Interactions," offers a conceptual framework to assess causal relationships in clinical genomics and, particularly, for evaluating the etiopathogenic significance of gene-disease associations and gene-environment interactions (i.e., a framework to assess the validity and significance of such environment-host-gene relationships in the etiology of human diseases). The framework includes a two-step approach that combines the causal criteria of Austin Bradford Hill with graphical models such as directed acyclic graphs (DAGs). The approach Geneletti and her colleagues propose thus helps, first, to untangle the web of interactions amongst several exposures and characteristics (environmental, clinical and genetic) and a disease, using criteria to assess causality that have long been used in clinical and epidemiological research. More generally, chapter 4 is an example of integrative research, i.e., research that integrates knowledge, data, methods, techniques, and reasoning from multiple disciplines, approaches and levels of analysis to generate knowledge that no discipline alone may achieve.

Because enzymes that catalyze acetylation are also transcriptional co-activators, which coordinate with transcription factors in regulating gene expression—underscoring the integration of transcription with metabolism—such enzymes present potential therapeutic targets. In chapter 5, "Lysine Acetylation in Transcriptional Programming and Metabolism," the overall goal authors Patel, Pathak and Mujtaba is to highlight the most recent advances in the field of acetylation biology that could spark new perspectives and illuminate novel research avenues.

In chapter 6, Delcuve, Khan and Davie take an in-depth look at the role histone deacetylation plays in epigenetic regulation. The scope of their review concerns emerging concepts regarding the roles of HDACs in modulating chromatin structure and function as revealed by studies with HDAC inhibitors. The zinc-dependent mammalian histone deacetylase (HDAC) family comprises 11 enzymes, which have specific and critical functions in development and tissue homeostasis. Mounting evidence points to a link between misregulated HDAC activity and many oncologic and nononcologic diseases. Thus the development of HDAC inhibitors for therapeutic treatment garners a lot of interest from academic researchers and biotechnology entrepreneurs. Numerous studies of HDAC inhibitor specificities and molecular mechanisms of action are ongoing, and in one of these studies, mass spectrometry was used to characterize the affinities and selectivities of HDAC inhibitors toward native HDAC multiprotein complexes in cell extracts. Such a novel approach reproduces *in vivo* molecular interactions more accurately than standard studies using purified proteins or protein domains as targets and could be very useful in the isolation of inhibitors with superior clinical efficacy and decreased toxicity compared to the ones presently tested or approved. HDAC inhibitor induced-transcriptional reprogramming, believed to contribute largely to their therapeutic benefits, is achieved through various and complex mechanisms not fully understood, including histone deacetylation, transcription factor or regulator (including HDAC1) deacetylation followed by chromatin remodeling and positive or negative outcome regarding transcription initiation. Although only a very low percentage of protein-coding genes are affected by the action of HDAC inhibitors, about 40% of noncoding microRNAs are upregulated or downregulated. Moreover, a whole new world of long noncoding RNAs is emerging, revealing a new class of potential targets for HDAC inhibition. HDAC inhibitors might also regulate transcription elongation and have been shown to impinge on alternative splicing.

In chapter 7, “Aberrant Epigenetic Silencing and Gene Expression,” Oyer and his colleagues have developed a system to directly test the hypothesis that a transient reduction in gene expression can sensitize a promoter to undergo epigenetic silencing. Their results demonstrate that this principle is correct. Additionally, they find that induction of silencing is dependent on histone deacetylase activity, but does not require DNA methylation. Silenced alleles readily reactivated spontaneously or after treatment of cells with inhibitors of histone deacetylation and/or DNA methylation, but re-silencing of reactivated alleles did not require a new round of Dox exposure. Inhibition of histone deacetylation inhibited both the induction of silencing and re-silencing, whereas inhibition of DNA methylation had no such effect

Blenn, Wyrsch and Althaus discuss in chapter 8, "RNAi Silencing in Poly (ADP-Ribose) Research," the potential of RNAi to manipulate the levels of PARPs and PARG, and consequently those of PAR and ADPR. They compare the results of their studies with those obtained after genetic or chemical disruption.

Chapter 9, "MicroRNA Deregulation in Rhabdomyosarcoma and Neuroblastoma," focuses on current knowledge about miRNAs deregulated in RMS and NB by epigenetic modifications. Romania et al highlight miRNAs' role in developmental pathways, highlighting RMS and NB tumorigenesis. They discuss the translational implications and challenges of miRNAs modulation in these pediatric tumors, explaing that gene expression control mediated by microRNAs and epigenetic remodeling of chromatin are interconnected processes often involved in feedback regulatory loops. These strictly guide proper tissue differentiation during embryonal development. Altered expression of microRNAs is one of the mechanisms leading to pathologic conditions, such as cancer, and several lines of evidence point to epigenetic alterations as responsible for aberrant microRNA expression in human cancers. Rhabdomyosarcoma and neuroblastoma are pediatric cancers derived from cells presenting features of skeletal muscle and neuronal precursors, respectively, blocked at different stages of differentiation. Consistently, tumor cells express tissue markers of origin but are unable to terminally differentiate. Several microRNAs playing a key role during tissue differentiation are often epigenetically downregulated in rhabdomyosarcoma and neuroblastoma and behave as tumor suppressors when re-expressed. Recently, inhibition of epigenetic modulators in adult tumors has provided encouraging results causing re-expression of anti-tumor master gene pathways. Thus, a similar approach could be used to correct the aberrant epigenetic regulation of microRNAs in rhabdomyosarcoma and neuroblastoma. Chapter 9 highlights the current insights on epigenetically deregulated microRNAs in rhabdomyosarcoma and neuroblastoma and their role in tumorigenesis and developmental pathways. The translational clinical implications and challenges regarding modulation of epigenetic chromatin remodeling/microRNAs interconnections are also discussed.

In chapter 10, "Epigenetic Effects of Environmental Chemicals," Singh and Shoei-Lung Li investigate the epigenetic effects on DNA methylation, histone modification, and expression of non-coding RNAs (including microRNAs) of environmental chemicals such as bisphenol A (BPA) and phthalates, expanding our understanding of the etiology of human complex diseases such as cancers and diabetes. Multiple lines of evidence from *in vitro* and *in vivo* models have established that epigenetic modifications caused by *in utero* exposure to environmental toxicants can induce alterations in gene expression that may persist throughout life. Epigenetics is an important mechanism in the ability of environmental chemicals to influence health and disease, and BPA and phthalates are epigenetically toxic. The epigenetic effect of BPA was clearly demonstrated in viable yellow mice by decreasing CpG methylation upstream of the Agouti gene, and the hypomethylating effect of BPA was prevented by maternal dietary supplementation with a methyl donor like folic acid or the phytoestrogen genistein. Histone H3 was found to be trimethylated at lysine 27 by BPA effect on EZH2 in a human breast cancer cell line and mice. BPA exposure of human placental cell lines has been shown to alter microRNA expression levels, and specifically, miR-146a was strongly induced by BPA treatment. In human breast cancer MCF7 cells, treatment with the phthalate BBP led to demethylation of estrogen receptor (ESR1) promoter-associated CpG islands, indicating that altered ESR1 mRNA expression by BBP is due to aberrant DNA methylation. Maternal exposure to phthalate DEHP was also shown to increase DNA methylation and expression levels of DNA methyltransferases in mouse testis. Further, some epigenetic effects of BPA and phthalates in female rats were found to be transgenerational. Finally, the authors describe the ways in which available new technologies for global analysis of epigenetic alterations will provide insight into the extent and patterns of alterations between human normal and diseased tissues. *In vitro* models such as human embryonic stem cells may be extremely useful in bettering the understanding of epigenetic effects on human development, health and disease, because the formation of embryoid bodies *in vitro* is very similar to the early stage of embryogenesis.

Clear cause-and-effect relationships are commonly established between genotype and the inherited risk of acquiring human and plant diseases and aberrant phenotypes. By contrast, few such cause-and-effect relationships are established linking a chromatin structure (that is, the epitype) with the transgenerational risk of acquiring a disease or abnormal phenotype. It is not entirely clear how epitypes are inherited from parent to offspring as populations evolve, even though epigenetics is proposed to be fundamental to evolution and the likelihood of acquiring many diseases. Chapter 11, “DNA Sequence and Epigenome-Induced Pathologies,” explores the hypothesis that, for transgenerationally inherited chromatin structures, “genotype predisposes epitype,” and that epitype functions as a modifier of gene expression within the classical central dogma of molecular biology. Evidence for the causal contribution of genotype to inherited epitypes and epigenetic risk comes primarily from two different kinds of studies discussed herein. The first and direct method of research proceeds by the examination of the transgenerational inheritance of epitype and the penetrance of phenotype among genetically related individuals. The second approach identifies epitypes that are duplicated (as DNA sequences are duplicated) and evolutionarily conserved among repeated patterns in the DNA sequence. The chapter summarizes particularly robust examples of these studies from humans, mice, Arabidopsis, and other organisms. The bulk of the data from both areas of research support the hypothesis that genotypes predispose the likelihood of displaying various epitypes, but for only a few classes of epitype. The authors’ analysis suggests that renewed efforts are needed in identifying polymorphic DNA sequences that determine variable nucleosome positioning and DNA methylation as the primary cause of inherited epigenome-induced pathologies. By contrast, there is very little evidence that DNA sequence directly determines the inherited positioning of numerous and diverse post-translational modifications of histone side chains within nucleosomes. The authors discuss the medical and scientific implications of these observations on future research and on the development of solutions to epigenetically induced disorders.

Chapter 12, “The Future of Epigenomics Research,” takes a look at next-generation sequencing that is bringing epigenomic studies to the forefront of current research. The power of massively parallel sequencing coupled to innovative molecular and computational techniques has allowed researchers to profile the epigenome at resolutions that were unimaginable only a few years ago. With early proof of concept studies published, the field is now moving into the next phase where the importance of method standardization and rigorous quality control are becoming paramount. In this chapter, Hirst and Marra describe methodologies that have been developed to profile the epigenome using next generation sequencing platforms. They will discuss these in terms of library preparation, sequence platforms and analysis techniques.

In summary, both nurture (environmental) and nature (genetic factors) play an important role in human disease etiology. Traditionally, these effects have been thought of as independent. In chapter 13, "Issues in Understanding Epigenetics and Disease," Liu et al indicate that this perspective is ill informed for non-mendelian complex disorders, which result as an interaction between genetics and environment. To understand health and disease, they affirm, we must study how nature and nurture interact. Recent advances in human genomics and high-throughput biotechnology make it possible to study large numbers of genetic markers and gene products simultaneously to explore their interactions with environment. This final chapter discusses design and analytic issues for gene-environment interaction studies in the "-omics" era, with a focus on environmental and genetic epidemiological studies. The authors present an expanded environmental genomic disease paradigm. They include in their discussion several study design issues for gene-environmental interaction studies, including confounding and selection bias, measurement of exposures and genotypes. They also discuss statistical issues in studying gene-environment interactions in different study designs, such as choices of statistical models, assumptions regarding biological factors, and power and sample size considerations, especially in genome-wide gene-environment studies. Future research directions are also discussed.

— **Kasirajan Ayyanathan, PhD**

CHAPTER 1

EPIGENETICS OF HOST-PATHOGEN INTERACTIONS: THE ROAD AHEAD AND THE ROAD BEHIND

ELENA GÓMEZ-DÍAZ, MIREIA JORDÀ, MIGUEL ANGEL PEINADO, and ANA RIVERO

WHAT IS EPIGENETICS?

Few areas in biology attract as much current attention and yet require as much presentation as the field of epigenetics. The term "epigenetics" was first used by Waddington to describe the process through which genotypes give rise to phenotypes during development [1]. Since then, there has been a burgeoning interest in the field of epigenetics that has been coupled with a diversification in the use of the term: epigenetics means different things to the different fields of biology, and even within a given field, different authors may use it in somewhat different contexts, generating a great deal of confusion in the process [2]. Broadly speaking, epigenetics refers to stimuli-triggered changes in gene expression due to processes that arise independent of changes in the underlying DNA sequence. Some of these processes have been elucidated and include DNA methylation [3], histone modifications and chromatin-remodeling proteins [4], and DNA silencing by noncoding RNAs (ncRNA) [5]. This general definition of "epigenetics" is, however, used in two broadly different contexts. For some authors, the term "epigenetics" includes all transient changes in gene expression that occur at the individual cell level, as well as those that are propagated during mitosis in multicellular organisms and remain stable at the time scale of an individual (Figure 1). For clarity, we refer to this as *epigenetic plasticity* (see [6]). A good example is the development of morphologically different castes of bees from genetically identical individuals through nutritionally triggered DNA methylation [7]. Yet for other authors, and most notably for evolutionary biologists, the term epigenetics refers exclusively to *epigenetic inheritance*: the stimuli-triggered variation in gene expression that is heritable across generations. Here, the epigenetic changes are generated in the germ line in multicellular organisms (either directly or indirectly, see Jablonka and Raz [8]) or maintained clonally in single-cell organisms (Figure 1) [8,9]. A classic example of transgenerational epigenetic inheritance involves a change in flower symmetry from bilateral to radial in *Linaria vulgaris*, which relates to different levels of methylation of the gene*Lcyc* [10]. In this review, we contend that both epigenetic plasticity and epigenetic inheritance are important in shaping host–pathogen interactions, and thus we use the term "epigenetics" to encompass both of these definitions.

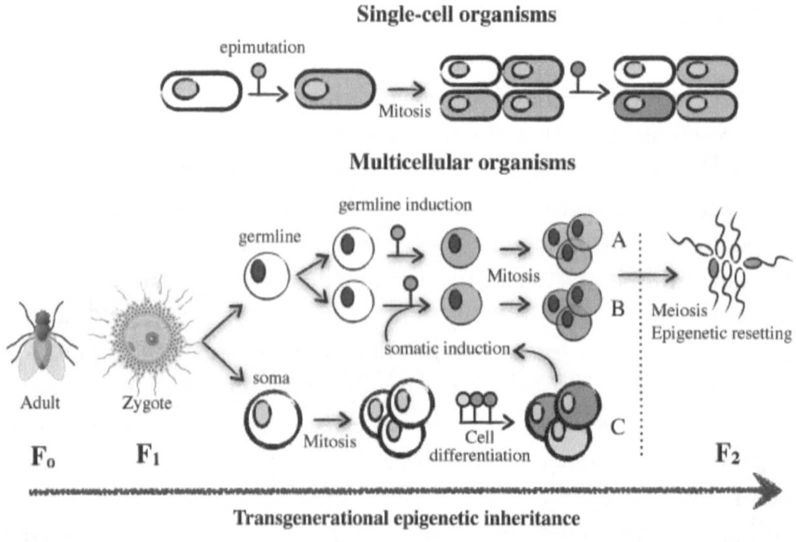

FIGURE 1. Mechanisms of epigenetic plasticity and inheritance. In single-cell organisms, epimutations induced by environmental stimuli (i.e., host) propagate in daughter cells by mitosis and result in transient or stable epigenetic states. In multicellular, sexually reproducing, organisms the zygote (F1) differentiates into germinal and somatic cells. Epimutations can be originated directly in the germline and propagated by mitosis ("germline induction") (A), or they can arise and propagate as a consequence of interactions with the soma ("somatic induction") (B). In the soma, after several rounds of cell divisions, epimutations tend to accumulate during cell and tissue differentiation processes (C). Only those epimutations generated in the germline that escape meiotic resetting during gametogenesis and oogenesis are expected to have transgenerational consequences (i.e., epigenetic inheritance) (F2).

Box 1. The Epigenetic Code

Specific combinations of epigenetic modifications constitute what has been called the epigenetic code, determining the functional (gene regulation, replication, repair, etc.) and structural features of each genomic region [68]. **Histone modifications**: A widely studied epigenetic mark is constituted by the set of *posttranslational modifications*(PTMs) on histones, which consist in the covalent addition of different chemical groups to particular residues, and that take place mostly in the tails of histones (see figure box). The association between different histone marks or variants and distinct chromatin and functional states (or *histone code* [69]) is well established. For instance, trimethylation of the histone 3 lysine 4 residue (H3K4me3) is usually linked to active genes, while trimethylation in lysine 9 residue (H3K9me3) is characteristic of repressed chromatin. **DNA methylation:** The DNA of most species is methylated and this modification takes place postreplicatively. In eukaryotes the modified base is 5-methylcytosine (5 mC) whereas in prokaryotes is mostly N6-methyladenine (6 mA) [70]. DNA methylation has a role in silencing gene expression and heterochromatin remodeling, among other functions [3]. DNA methylation patterns are dynamic and have changed several times through the tree of life, exhibiting a considerable structural, functional,and mechanistic diversity [71]. Hence, while in plants and vertebrates, DNA methylation occurs widely at CpG (C—phosphate—G) dinucleotide sites, regions of DNA where a cytosine nucleotide occurs next to a guanine nucleotide in the linear sequence of bases, and appear preferentially associated with transposons and silenced DNA; in invertebrates, DNA methylation is mainly found in gene bodies, but its regulatory function is only partially understood [72]. Interestingly, DNA methylation is not ubiquitous across the tree of life. Several species seem to have undergone loss of DNA methylation to a large degree, including model-species such as the nematode *C. elegans*, the insect *D. melanogaster*, and the yeast *S. cerevisiae*. **RNA-mediated silencing**: A variety of noncoding RNAs (ncRNA) have been shown to act in concert with the cell's epigenetic machinery, for example by establishing DNA methylation and by regulating histone modifiers [5]. Among those, the best characterized are the so-called microRNAs (miRNA), small ncRNAs of 19 to 24 nucleotides that bind target messenger RNAs and induce their translational repression, cleavage, or accelerated decay [73]. Yet the nature and function of this class of molecules are poorly understood, as well as the degree to which they contribute to epigenetic phenomena.

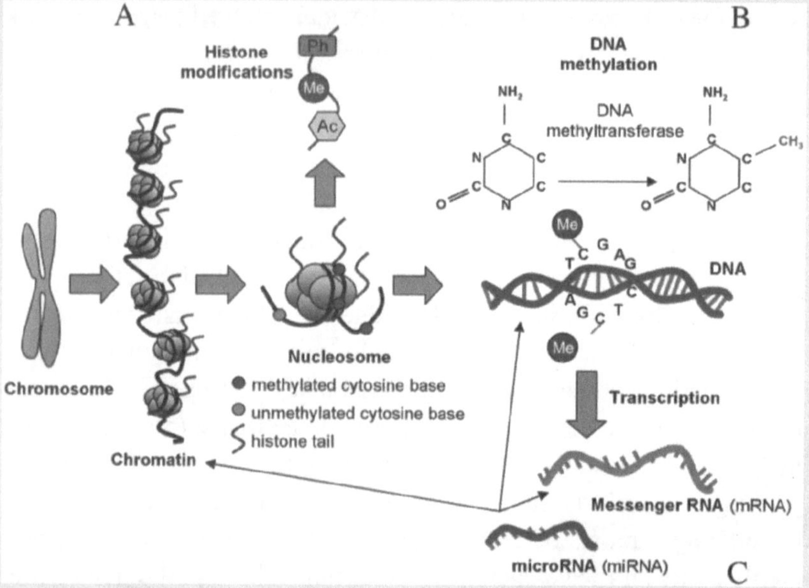

FIGURE BOX 1. Types of epigenetic modifications. (A) Histones can undergo phosphorylation (Ph), methylation (Me), and acetylation (Ac), among other chemical modifications. These modifications are involved in chromatin remodeling and transcriptional regulation. (B) DNA molecules are methylated by the addition of a methyl group to carbon position 5 on cytosine bases, a reaction catalyzed by DNA methyltransferase enzymes, which maintains repressed gene activity. (C) mRNA is translated into a protein product, but this process can be repressed by binding of microRNAs (miRNA), a class of noncoding RNA (ncRNA). Figure adapted with permission from [45].

EPIGENETICS OF HOST–PATHOGEN INTERACTIONS

In recent years, a plethora of papers on the role of epigenetic phenomena on gene expression and phenotype have brought about enormous progress in other fields (such as cancer epigenetics [11]) thanks in part to the significant advances of epigenetic technologies (see Box 2). Conversely, we still know comparatively little about the extent and significance of epigenetic variation in host–pathogen interactions.

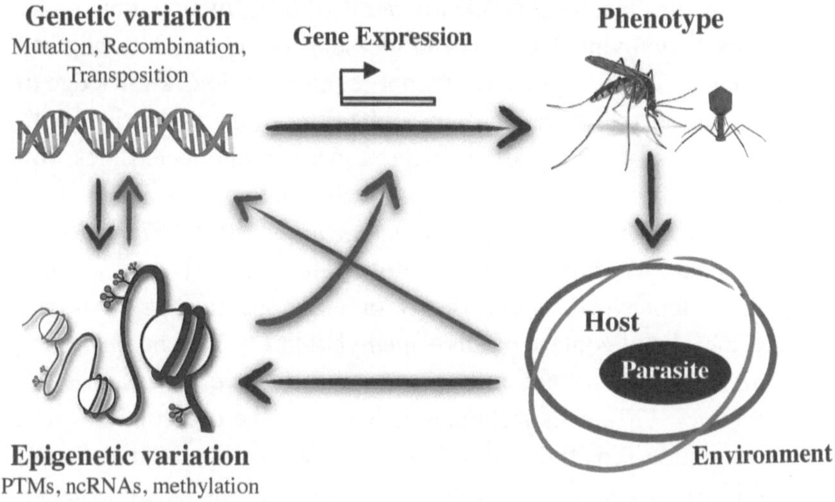

FIGURE 2. Schematic representation of the interrelations between epigenetic variation, phenotypic variation, and host–pathogen interactions. The infection phenotype, which varies between host and pathogen phenotypes and is environmentally dependent, can induce changes at both the genomic and epigenomic levels. These changes can in turn alter gene expression patterns. Apart from these direct effects of epigenetic variation on host and pathogen phenotypes, epigenetic variation can also have indirect, and transgenerational, phenotypic effects by influencing the probability of mutation, transposition, and/or recombination of the DNA sequence, as well as the predisposition of a gene with a particular epigenetic mark to be selected. See text for further explanation. Arrows indicate action routes with potential inherited effects (see Figure 1).

Box 2. Methods of Epigenetic Analysis

Over the last decade, numerous techniques have been developed to analyze epigenetic marks at both genome-wide and sequence-specific levels. Here we summarize novel and cutting-edge methodologies that due to their versatile and straightforward nature can be cross-applied to host–parasite studies. A more comprehensive list of available technologies may be found elsewhere [11,74].

DNA methylation

As a first step in any epigenetic study, **global DNA methylation analyses** allow the detection and identification of DNA methylation (either C and/or A methylnucleotides) and measure its frequency throughout the genome. These approaches do not require previous knowledge of the genome of reference, and most rely on a prior enzymatic/chemical hydrolysis of DNA to obtain the 2′-deoxymononucleosides, followed by the subsequent separation by chromatographic means such as *High Performance Liquid Chromatography* (HPLC) [75] or *High Performance Capillary Electrophoresis*(HPCE) [76], and a final detection step by UV spectroscopy or mass spectrometry. Alternatively, the global content of DNA methylation can also be quantified by enzymatic approaches such as the *Luminometric Methylation Assay* (LUMA) [77]. This technique is based on the digestion of DNA by methylation-sensitive and -insensitive isoschizomers (HpaII/MspI) and followed by pyrosequencing [78] to measure the extent of endonucleases cleavage.

Once the type of DNA methylation is determined, the next step is to study the **distribution and extent of DNA methylation**. The majority of methods are based on three strategies: DNA digestion by methylation-sensitive restriction enzymes, DNA bisulphite conversion, and affinity enrichment of methylated DNA using specific antibodies. The combination of these techniques with different molecular and analytical procedures has resulted in a plethora of approaches for determining DNA methylation patterns both at the specific and the genomic scales. At the scale of specific sequences, the **bisulphite sequencing** has become the gold-standard in mapping m5C sites at single base-pair resolution [79]. Following the bisulphite DNA treatment, cytosines in single-stranded DNA are deaminated to give uracil. After PCR amplification and DNA sequencing using primers that do not contain any CpG site, nonmethylated cytosines are recognized as thymines, while methylated cytosines remain as cytosines. This way, any cytosine that remains in bisulphite-treated DNA must have been methylated. But in recent years there have been major advances at the level of whole methylomes, and numerous techniques have been developed that now allow the study of

DNA methylation at a genome-wide scale. The **Amplification of Inter-Methylated Sites** (AIMS) [80] is based on the differential enzymatic digestion of genomic DNA with methylation-sensitive and -insensitive isoschizomers (SmaI/XmaI) followed by the ligation of specific adapters and the amplification by PCR of the methylated sequences. Amplicons are resolved in denaturing polyacrylamide-sequencing gels, resulting in readable fingerprints that represent the organismal cell's DNA methylation profile. It has been widely applied to study DNA methylation in cancer [81], and more recently, to the discovery of DNA methylation in a social insect (*Apis mellifera*) [82]. Another straightforward approach is **Methylated DNA Immunoprecipitation** (MeDIP) [83], which is based on the isolation of methylated DNA fragments using an antibody specific for 5-methylcytosines. The utility of this technique depends upon the quality of the available antibodies, which at present limits the MeDIP analysis to 5 mC. Among the newest genome-wide technologies that can be applied to host–parasite studies, **microarray technology** provides a good resolution DNA methylation profiling, but its use is restricted to the availability of specific probes. In addition, in nonmodel organisms, a custom array must be designed. In recent years, fast advances in **Next Generation Sequencing** (NGS) have been successfully incorporated to analyze DNA methylation in a cost-effective manner, particularly when combined with enrichment techniques like the MeDIP-seq [84]. Nowadays, complete methylomes can be obtained at single-base resolution by sequencing bisulphate-converted whole genomes [85]. This approach requires, however, complex bioinformatic analysis because bisulphite conversion significantly reduces the complexity of the genome by converting Cs into Ts, thus complicating the alignment of short reads to reference genomes. More recently, several new technologies appear to be able to detect different DNA modifications directly without the bisulphite transformation. These technologies include the nanopore-based methods [86] and single molecule real time (SMRT) DNA sequencing [87].

Modifications and variants of histones

The identification and quantification of the *posttranslational modifications* (PTMs) and histone variants is an essential first characterization step, especially in nonorganisms models. **Mass spectrometry** is the gold standard in terms of accuracy [88]. However, most epigenetic research in this field focuses on detecting the association of individual proteins and histones with specific genomic regions. At present, the most powerful technique is **Chromatin ImmunoPrecipitation** (ChIP) [89]. After cross-linking DNA-binding proteins to DNA with formaldehyde in vivo, the chromatin is isolated and the DNA along with its associated proteins are sheared into small fragments. The DN-binding protein of interest is then precipitated using specific antibodies to isolate the complex, and as a final step, the cross-link is reverted to release the DNA. This method also relies on the availability and quality of antibodies. The immunoprecipitated DNA can be then analyzed by conventional or real-time PCR (ChIP-PCR) [90]. For genome-wide analyses, ChIP is followed by microarray hybridization (ChIP-on-chip) [91] or next-generation sequencing (ChIP-seq) [92]. ChIP-seq has become the state-of-the-art technology for mapping protein–DNA interactions in a genome-wide fashion, but data analysis is time-consuming and its application to nonmodel organisms is still limited.

Host–pathogen interactions are amongst the most plastic and dynamic systems in nature. To cope with the selective constraints imposed by their hosts, many pathogens have evolved an unparalleled level of phenotypic plasticity in their life history traits [12]. Likewise, the host phenotype is drastically and rapidly altered by the presence of a pathogen, and in some cases, the parasitized phenotype is inherited across host generations (see [13] for a review). In addition, co-adaptations between hosts and pathogens often occur over such short evolutionary time scales as to call into question the sole role of genetic modifications (i.e., mutation and/or recombination) as an underlying mechanism [14]. In this sense, epigenetic modifications may provide an accessory source of fast-acting, reversible, and readily available phenotypic variation that can be directly shaped by both host and pathogen selection pressures (Figure 2) [9,14].

We describe in this chapter recent examples of host–pathogen studies where epigenetic processes have already been shown to play a role and which can be broadly classified into (1) pathogen plasticity and (2) pathogen-induced alterations of the host.

PATHOGEN PLASTICITY

One of the most notorious aspects of pathogens is the morphological and developmental plasticity they exhibit, which is intimately linked to their survival and transmission in the host. Complex life-history transitions that occur in response to the changing host environment require rapid and profound alterations of their gene expression profiles. Take, for example, the malaria parasites in the genus *Plasmodium*. In the vertebrate host, the parasite has distinct hepatocytic and erythrocytic stages, and it forms sexually differentiated gametocytes in the blood that are taken up by the mosquito, where these gametocytes mate, then migrate through the midgut to form oocysts and from there to the salivary glands as sporozoites. Previous studies have revealed distinct gene expression profiles in all of these phases (reviewed in [15]), suggesting that developmental switches are transcriptionally regulated. However, apicomplexan parasites such as *Plasmodium* are notoriously poor in transcription factors [16]. In contrast, these parasites contain a rich repertoire of histone variants, chromatin and histone modifying enzymes, and RNA-mediated silencing mechanisms [17,18]. In *Toxoplasma gondii*, histone acetylation has been shown to be responsible for the switch between the replicative and nonreplicative stages of the pathogen [19,20]. Similar mechanisms of epigenetic regulation have been characterized in other protists. Although less studied, *Trypanosoma brucei* is the only Apicomplexa where DNA methylation has been detected, but the significance of these epigenetic modifications in parasite cell-cycle regulation remains unexplored [18]. More recently, DNA methylation has been shown to be responsible for the transition between the yeast and hyphal forms of the polymorphic yeast *Candida albicans* [21].

The second striking aspect of pathogen plasticity concerns their ability to alter the expression of genes linked to virulence processes, which allows them to colonize, replicate, and/or disseminate within the host. Within the Apicomplexa, *Plasmodium falciparum* switches its variant surface proteins during its erythrocytic stage to avoid the host's immune system (antigenic variation). These surface proteins are encoded by highly polymorphic gene families (*var*, *rif*, *stevor*, and *pfmc-2tm*, among others). In the case of the *var* family, the ability of the parasite to express only one of the 60 genes that encode for these proteins (called PfEMP1) is epigenetically regulated through histone modifications [22]. In a recent study, Rovira-Graells et al. [23] have reported a more general association between these histone and chromatin marks and clonally variant expression, extending previous results on *Plasmodium var* genes to all but two of the 28 variantly expressed gene families. In addition, recent work has shown that the epigenetic state of the parasite is maintained during several rounds of cell division [24]. Epigenetic control of virulence factors is well demonstrated in several microbial pathogens. In *Entamoeba histolitica*, for example, histone methylation and demethylation regulate the expression of the amoebapore protein (a protein responsible for the cytotoxicity of the pathogen [25]). DNA methylation is also an essential regulatory mechanism of virulence in several pathogenic bacteria [26,27]. In *Salmonella enterica*, for example, lack of *Dam* (DNA adenine methyltransferase) methylation causes, amongst other things, envelope instability, reduced motility, and an impaired ability to invade the intestinal epithelium [28].

Given the importance of epigenetics for pathogen biology, understanding how the host environment cues the epigenetic transition between the replicative and transmission stages and the virulence factors of morbid and deadly parasites such as *Plasmodium* is not only an academic exercise, but it will also provide novel targets for drug development; an option that has been termed "*epigenetic therapy*" is currently being tested in clinical trials for other (noninfectious) diseases. These prevention and treatment strategies translated to the field of host–parasite interactions could be aimed at arresting the developmental switches of parasites within the host or at blocking or limiting their virulence. This could be achieved by using chemical inhibitors, gene knockout, and RNA interference (RNAi) approaches, designed to target the epigenetic machinery of the parasite such as the DNA methyltransferases or the chromatin and histone modifying enzymes (see Table 1, [29]).

PATHOGEN-INDUCED ALTERATIONS OF THE HOST

Pathogen-induced alterations of host physiology, morphology, and behavior are widely documented in the scientific literature. Perhaps the most fascinating examples of these changes are those that have been shown to be the result of a manipulative strategy of the pathogen aimed at maximizing its survival and transmission. Although some of the mechanisms underlying such pathogen manipulation have been unraveled [30–32], by and large, we know startlingly little of the strategies used by pathogens to achieve this end. In the last few years, however, evidence has accumulated that histone modifications and chromatin remodeling regulate gene expression and are thus key targets for pathogen manipulation during an infection [33]. One such obvious target is the host's immune system. In recent years, the epigenetic modulation of host's transcriptional program linked to host defense genes has emerged as a relatively common occurrence of pathogenic viral and bacterial infections [33,34]. Bacteria are the hallmark of epigenetic studies on microbes and provide several pioneer examples on infection-induced host gene reprogramming [32]. A diverse array of bacterial effectors has been identified that either mimic or inhibit the host cellular machinery, thus facilitating the pathogen's life-cycle. MAPK (mitogen-activated protein kinase), Interferon (IFN), and transcription factor NF-κB signaling pathways are common targets of bacterial-induced post-translational modifications, acetylation, ubiquitylation, and phosphorylation on histones and chromatin-associated proteins [35]. Within the alveolar macrophages, *Mycobacterium tuberculosis*, for example, inhibits interferon-γ-induced expression of several immune genes through histone acetylation [36], which explains the persistence of long-term chronic tuberculosis infections in some patients. This mechanism is not restricted to bacteria but appears rather ubiquitous among intracellular pathogens. Influenza viruses go a step further at circumventing host immune defenses. In a recent study, Marazzi et al. [37] report an influenza protein called NS1 that contains an amino-acid sequence (ARTK) very similar to the host's H3 histone tail. The authors provide compelling evidence of how using this histone mimic sequence, the viral NS1 protein hijacks a host transcription elongation factor (hPAF1), selectively suppressing the cell's production of antiviral proteins. This work is a good example of how a molecule of pathogen origin can directly induce an epigenetic modification in the host. More studies are needed that establish causative relationships between the pathogen infection and host epigenetic modifications such as DNA methylation and posttranslational histone modifications. Indeed, most of the evidence currently available is correlational (but see Table 1), and cases where proteins of pathogen origin have been shown to interact directly with the host epigenetic machinery are still scarce.

An additional characteristic of many pathogens is their ability to manipulate the reproductive biology of their hosts. The endobacteria *Wolbachia pipientis* is the archetypal example of such reproductive manipulations. *Wolbachia* is the most common parasitic microorganism in insects. Its maternal inheritance has selected for a variety of phenotypes associated with manipulating the reproduction of its hosts: forcing asexuality, feminizing hosts, killing males, and inducing incompatibility between infected males and uninfected (or differently infected) females [38]. Negri et al. [39] have provided the first evidence that a feminizing strain of *Wolbachia* interferes with the genetic imprinting of its host (the leafhopper *Zyginidia pullula*) by altering the host's methylation pattern. Recently, the widespread existence of putative DNA-methyltransferases in the prophage of the *Wolbachia* infecting several *Drosophila* species [40] has raised the possibility that this may be a widespread mechanism of epigenetic interference in this endosymbiotic bacteria. The link is, however, unclear since these enzymes have been identified as adenine methyltransferases, a family of prokaryotic enzymes that methylate the amino group at the C-6 position of adenines, whereas in the example reported above, genetic imprinting of the invertebrate host seems to occur at the C-5 carbon of CpG cytosines.

Not all modifications that take place in the infected host are, however, adaptive for the pathogen. Some of them are adaptive strategies of the host aimed to compensate or minimize the effects of the infection. In vertebrates, invertebrates, and plants, individuals that have recovered from certain infectious diseases are protected against later infection with those same diseases (immune priming). While the mechanistic basis of immune priming in invertebrates is still unresolved, in vertebrates, histone modifications may be associated with immune memory following a viral infection in CD8 T cells (reviewed in [41]). Histone modifications, DNA methylation, and other chromatin remodeling mechanisms, including deposition of histone variants and ATP-dependent chromatin remodelers, also seem to serve as a memory for priming in plant immunity [42,43]. In some cases, acquired immunity can be passed from mother to offspring, endowing the offspring with improved defense against infection (transgenerational immune priming). For instance, a very recent paper has shown that in *Arabidopsis thaliana*, immune priming to *Pseudomonas syringae* is transmitted between plant generations through the hypomethylation of defense-related genes [44].

FUTURE DIRECTIONS

In this chapter, we have concentrated our attention on the current evidence available for the role of epigenetic mechanisms in pathogens' life cycle and pathogen-induced modification of host phenotype. However, epigenetics not only represents a paradigm shift in our understanding of host and pathogen phenotypic plasticity. We believe that in the next few years, perhaps the most exciting developments in the field of epigenetics will come by linking epigenetic variation and inheritance to the epidemiology and evolution of infectious diseases.

Epigenetic epidemiology has recently emerged as a promising area for future research on infectious diseases [29,45]. In recent years, disease association studies based on epigenomic mapping have arisen as a powerful tool for disease risk prediction in humans. But these studies typically face the "chicken-and-egg" causality problem: there is an association between a particular disease phenotype and the epigenome, but it is not easy to establish whether it is the disease which is causing the epigenetic changes or whether the epigenetic changes are the ones causing the disease pathogenesis [29]. New epidemiological approaches are, however, being developed in epigenetic disease studies to control for such cause–effect relationships [46]. However, the reversible and context-dependent nature of epigenetic changes poses serious caveats to epidemiological studies. For example, many epigenetic changes linked to disease risk can be lost after one generation, change from tissue to tissue, or be differentially expressed in an age-dependent, sex-, and parent-of-origin-specific manner [47–49]. To overcome these difficulties, epigenetic studies of disease must be accompanied by comprehensive longitudinal (multistage and multi-individual) and transgenerational data. Although there is a lot of effort to bring epigenetics into epidemiological research in several noninfectious human diseases (see examples reviewed by [50,51]), we still know very little about the consequences of epigenetic processes in the emergence and epidemiology of infectious diseases. Therefore, a comprehensive survey of epigenetic determinants of pathogenesis coupled with population-level epigenetic diversity studies in host–pathogen systems is needed before any disease prediction and prevention strategies can become a reality [52].

The second area of research is the role of epigenetic variation in *host and pathogen coevolution and evolution*. Since the incorporation of epigenetic inheritance and epigenetic plasticity mechanisms to evolutionary models and empirical studies of host–pathogen interactions is still lacking, our discussion is necessarily speculative. However, we envisage several areas that are ripe for future research. Models of host–pathogen co-evolution assume the presence of genetic variation for host resistance and pathogen infectivity, as well as genotype-specific interactions [53]. Mechanisms of genetic variation alone are, however, often unsatisfactory to explain the compatibility between host and pathogen phenotypes [54], and nongenetic inheritance mechanisms may have an important role to play (e.g., [55–57]). In addition, host–pathogen co-evolutionary interactions are often context-dependent (i.e., spatially and temporally variable), and the output of infection often depends upon several environmental factors such as temperature or nutrition [58]. Given the prominent role of epigenetic processes in environmentally induced phenotypic plasticity and adaptation [59], the fact that genotype-by-environment interactions in host–pathogen systems are epigenetically regulated seems a reasonable assumption. The work by Laine et al. [60] on a fungal pathogen and its host plant has, for example, demonstrated a temperature-dependent effect on pathogen performance on local versus foreign hosts. Multiple cases of environmentally triggered co-adaptations have been reported in several other host–pathogen systems (reviewed by [61]), which we contend will provide the raw material for future epigenetic studies.

A further unresolved matter is to establish the extent, nature, and significance of *epigenetic inheritance* in host–pathogen interactions. For example, transgenerational immune priming in invertebrates [62–64] and plants [42] is likely to have an epigenetic component, but the actual mechanism of inheritance is not known. Other transgenerational infection effects on host behavior and physiology, often so-called maternal effects, still need to be investigated (reviewed by [13]). Several authors (Bonduariansky and Day [9], Bossdorf et al. [65], Ho and Burggren [66]) have provided recommendations for testing epigenetic inheritance experimentally, which could be cross-applied to host–pathogen studies. In addition, there is an urgent need for epigenetic studies to develop solid theoretical evolutionary models [67]. Bonduriansky and Day [9] have suggested that epigenetic inheritance allows us to overcome three major limitations of genetic inheritance on phenotypic evolution: (1) It allows for traits acquired during the lifetime of an individual to be directly transmitted to the offspring, (2) it allows the transmission of favorable trait combinations across generations (genetic recombination tends to break such combinations), and (3) it provides an additional source of phenotypic variation for selection to act upon. At present, however, we know startlingly little about how these phenomena may impact the evolution of host–pathogen interactions. In this sense, the collaboration between molecular epigeneticists, functional and experimental parasitologists, and theoretical evolutionary biologists is needed to extend the current gene-based view of host–pathogen interactions into a more integrated one that includes both genetic and epigenetic dimensions.

CONCLUDING REMARKS

In recent years there has been an explosion in the number of epigenetics papers across biological disciplines, a progression that has been accompanied by technological breakthroughs that now make it possible to undertake sophisticated epigenomic studies across a range of organisms (Box 2). However, studies on the complex and multifaceted co-evolutionary interactions between hosts and pathogens have received comparatively little attention, and this in spite of their potentially evolutionary and epidemiological implications.

In this review, we have concentrated our attention on the current evidence available for those few cases in which an epigenetic mechanism has been described, but we lack evidence on the evolutionary and epidemiological significance of these changes. Conversely, there are many pathogen and host traits of key epidemiological importance that may be epigenetically controlled, some of which may have transgenerational consequences and whose mechanistic basis would merit further investigation.

In conclusion, the future is bright for the epigenetics of host–pathogen interactions. We are confident that in the next few years, cutting-edge epigenomic techniques combined with experimental (whole organism), functional, and theoretical (modeling) approaches will provide fascinating insights into the interrelations between genetic, epigenetic, and phenotypic variation in the complex world of host–pathogen relationships.

Supporting Information

Experimental approach to detect transgenerational epigenetic and phenotypic changes of infection in a model study involving mosquitoes. Starting from isogenic lines and controlled environmental conditions, female mosquitoes are experimentally infected for successive generations to detect adaptive traits in response to a continuous selection pressure (i.e., infection). Phenotype (behavior, immune response, and physiology), epigenotype, and fitness (i.e., fecundity, longevity, and survival) are then quantified and statistically compared. In F1, two groups of females, either infected or noninfected, are backcrossed with noninfected mosquito control males (NI(C)). If the descendants of infected (I×NI(C)) versus noninfected lines (NI(C)×NI) are phenotypically different but show significant divergence in epigenetic profiles, gene or protein expression—in spite of being still identical at the DNA level—this will be evidence for epigenetically based phenotypic change. In subsequent generations, the comparison of infected versus noninfected mosquito groups that descend of infected mosquito females will allow us to test transient versus stable changes (i.e., adaptive traits) as well as cumulative effects of infection (we may expect them to be greater in V than in III). In addition, differences between the descendants of IV–V in Fx will be indicative of maternal effects.

REFERENCES

1. Waddington CH. The epigenome. *Endeavor* **1942**, 1: 18–20.
2. Jablonka E, Lamb MJ. The changing concept of epigenetics. In: VanSpeybroeck L, VandeVijver G, DeWaele D, editors. *From epigenesis to epigenetics: the genome in context*. New York: New York Academy of *Sciences*, **2002**, pp. 82–96.
3. Suzuki MM, Bird A. DNA methylation landscapes: provocative insights from epigenomics. *Nature Reviews in Genetics* **2008**, 9: 465–476.
4. Bannister AJ, Kouzarides T. Regulation of chromatin by histone modifications. *Cell Research* **2011**, 21: 381–395.
5. Storz G. An expanding universe of noncoding RNAs. *Science* **2002**, 296: 1260–1263
6. Moczek AP, Snell-Rood EC. The basis of bee-ing different: the role of gene silencing in plasticity. *Evolution & Development* **2008**, 10: 511–513.
7. Kucharski R, Maleszka J, Foret S, Maleszka R. Nutritional control of reproductive status in honeybees via DNA methylation. *Science* **2008**, 319: 1827–1830.
8. Jablonka E, Raz G. Transgenerational epigenetic inheritance: prevalence, mechanisms, and implications for the study of heredity and evolution. *The Quarterly Review of Biology* **2009**, 84: 131–176.
9. Bonduriansky R, Day T. Nongenetic inheritance and its evolutionary implications. *Annual Review of Ecology, Evolution, and Systematics* **2009**, 40: 103–125.
10. Cubas P, Vincent C, Coen E. An epigenetic mutation responsible for natural variation in floral symmetry. *Nature* **1999,** 401: 157–161.
11. Esteller M. Cancer epigenomics: DNA methylomes and histone-modification maps. *Nature Reviews in Genetics* **2007**, 8: 286–298.
12. Reece SE, Ricardo SR, Daniel HN. Plastic parasites: sophisticated strategies for survival and reproduction? *Evolutionary Applications* **2009**, 2: 11–23.
13. Poulin R, Thomas F. Epigenetic effects of infection on the phenotype of host offspring: parasites reaching across host generations. *Oikos* **2008**, 117: 331–335.
14. Rando OJ, Verstrepen KJ. Timescales of genetic and epigenetic inheritance. *Cell* **2007**, 128: 655–668.
15. Merrick CJ, Duraisingh MT. Epigenetics in *Plasmodium*: what do we really know? *Eukaryotic Cell* **2010**, 9: 1150–1158.
16. Meissner M, Soldati D. The transcription machinery and the molecular toolbox to control gene expression in *Toxoplasma gondii* and other protozoan parasites. *Microbes and Infection* **2005**, 7: 1376–1384.
17. Hakimi M-A, Deitsch KW. Epigenetics in Apicomplexa: control of gene expression during cell cycle progression, differentiation and antigenic variation. *Current Opinion in Microbiology* **2007**, 10: 357–362.
18. Croken MM, Nardelli SC, Kim K. Chromatin modifications, epigenetics, and how protozoan parasites regulate their lives. *Trends in Parasitology* **2012**, 28: 202–213.
19. Bougdour A, Maubon D, Baldacci P, Ortet P, Bastien O, et al. Drug inhibition of HDAC3 and epigenetic control of differentiation in Apicomplexa parasites. *Journal of Experimental Medicine* **2009**, 206: 953–966.

20. Dixon SE, Stilger KL, Elias EV, Naguleswaran A, Sullivan Jr WJ. A decade of epigenetic research in *Toxoplasma gondii*. *Molecular and Biochemical Parasitology* **2010**, 173: 1–9.
21. Mishra PK, Baum M, Carbon J. DNA methylation regulates phenotype-dependent transcriptional activity in *Candida albicans*. *Proc Natl Acad Sci* **2011**, 108: 11965–11970.
22. Chookajorn T, Dzikowski R, Frank M, Li F, Jiwani AZ, et al.. Epigenetic memory at malaria virulence genes. *Proc Natl Acad Sci* **2007**, 104: 899–902.
23. Rovira-Graells N, Gupta AP, Planet E, Crowley VM, Mok S, et al. Transcriptional variation in the malaria parasite *Plasmodium falciparum*. *Genome Research* **2012**, 22(5): 925–938.
24. Volz JC, Bártfai R, Petter M, Langer C, Josling GA, et al.. PfSET10, a *Plasmodium falciparum* methyltransferase, maintains the active *var* gene in a poised state during parasite division. *Cell Host & Microbe* **2012**, 11: 7–18.
25. Huguenin M, Bracha R, Chookajorn T, Mirelman D. Epigenetic transcriptional gene silencing in *Entamoeba histolytica*: insight into histone and chromatin modifications. *Parasitology* **2010**, 137: 619–627.
26. Løbner-Olesen A, Skovgaard O, Marinus MG. Dam methylation: coordinating cellular processes. *Current Opinion in Microbiology* **2005**, 8: 154–160.
27. Heithoff DM, Sinsheimer RL, Low DA, Mahan MJ. An essential role for DNA adenine methylation in bacterial virulence. *Science* **1999,** 284: 967–970.
28. Marinus MG, Casadesus J. Roles of DNA adenine methylation in host–pathogen interactions: mismatch repair, transcriptional regulation, and more. *FEMS Microbiology Reviews* **2009**, 33: 488–503.
29. Handel A, Ebers G, Ramagopalan S. Epigenetics: molecular mechanisms and implications for disease. *Trends in Molecular Medicine* **2010**, 16: 7–23.
30. Lefèvre T, Lebarbenchon C, Gauthier-Clerc M, Missé D, Poulin R, et al. The ecological significance of manipulative parasites. *Trends in Ecology & Evolution* **2009**, 24: 41–48.
31. Hurd H. Chapter 4 evolutionary drivers of parasite-induced changes in insect life-history traits: from theory to underlying mechanisms. In: Joanne PW, editor. *Advances in parasitology*. Academic Press, **2009**, pp. 85–110.
32. Bhavsar A, Guttman J, Finlay B. Manipulation of host-cell pathways by bacterial pathogens. *Nature* **2007**, 449: 827–861.
33. Hamon MA, Cossart P. Histone modifications and chromatin remodeling during bacterial infections. *Cell Host Microbe* **2008**, 4(2,): 100–109.
34. Paschos K, Allday MJ. Epigenetic reprogramming of host genes in viral and microbial pathogenesis. *Trends in Microbiology* **2010**, 18: 439–447.
35. Ribet D, Cossart P. Post-translational modifications in host cells during bacterial infection. *FEBS Letters* **2010**, 584: 2748–2806.
36. Pennini ME, Pai RK, Schultz DC, Boom WH, Harding CV. *Mycobacterium tuberculosis* 19-kDa lipoprotein inhibits IFN-γ-induced chromatin remodeling of MHC2TA by TLR2 and MAPK signaling. *The Journal of Immunology* **2006**, 176: 4323–4330.
37. Marazzi I, Ho JSY, Kim J, Manicassamy B, Dewell S, et al..(**2012**,) Suppression of the antiviral response by an influenza histone mimic. *Nature* 483: 428–433.

38. Werren JH, Baldo L, Clark ME. *Wolbachia*: master manipulators of invertebrate biology. *Nature Review in Microbiology* **2008**, 6: 741–751.
39. Negri I, Franchini A, Gonella E, Daffonchio D, Mazzoglio PJ, et al. Unravelling the *Wolbachia* evolutionary role: the reprogramming of the host genomic imprinting. *Proc Biol Sci* **2009**, 276(1666): 2485–2491.
40. Saridaki A, Sapountzis P, Harris HL, Batista PD, Biliske JA, et al. *Wolbachia*prophage DNA adenine methyltransferase genes in different *Drosophila-Wolbachia*associations. *PLoS ONE* **2011**, 6: e19708.
41. Youngblood B, Davis CW, Ahmed R. Making memories that last a lifetime: heritable functions of self-renewing memory CD8 T cells. *International Immunology* **2010**, 22: 797–803.
42. Conrath U. Molecular aspects of defence priming. *Trends in Plant Science* **2011**, 16: 524–531.
43. Boyko A, Kovalchuk I. Genetic and epigenetic effects of plant–pathogen interactions: an evolutionary perspective. *Molecular Plant* **2011**, 4: 1014–1023.
44. Luna E, Bruce TJA, Roberts MR, Flors V, Ton J. Next-generation systemic acquired resistance. *Plant Physiology* **2012**, 158: 844–853.
45. Relton CL, Davey Smith G. Epigenetic epidemiology of common complex disease: prospects for prediction, prevention, and treatment. *PLoS Med* **2010**, 7: e1000356.
46. Relton CL, Davey Smith G. Two-step epigenetic Mendelian randomization: a strategy for establishing the causal role of epigenetic processes in pathways to disease. *International Journal of Epidemiology* **2012**, 41: 161–176.
47. Fraga MF, Esteller M. Epigenetics and aging: the targets and the marks. *Trends in Genetics* **2007**, 23: 413–418.
48. Hemberger M, Dean W, Reik W. Epigenetic dynamics of stem cells and cell lineage commitment: digging Waddington's canal. *Nature Reviews in Molecular Cell Biology* **2009**, 10: 526–537. doi: 10. 1038/nrm2727.
49. Ferguson-Smith AC, Surani MA. Imprinting and the epigenetic asymmetry between parental genomes. *Science* **2001**, 293: 1086–1089.
50. Relton CL, Davey Smith G. Is epidemiology ready for epigenetics? International *Journal of Epidemiology* **2012**, 41: 5–9.
51. Baccarelli A, Ghosh S. Environmental exposures, epigenetics and cardiovascular disease. *Current Opinion in Clinical Nutrition & Metabolic Care* **2012**, 15: 323–329.
52. Richards E. Population epigenetics. *Current Opinion in Genetics & Development* **2008**, 18: 221–227.
53. Carius HJ, Little TJ, Ebert D. Genetic variation in a host-parasite association: potential for coevolution and frequency-dependent selection. *Evolution* **2001**, 55: 1136–1145.
54. Lambrechts L, Fellous S, Koella JC. Coevolutionary interactions between host and parasite genotypes. *Trends in Parasitology* **2006**, 22: 12–16.
55. Stjernman M, Little TJ.( ) Genetic variation for maternal effects on parasite susceptibility. *Journal of Evolutionary Biology* **2011**,24: 2357–2363.
56. Cosseau C, Azzi A, Rognon A, Boissier J, Gourbière S, et al. Epigenetic and phenotypic variability in populations of *Schistosoma mansoni*—a possible kick-off for adaptive host/parasite evolution. *Oikos* **2010**, 119: 669–678.
57. Ebert D. Host–parasite coevolution: insights from the daphnia–parasite model system. *Current Opinion in Microbiology* **2008**, 11: 290–301.

58. Vale PF, Wilson AJ, Best A, Boots M, Little TJ. Epidemiological, evolutionary, and coevolutionary implications of context-dependent parasitism. *The American Naturalist* **2011**, 177: 510–521. doi: 10. 1086/659002.
59. Feil R, Fraga MF. Epigenetics and the environment: emerging patterns and implications. *Nature Reviews in Genetics* **2012**, 13: 97–109.
60. Laine A-L. Temperature-mediated patterns of local adaptation in a natural plant–pathogen metapopulation. *Ecology Letters* **2008**, 11: 327–337.
61. Wolinska J, King KC. Environment can alter selection in host–parasite interactions. *Trends in Parasitology* **2009**, 25: 236–244.
62. Moret Y, Schmid-Hempel P. Entomology—immune defence in bumble-bee offspring. *Nature* **2001**, 414: 506–506.
63. Little TJ, O'Connor B, Colegrave N, Watt K, Read AF. Maternal transfer of strain-specific immunity in an invertebrate. *Current Biology* **2003**, 13: 489–492.
64. Kurtz J, Franz K. Innate defence: evidence for memory in invertebrate immunity. *Nature* **2003**, 425: 37–38.
65. Bossdorf O, Richards CL, Pigliucci M. Epigenetics for ecologists. *Ecology Letters* **2008**, 11: 106–115.
66. Ho D, Burggren W. Epigenetics and transgenerational transfer: a physiological perspective. *The Journal of Experimental Biology* **2010**, 213: 3–19.
67. Hunter B, Hollister J, Bomblies K. Epigenetic inheritance: what news for evolution? *Current Biology* **2012**, 22: R54–R56.
68. Kouzarides T. Chromatin modifications and their function. *Cell* **2007**, 128: 693–705.
69. Jenuwein T, Allis CD. Translating the histone code. *Science* **2001**, 293: 1074–1080.
70. Wion D, Casadesus J. N6-methyl-adenine: an epigenetic signal for DNA-protein interactions. *Nature Reviews in Microbiology* **2006**, 4: 183–192.
71. Jeltsch A. Phylogeny of methylomes. *Science* **2010**, 328: 837–838.
72. Zemach A, McDaniel IE, Silva P, Zilberman D. Genome-wide evolutionary analysis of eukaryotic DNA methylation. *Science* **2010**, 328: 916–919.
73. Guil S, Esteller M.( ) DNA methylomes, histone codes and miRNAs: tying it all together. *The International Journal of Biochemistry & Cell Biology* **2009**,41: 87–95.
74. Laird PW. Principles and challenges of genomewide DNA methylation analysis. *Nature Reviews in Genetics* **2010**, 11: 191–203.
75. Eick D, Fritz HJ, Doerfler W. Quantitative determination of 5-methylcytosine in DNA by reverse-phase high-performance liquid chromatography. *Analytical Biochemistry* **1983,** 135: 165–171.
76. Fraga MF, Rodriguez R, Canal MJ. Rapid quantification of DNA methylation by high performance capillary electrophoresis. *Electrophoresis* **2000**, 21: 2990–2994.
77. Karimi M, Johansson S, Stach D, Corcoran M, Grander D, et al. LUMA.(LUminometric Methylation Assay,)–a high throughput method to the analysis of genomic DNA methylation. *Experimental Cell Research* **2006**, 312: 1989–1995.
78. Ronaghi M, Uhlen M, Nyren P. A sequencing method based on real-time pyrophosphate. *Science* **1998,** 281: 363, 365.
79. Clark SJ, Harrison J, Paul CL, Frommer M. High sensitivity mapping of methylated cytosines. *Nucleic Acids Research* **1994,** 22: 2990–2997.
80. Jordà M, Rodríguez J, Frigola J, Peinado MA. Analysis of DNA Methylation by Amplification of Intermethylated Sites.(AIMS,). In: Tost J, editor. *DNA methylation: methods and protocols*, second edition. Humana Press, **2009**, pp. 107–116.

81. Frigola J, Sole X, Paz MF, Moreno V, Esteller M, et al. Differential DNA hypermethylation and hypomethylation signatures in colorectal cancer. *Human Molecular Genetics* **2005**, 14: 319–326.
82. Wang Y, Jorda M, Jones PL, Maleszka R, Ling X, et al. Functional CpG methylation system in a social insect. *Science* **2006**, 314: 645–647.
83. Weber M, Davies JJ, Wittig D, Oakeley EJ, Haase M, et al., Chromosome-wide and promoter-specific analyses identify sites of differential DNA methylation in normal and transformed human cells. *Nature Genetics* **2005**, 37: 853–862.
84. Down TA, Rakyan VK, Turner DJ, Flicek P, Li H, et al. A Bayesian deconvolution strategy for immunoprecipitation-based DNA methylome analysis. *Nature Biotechnology* **2008**, 26: 779–785.
85. Lister R, Pelizzola M, Dowen RH, Hawkins RD, Hon G, et al. Human DNA methylomes at base resolution show widespread epigenomic differences. *Nature* **2009**, 462: 315–322.
86. Clarke J, Wu HC, Jayasinghe L, Patel A, Reid S, et al. Continuous base identification for single-molecule nanopore DNA sequencing. *Nature Nanotechnology* **2009**, 4: 265–270.
87. Flusberg BA, Webster DR, Lee JH, Travers KJ, Olivares EC, et al. Direct detection of DNA methylation during single-molecule, real-time sequencing. *Nature Methods* **2010**, 7: 461–465.
88. Fraga MF, Ballestar E, Villar-Garea A, Boix-Chornet M, Espada J, et al. Loss of acetylation at Lys16 and trimethylation at Lys20 of histone H4 is a common hallmark of human cancer. *Nature Genetics* **2005**, 37: 391–400.
89. Carey MF, Peterson CL, Smale ST. Chromatin immunoprecipitation (ChIP). Cold Spring Harbor Protocols **2009**: pdb prot5279.
90. Solomon MJ, Larsen PL, Varshavsky A. Mapping protein-DNA interactions in vivo with formaldehyde: evidence that histone H4 is retained on a highly transcribed gene. *Cell* **1988,** 53: 937–947.
91. Kurdistani SK, Tavazoie S, Grunstein M. Mapping global histone acetylation patterns to gene expression. *Cell* **2004**, 117: 721–733.
92. Barski A, Cuddapah S, Cui K, Roh TY, Schones DE, et al. High-resolution profiling of histone methylations in the human genome. *Cell* **2007**, 129: 823–837.
93. Cui L, Miao J. Chromatin-mediated epigenetic regulation in the malaria parasite*Plasmodium falciparum*. Eukaryotic *Cell* **2010**, 9: 1138–1149.
94. Chookajorn T, Ponsuwanna P, Cui L. Mutually exclusive var gene expression in the malaria parasite: multiple layers of regulation. *Trends in Parasitology* **2008**, 24: 455–461.
95. Sonda S, Morf L, Bottova I, Baetschmann H, Rehrauer H, et al. Epigenetic mechanisms regulate stage differentiation in the minimized protozoan *Giardia lamblia. Molecular Microbiology* **2010**, 76: 48–67.
96. Geyer KK, Rodriguez Lopez CM, Chalmers IW, Munshi SE, Truscott M, et al. Cytosine methylation regulates oviposition in the pathogenic blood fluke *Schistosoma mansoni. Nature Communications* **2011**, 2: 424.
97. Fernandez AF, Rosales C, Lopez-Nieva P, Graña O, Ballestar E, et al. The dynamic DNA methylomes of double-stranded DNA viruses associated with human cancer. *Genome Research* **2009**, 19: 438–451.

98. Fonseca GJ, Thillainadesan G, Yousef AF, Ablack JN, Mossman KL, et al. Adenovirus evasion of interferon-mediated innate immunity by direct antagonism of a cellular histone posttranslational modification. *Cell Host & Microbe* **2012**, 11: 597–606.
99. Lang C, Hildebrandt A, Brand F, Opitz L, Dihazi H, et al. Impaired chromatin remodelling at STAT1-regulated promoters leads to global unresponsiveness of*Toxoplasma gondii* infected macrophages to IFN-γ. *PLoS Pathog* **2012**, 8: e1002483.
100. Garcia-Garcia JC, Barat NC, Trembley SJ, Dumler JS. Epigenetic silencing of host cell defense genes enhances intracellular survival of the rickettsial pathogen*Anaplasma phagocytophilum*. *PLoS Pathog* **2009**, 5: e1000488.

CHAPTER 2

A CONCISE REVIEW ON EPIGENETIC REGULATION: INSIGHT INTO MOLECULAR MECHANISMS

SHAHRAM GOLBABAPOUR, MAHMOOD AMEEN ABDULLA, and MARYAM HAJREZAEI

INTRODUCTION

The transition from the differentiated somatic cell to the embryonic stage through somatic cell nuclear transfer (SCNT) requires activation energy to efficiently reprogram the resultant zygote to a proper pluripotent state [1,2]. SCNT is a tool to clone nuclear material into the enucleated cytoplasm of an unfertilized oocyte and thereby create genetically identical animals (Figure 1). SCNT not only benefits agricultural applications, but has the potential for great advances in the field of medicine. In addition, SCNT has paved the way to better understand the changes in cell differentiation and reprogramming. Despite many investigations that have been done by numerous laboratories, the efficiency (*i.e.,* the ability to create a live born animal per nuclear transfer) by this technique is still below 5% and several abnormalities have been reported [3]. One of the main reasons for these abnormalities is the failure in reprogramming/remodeling of differentiated cells to the stage that will evolve to a normal neonate. In the other words, programs involved in differentiated cells should be replaced with totipotency to ensure nuclear cloning and production of healthy offspring. Gene regulatory pathways are the critical network that could redefine SCNT. Clones, on the other hand, have to change expression profiles to embryo-specific, global rearrangement of chromatin structure. As a result, the cloning study is a way to understand epigenetic mechanisms and reprogram differentiated nuclei. Epigenetic modifications in the donor cells remodel the gene expression profile to the extent that is similar to the normal embryo. However, the epigenetic mechanisms that are responsible for the transformation from a differentiated somatic cell into a pluripotent state remain mysterious.

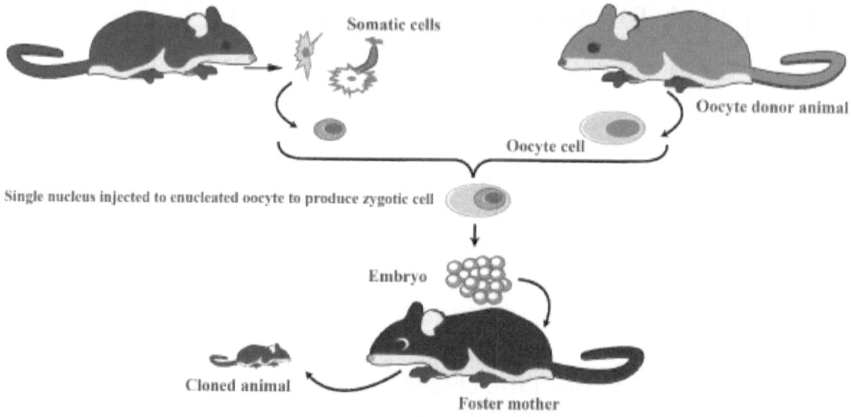

FIGURE 1. The schematic method used to create a cloned animal. A nucleus is taken from a somatic cell (nucleus donor animal) and injected into enucleated Oocyte (Oocyte donor animal). The zygotic cell begins dividing and the resultant blastocyst (embryo) transfers to a foster mother to develop the cloned animal.

TRANSITION TO PLURIPOTENCY

SCNT provides new insight into gene manipulation to achieve defined purposes. This technique is to reprogram the differentiated somatic cell to a pluripotent state by transferring the nucleus of a somatic cell into an enucleated oocyte and produce a zygote, which results in a live offspring. In mammals, genomes of differentiated cell have to reprogram to a totipotent state to establish SCNT during pre-implantation. Consequently, the development of a zygote initiates and follows with blastocyst and the subsequent embryonic stages. Cloned embryos derived from less differentiated cells (as nucleus donors), such as embryonic stem cells, show better implantation than those derived from more differentiated somatic cells probably due to minimum or no reprogramming requirement [4]. It was shown that the efficiency of bovine SCNT is relatively higher than the other experienced species (see review [5]) and pregnancy in *Bos taurus* is very similar to that of human in terms of length and development.

Generation of induced pluripotent stem cell is to transport defined regulatory signals, influence the epigenetic state and change it to another state (plasticity), which emphasizes the mutual reliance between cell identity and epigenetic states [6,7]. This is, especially true during early embryo development and gametogenesis [1]. Pluripotent stem cells are driven from somatic cells that are introduced by specific reprogramming factors through either cell fusion or delivery of defined biochemical and/or chemical factors, which are also categorized as a reprogramming approach. The fusion technique produces hybrid cells from differentiated somatic cells by nuclear reprogramming through the reactivation of embryo-specific genes, whose expressions are suppressed in somatic cells [8]. In 2006, a four-gene set was introduced to reprogram somatic cells to a pluripotent state [7]. Hybrid cells produced by fusion technique show a pluripotent state by expression of the pluripotent markers such as *OCT4* [9]. Moreover, a number of other genes such as *Nanong*, *Sox2*, *Lin28*, *Klf4*, *c-Myc* and *AID* have been correlated to the pluripotent state of a cell. The expressions of these genes result in cell reprogramming [7,10–13]. Based on these evidences, identification of embryo-specific genes is crucial to defining their expression profiles during embryogenesis, functions during different stages of embryogenesis and the development of placenta. These regulations, actually, are defined epigenetic regulation for which molecular signals modulate the modifications.

Several morphological abnormalities such as hydroallantois, placentomegaly, cardiomegaly, enlarged umbilical cord, abdominal ascites and placental dysfunctions [14,15], have been observed in the cloned offsprings. Large offspring syndrome (LOS) is a developmental disorder mostly seen in SCNT driven embryos. This syndrome in addition to the failure in the development of embryo and placenta and other abnormalities is attributed to inappropriate and/or inadequate somatic nuclear reprogramming events. Significant increase in genomic methylation in liver of cloned bovine fetuses is attributed to fetal overgrowth [16]. LOS and failure in the normal development of an embryo that are seen in cloned animals could be due to abnormal epigenetic patterns [17]. In fact, assisted reproductive techniques appear to be accompanied by several anomalies, especially in the second half of the gestation [14,18–20].

MOLECULAR SIGNALS IN EPIGENETIC REGULATION

Cells' information is inherited to the next generation through genetic and epigenetic routes. Genetic information is encoded in the DNA sequence while, epigenetic information is defined basically by DNA modification (DNA methylation) and chromatin modifications (methylation, phosphorylation, acetylation and ubiquitination of histone cores). Combination of these modifications characterizes the chromatin configuration and the accessibility of genes to the transcription machinery and consequently, transcriptional regulation of the expression of genes. Cheng [21] introduced three categories by which transcriptional function is generally initiated and controlled: First, general intrinsic promoter and transcriptional machinery [22–24], second, specific transcriptional regulatory factors [25–27] and, third, the configuration and accessibility of chromatin structure and DNA to the transcriptional machinery through posttranslational modifications of histone and post replicational modification of DNA [27–29].

MAIN EPIGENETIC REGULATORY MECHANISMS

Complex epigenetic regulation comprises several molecular signals that direct the expression of genes based on environmental changes and developmental status. Transcription factors, non-coding RNAs (ncRNAs) [30], DNA methylation, histone modification and chromatin remodeling are such epigenetic signals that mediate accessibility and expression of genes as needed. Transcription mainly defines a self-propagating state mediated by *cis*-acting and/or *trans*-acting regulatory mechanisms [31], and are able to establish epigenetic states through *cis*-acting [32] and non-coding polycomb domains [31,33]. Reinforcement of epigenetic states happens through mutual relationship between DNA methylation and histone modifications [34]. DNA methylation postulates a reinforcing signal for other regulatory mechanisms whose functions are not that much strong [31].

DNA methylation and histone modification are two important mechanisms for modulating the chromatin structure and regulating the expressions of the genes (for review, see [35]). Epigenetic regulation is a complex phenomenon that consists of a variety of different processes [21] such as imprinting [36], X chromosome inactivation [37] and gene silencing [38,39]. In addition it encompasses the development of an embryo [40–43] and placenta [44–46], nuclear reprogramming in SCNT embryos [3,6] and carcinogenesis [47,48].

TRANSCRIPTIONAL REGULATION

The transcriptional regulation of genes is mainly directed by different strategies. These include the state of genomic methylation [21], chromatin configuration [49,50], chromatin structural variations (euchromatin and heterochromatin) [51,52], and chromatin modifications [53]. Chromatin modification in turn is influenced by methylation, acetylation and phosphorylation, as well as polycomb proteins [54] and matrix attached region [55]. Transcriptional regulation is mostly controlled by the methylation pattern of the genome. DNA methylation on specific CpG dinucleotide (CpG) located in a cluster (CpG islands) is the regulatory mechanism by which expression of gene is either activated or suppressed (for review see [56]). Moreover, chemical modifications of chromatin histone cores are mediated by DNA methylation of CpG islands [57]. These modifications have a mutual relationship with each other [58]. Germ cells and embryonic cells during early development are two epigenetic sites where methylation patterns erase, establish and reestablish [59].

EPIGENETIC REPROGRAMMING DURING EMBRYOGENESIS

In mammals, epigenetic reprogramming in germ cells and during preimplantation, especially its effects on imprinting genes, predominantly establishes developmental stages [60]. The DNA methylation patterns characterize developmental status during cell differentiation. In the concept of epigenomics, molecular signals are responsible to establish the proper expression of embryo-specific genes, mainly during gametogenesis and embryogenesis. Therefore, the main issue for a successful SCNT is the establishment of these modifications, occurring during embryogenesis, which should be similar and ideally identical to its normal embryo counterpart. However, undoubtedly, several lessons are still to be learnt regarding epigenetic modifications during gametogenesis.

EPIGENETIC FEATURES OF DNA METHYLATION

As mentioned before, DNA methylation and histone modification are the main epigenetic factors, by which gene expression could be regulated, and have important roles in nuclear reprogramming during embryogenesis. DNA methylation is a heritable epigenetic marker by which expression of a gene may be regulated through alteration in the local chromatin structure that mostly happen within the CpG islands and imprinted genes at cytosine carbon 5 within palindromic dinucleotide 5′-CpG-3′ and differentially methylated domains (DMDs) respectively (see review [60]). Cytosine residue of CpG is the site for DNA methylation by which gene expression is regulated. Generally, DNA methylation at CpG sequences suppresses the expression of the methylated gene [61]. CpG islands are usually located within repetitive elements such as centromic repeats, satellite sequences and ribosomal RNA genes [62,63]. DNA methylation can be varied in terms of patterns and the level of global/regional DNA methylation, is specific to developmental stages [64] and origin of tissue [16,65]. In fact, the mature parental gametes at fertilization are significantly methylated. For instance, DNA of sperm, in comparison with that of the oocyte, is more methylated [66,67] and, undergoes demethylation after fertilization [68–70]. However, imprinted genes and some retrotransposons mostly remain methylated. In mouse, hypermethylation pattern in repetitive regions and heterochromatin region has been observed; whilst in gene-specific region of DNA hypomethylation is predominant [17,71,72]. Abnormal DNA methylation of various repetitive elements in cloned blastocysts was reported for the first time by Kang and coworkers [73].

Methylation of imprints, monoallelic expressed genes [74,75], on the other hand, is a maintained (not *de novo*) and highly conserved event [76,77]. Recently, in a human study, comparison between embryonic stem cells and differentiated cells illustrated that there are a number of methylated cytosine in non-CpG regions of the embryonic stem cells [78].

DNA METHYLATION SIGNALS

DNA methylation is under the control of two types of signals: *cis*-acting signals and *trans*-acting signals. *IGF2R, SNRPN, H19* and *RASGRF1* are genes regulated by *cis*-acting signals (see review [79]). Global DNA methylation takes place especially after fertilization and with different rate of demethylation that is specific to either parental genome [61]. Cell cycle observations reveal that paternal demethylation generally happens during the first cell cycle but maternal alleles take a few cycles to be demethylated [80]. After fertilization, imprinting control regions (ICRs) methylation is established in a sex-dependent manner [61]. Despite the maintained methylation pattern in somatic cells, methylation pattern in germ cells needs to be appropriately reestablished to provide a methylation pattern that is heritable to the next generation. This suggests that methylation is modulated in a sex-dependent manner [81]. Although hypomethylation of female germ line seems not to be correlated to the sex chromosomes, their regulation is thought to be associated with genital ridge. However, in the male germ line it is regulated by both mechanisms [81,82].

DNA METHYLATION ANALYSIS

Epigenetic studies are strongly involved in DNA methylation. Analysis of the methylation patterns is the main approach in different studies that focuses on gene regulation. The cytosine 5 methylation in the context of CpGs mostly takes place within CpGs islands at the promoter region of genes and this leads to suppression of the expression of the gene. Several studies correlate aberrant methylation pattern of DNA to developmental failure during embryogenesis [83–85] and placentation [86,87] as well as several diseases and disorders [88–91] (for review see [92]). DNA methylation techniques cover a wide range of analysis from gene specific, locus specific to entire genome analysis using proper methods categorized in four groups based on DNA methylation analysis techniques [93]: in the first category cytosine residues are converted to uracil by a bisulfite conversion, the second category is based on methylation-sensitive restriction endonuclease, enrichment based methods and the last is the capturing method based on the affinity to retain methylated DNA [94,95].

REGULATORY FACTORS IN DNA METHYLATION

As mentioned before, DNA methylation is a common epigenetic modification taking place by enzymatic reactions to be added to the cytosines at CpG, mostly known as repetitive elements and imprinted genes [79]. Generally, DNA methylation is classified either as *de novo* methylation or maintained methylation. Therefore, there are two classes of enzymatic: *de novo* methyltransferases and maintenance methyltransferases [96,97]. In mammals, DNA methylation occurs by the addition of a methyl group from S-adenosylmethionine to cytosine using DNA methyltransferases (DNMTs). DNMTs are *trans*-acting factors targeting DNA sites for methylation using *cis*-acting signals. There are a number of mammalian DNMTs (see Table 1) that have been identified since 1980s [98] (for review see [99]).

TABLE 1. Types of DNA methyltransferases and their epigenetic functions.

DNMT # types [100]	Functions
DNMT1	Maintaining methylation pattern [21,101,102] Essential for chromosome replication and repair [21,103,104] Essential for de novo methylation [105]
DNMT2	Effective in DNA and RNA methylation (for review see [106])

TABLE 1 continued...

DNMT3a	Establishment of de novo methylation pattern [107,108] especially during gametogenesis [109] Maintaining methylation pattern [101]
DNMT3b	Establishment of de novo methylation [107,108]
DNMT3L	Essential for de novo methylation [110] Enhance de novo methylation activity of DNMT3a [111] and DNMT3b [112] Establishment of de novo methylation pattern especially during gametogenesis [113]

#DNA methyltransferase

DNA METHYLTRANSFERASES

DNA methylation at cytosine 5 nucleotide is catalyzed by DNMTs. This family of enzyme is vitally important in epigenetic regulation, which modulates the expression of genes especially imprinted ones as well as X chromosome inactivation [114,115]. There are five main DNMTs that are important in *de novo* and/or maintenance DNA methylation: DNMT1, DNMT10, DNMT3a, DNMT3b and DNMT3L [99]. DNMT1, DNMT2 and DNMT3 are mostly characterized DNMTs that can categorize either maintenance or *de novo* DNMT. DNMT1 is a maintenance DNMT that methylates both imprints and non-imprints genes. DNMT2 seems to have a regulatory role in DNA methylation but the mechanism and its role in methylation maintenance or *de novo* remains unclear. DNMT3 as a *de novo* DNMT (DNMT3a) is a key factor in imprints' methylation. Its isoforms are suggested to have roles in global DNA methylation in germ cells [116].

DNMT3L is defined as an imprints' regulatory candidate for DNA methylation by regulating NMT3a/b [117]. The expression of *DNMT1* gene has a positive correlation with DNA methylation status on the satellite I region. Consequently, it has been shown that *in vitro* development of bovine SCNT embryos to the blastocysts state can be enhanced through down regulation of DNMT1 [118]. It is also shown that the DNMT is responsible for ICRs methylation [109]. Moreover, transcriptional analysis on the pS2/TFF1 during cell cycles reveals that DNMTs carry out two distinct actions, namely methylation and demethylation of CpGs [119].

In comparison to the male germ line in which the establishment of ICR methylation of an imprinted gene, *H19*, is regulated by DNMT3a and DNMTL [109,113], DNMT3L is the *de novo* methylating regulatory factor for the female germ line [109]. In the female germ line, DNMT3L establishes the methylation of ICRs that selectively interact with histone H3 [120]. This evidence in addition to DNMT3L's stimulating role for DNMT3a and DNMT3b [117] shows its potential in chromatin mark-specific recognition and methylation establishment [61]. Promoter methylation-mediated DNMTs show down regulation of DNMT1 and up regulation of DNMT3L in the human placenta and brings strength to the capability of DNMT family in the establishment of *de novo* DNA methylation in extraembryonic tissue [46]. A novel DNMT3b splice variant, DNMT3B3Δ5, is highly expressed in pluripotent embryonic stem cell and in contrast, is repressed during differentiation [121].

In a human study on DNMTS, global hypomethylation is shown to be induced by significant reduction in the expression of *DNMT3A*, *DNMT3b* and *DNMT1* using microRNA-29b [122]. DNMT3L by itself has no methyltransferase activity; however, its association with the DNMT3 family seems essential for *de novo* methylation in mice [123]. There are some evidence on the activity of DNMT1 in the establishment of methylation at non-CpG regions [55] and CpG islands [21,124,125]. Histone modification and CpG spacing are able to direct DMR methylation of imprints [21]. Crystallographic analysis showed that *de novo* DNA methylation might be controlled by specific histone modifications in that a heterotetramer structure, assembled from DNMT3a and DNMTL, provides two active sites of CpG that are 8–19 base-pair distance from each other [126–128]. A study on chromosome 21 also reinforces the crucial role of CpG spacing in DNA methylation [129]. Active demethylation in mammalian genome seems promising; however, there has been no report of an enzyme that can catalyze this reaction [130]. Some studies have emphasized an active demethylation process, independent from DNA replication [131,132]. In fact, demethylation of the paternal alleles is an active event that happens rapidly after fertilization.

The maternal genome, however, demethylates during first cell cycles in which demethylation mostly appears to be an inactive process. In addition to DNA methylation and histone modification, ncRNAs and regulatory proteins are the most studied epigenetic mechanisms that modulate epigenetic reprogramming. Small interfering RNAs (siRNAs) transfection is a technique to silence DNMT mRNA and modify the DNA methylation pattern in cells [6]. In a recent study, To examine the efficacy of the technique in SCNT embryos, DNMT1 RNA was silenced using siRNA in SCNT bovine embryo which demonstrated the capability in nucleus reprogramming through inducing DNA methylation [118]. In the expression of *H19* in the male germ line, DNMT3a and DNMTL are counterparts and reached their maximum [133,134] on embryonic day 13 while there is no methylation on the *H19* ICR, in mouse [135]. In addition, these enzymes seem not to be specific for DNA binding [99], suggesting direct/indirect interactions with specific chromatin modifications [61,136,137].

EPIGENETIC FEATURES OF ncRNAS

The cluster-oriented imprinted genes are laid in ~1Mb length base pair, containing parental expressed genes, ncRNA sequences that regulate the nearby imprinted genes [138–141]. ncRNAs are mostly placed in clusters and regulated by ICRs [142]. The *GNAS* and *KCNQ1* are examples of such imprints; containing ncRNAs that mediate the gene expression [143,144]. ncRNAs are divided into two groups, small ncRNAs and long ncRNAs. Small ncRNAs attach chromatin modifiers to specific genome sequence [145] and may interact with either RNA, single-stranded DNA or double-stranded DNA [1,146]. Long ncRNAs have complex tertiary structure and act globally to bridge chromatin modifiers to the genome [147]. But there are some evidences for local function of long ncRNAs, which is considered to function in *cis*-acting regulation of parental imprinted gene and inactivation of chromosome X [1].

In mammalian transcription of ncRNA genes is an important feature. ncRNAs usually are classified based on their mature length, location and orientation according to the nearest protein-coding gene, and their function which could be *cis* or *trans* [148–150]. Macro RNAs, such as inactive X-specific transcript (Xist) and X (inactive)-specific transcript, antisense (Tsix), are categorized as *cis*-acting ncRNAs that usually locate within clusters of imprinted genes. On the other hand, short ncRNAs such as short interfering RNAs, micro RNAs, piwi-interacting RNAs and short nucleolar RNAs are categorized as *trans*-acting ncRNAs (for review see [151]) [148]. Koerner (2009) concluded that chromosomes express macro ncRNA usually do not express imprinted mRNA genes and the expression of imprinted macro ncRNAs may be regulated by an unmethylated imprint control element [148]. Moreover, there are a number of evidences that show *trans*-acting regulators for imprinted small ncRNAs such as *Snurf-SNRPN* and *Dlk1-Gtl2* [152,153]. It has been shown that ncRNAs have a critical role during development. For instance, two ncRNAs, *Dicer1* and *Dgcr8*, show developmental impact in mice [154,155]. Moreover, studies on effects of ncRNAs during animal embryogenesis show their specific and crucial role during embryonic development (for review see [156]). Micro ncRNAs, specifically, show precise control on expression of imprinted genes during development [157]. For instances, *miR-15* and *miR-16* are important in early embryonic development [158], *miR-1*, *miR-133* and *miR-206* in development of skeletal and heart muscle [159,160], *miR-124* in neuronal development [161] (for review see [162]). During placentation, ncRNAs such as *KCNQ1OT1*, a long ncRNA, also illustrate a leading role in imprinted genes regulation [163–165]. Regulation of ncRNAs is an important silencing mechanism in plancenta (for review see [148]). It was shown that the repression of imprinted genes during gestation is directly regulated by micro RNAs during placentation and embryogenesis [166,167]. It seems that ncRNAs targets placental histone methyltransferases by ncRNAs through chromatin modification [148].

EPIGENETIC FEATURES OF SMALL RNAS

Small RNAs (terminologically different from ncRNAs), generated by activity of RNaseIII enzymes (reviewed in [168]), have variety of biological functions such as heterochromatin formation, mRNA inactivation and transcriptional regulation [169,170]. Generally, their bioactivity is due to their association with Argonaute (Ago)-family proteins [171]. MicroRNAs (miRNAs), endogenous small interfering RNAs and Piwi-interacting RNAs (piRNAs) are classes of small RNAs. In mammalians, small RNA-associated Ago proteins are mostly classified into Piwi subfamily and Ago subfamily (for review see [171]). miRNAs to do their biological activity, which is post translational regulation by acting on mRNAs, needs to be bound by Ago subfamily proteins (for review see [170]). Moreover, regulation of most miRNAs may control by developmental signaling [172]. piRNAs are mostly bound by Piwi subfamily proteins and have a critical role during gametogenesis [173] in germ line [174]. This subfamily protein has shown to have a critical role in regulation of germline stem cells [175].

EPIGENETIC FEATURES OF CHROMATIN MODIFICATIONS

Chromatin structure is crucial for gene regulation/expression, which is carried out by exploiting recruitment of protein complexes [29]. Euchromatin structure of embryonic stem cells is a predominant chromatic structure that allows for global gene expression accessibility [176] and facilitates reprogramming to the pluripotent state. It is not surprising that histone modifications might in turn influence the global gene expression by modulating chromatin configuration [177]. Covalent modification of the core histone has a critical role in the regulation of gene expression through acetylation and methylation. Chromatin modification and their function are important especially for gene regulation. Kouzarides (2007) reviewed a number of chromatin modifications characterized by mass spectrometry (for nucleosomal modification) and specific antibodies (for global histone modification) [178]. Cellular condition is the key element for such modifications and these chromatin modifications, as a dynamic procedure, are mediated by a number of histone-modifying enzymes that can fascinate unravels chromatin, recruitment of nonhistone proteins and transcriptional regulation (for review see [178]). Chromatin modifications, to regulate gene expression, are mostly implied be a number of chromatin modifications such as acetylation/deacetylation, phosphorylation, lysine/arginine methylation, deimination, ubiquitylation/

deubiquitylation, sumoylatio, ADP ribosylation and proline isomerization which are properly reviewed by Kouzarides (2007) [178].

Histone acetylation is the main type of histone modification during oogenesis, and it is shown that histone acetylation is critical in epigenetic reprogramming [179,180]. For instance, *in vitro* study on acethylation of histones in cloned porcine blastocyst showed that increase the level of acetylation may enhance the embryonic development [181,182]. Generally, hyperacetylation of histone H3 andH4 improve the accessibility of nucleosome to transcriptional machinery [183]. The level of histone acetylation may correlate with the regulation of the expression of genes because more histone acetylation the more expression of a given gene, and vice versa [180]. As mentioned before, histone modifications and DNA methylation are cooperative. Histone modification is able to direct DNA methylation as shown in H3 in Neurospora crassa [184,185] (for review see [186]). Results from recent studies [187,188] have recapitulated that some chromatin modifiers directly act in a *cis*-acting manner [31]. However, a study on the relationship between DNA methylation and histone methylation suggests that they act mostly independently [189]. The affinity of UHRF1 binding protein to the nucleosomal H3K9me3 increases if CpG islands at the nucleosome are methylated but in contrast, in the absence of DNA methylation KDM2A binds to nucleosome having H3J9me3 [190]. Two epigenetic markers, H3K27me3 and CpG DNA methylation, at the *RASGRF1* locus, are interdependent and antagonistic so they are more likely to exclude each other at the same loci [191]. The SNF2 family is an ATP-dependent remodeling complex. In this family, LSH has a role in establishment of normal DNA methylation. A null mutation in Hells gene, codes for LSH that results in the reduction or loss of methylation. Besides, this study suggests the importance of LSH in *de novo* methylation during embryogenesis [192]. Although histone methylation at H3K4 is able to control methylation at DMR of imprinted genes in an allele-specific manner [193,194], it seems to have preventive influence in terms of *de novo* methylation in mammalian somatic cells and may require low promoter methylation [195,196]. Furthermore, mutation in genes, coding for histone methyltransferase such as EZH2 and G9a [197,198] and histone deacetylases like HDAC1

[199] leads to premature death of mammalian embryos typically in less than ten days from fertilization.

GENERAL FEATURES OF IMPRINTED GENES AND THEIR REGULATION

After fertilization, a mammalian zygote undergoes proliferation and development. Although there are many active parental genes, involved in a normal embryo development, there are a few genes with bias regulation and transcription, referred to as imprinted genes [61]. Imprinting genes are important for normal embryonic development in mammals. Imprinting genes are selectively (on bias) expressed from a single parental allele [200] and conserved in their molecular structures and epigenomics [75,201]. These genes, essential for normal development, are expressed in a parent-specific manner, regulated by complex epigenetic mechanisms (e.g., DNA methylation, post-translational histone modification) using epigenetic markers (e.g., DNA methylation) [61]. The conflicting interests of parental, imprinted genes are hypothesized as maternally and paternally expressed imprints suppress and enhance the fetal growth respectively (see review [60]). In the nucleus, imprinted genes are mostly placed in a cluster orientation but some are identified as isolated ones [72] such as *Nap1l5*, *Nnat*, *Inpp5f_v2* [202–206] and *Gatm*, *Dcn* and *Htr2a* (for review see [207]). Imprints that are placed within CpG rich region are mostly in clusters, controlled by imprinting the control regions through DNA methylation and histone modifications [58,208,209]. Regulations of imprinted genes are generally proceeded through DNA methylation, post translational histone modification and ncRNAs [210]. In addition, active imprinted genes (expressed allele) contains the allele-discriminating signal (ADS) and the *de novo* methylation signal (DNS) that are necessary for establishing or maintaining methylation [211,212]. For instance, SNRPN is a paternally expressed imprint whose regulation is similar to that of *IGF2r* [212]. Human SNRPB contains two DNS signals; an ADS signal and a signal to maintain paternal imprint (MPI) [213].

METHYLATION OF IMPRINTED GENES AND ITS ABNORMALITIES IN CLONED ANIMALS

Short regions of DNA, described as differentially methylated regions (DMRs), are marked by methylation in a parental specific manner and therefore the expressions of such genes are monoallelic. Regulation of clusters of imprinted genes and their activities are mostly controlled by differentially methylated ICRs. In fact, ICRs are DMRs that obtain methylation on one allele (bios) and regulate clustered imprinted genes [138]. In the other words, those DMRs that have a critical role in maintaining imprinting are known as ICRs [81]. CpG spacing suggests a potential influence on ICRs recognition and DMRs methylation in imprints [126]. Moreover, the transcriptional system, especially those traversing ICRs, are considered a common requirement to open chromatin domains, and make targets available for methylation specially in the germ line [32]. Besides, an *in vivo* study in a mouse model illustrated a novel *cis*-acting function for the *H19* ICR [214]. This study shows changes in the size and CpG density that coincide with biallelic expression of the *H19* without any detectable alteration in the methylation pattern. The researchers concluded that, in addition to CTCF sites, there are sequences within the ICR that are essential for its regulatory function. Moreover, the ICR size and CpG density are of determinant elements.

Maternal alleles are dramatically more exposed to ICRs methylation than paternal ones [61,138]. Maternal alleles are mostly methylated on promoters of antisense transcripts but those of paternal alleles are placed between genes (non-promoter regions), suggesting that parental imprinting methylation acts differently [215]. In general, there is higher degree of methylation of the maternal ICRs allele in comparison with that of paternal allele [61]. The *H19* is an example of imprinted genes whose preference is to be expressed from maternal allele. Methylation of the DMDs of *H19*, maternally expressed imprinted gene, is needed for maintenance methylation [141].

DMRs of imprints, mostly, epigenetically signal for monoallelic expression of the gene. *IGF2* encodes a fetal growth-factor and is predominantly expressed from the paternal allele, while *H19* is expressed from the maternal allele and encodes a transcript which may reduce cellular proliferation. In mouse, *IGF2* has a few identified DMRs named DMR0, DMR1, DMR2 and DMR3 among which the first two DMRs are positioned upstream and DMR2 within the *IGF2* gene [216–218]. Recently, an intragenic regulatory DMR has been reported within the last exon of the *IGF2* gene [81]. The comparison between methylation patterns of *IGF2* DMR from parthenogenetic and androgenetic blastocysts on one hand and that evolved from a normal zygote suggests that in normal embryos, paternal allele significantly contributes in the DNA methylation at the locus [81]. Methylation on DMDs of imprints are initially established during gametogenesis and prior to parental pronucleus fusion in the zygote [61,219]. After fertilization, the intergenic DMR of bovine *IGF2* undergoes demethylation followed by low-level remethylation before blastocyst stage, which in turn precedes implantation by which the DMR is heavily remethylated. The study speculates that global methylation pattern of SCNT blastocysts is reprogrammed and maintained in a sex specific manner, similar to its normal counterpart. A recent study shows that, except for *RASGRF1* DMR (paternally expressed imprinted gene), methylation of the most imprinted genes during mouse embryonic cleavage stages (preimplantation phase) are mainly controlled by maternal and zygotic DNA methyltransferase 1 (DNMT1) protein family [220].

Abnormalities at imprinted loci have been observed in cloned mammals. In cloned cattle abnormal imprinted gene profiles have been observed especially in the expression of *IGF2* and *H19* [221]. In the *Bos taurus* model, the *IGF2* and *H19* (*IGF2*/*H19*), a conserved cluster of imprinted gene, showed significant variations from the normal pattern, mostly hypomethylation, associated with abnormal expressions of the *H19* (but not *IGF2*) from both alleles in methylation pattern of DMRs [17,222]. Moreover, methylation pattern, which mostly occurs in early embryogenesis is dependent on developmental stage and specific to different tissues, as was studied in *IGF2*/*H19* [218].

Super ovulation, also, can cause abnormal imprinting patterns in oocytes [223] that might be attributed to the reduced expression of imprinted parental alleles, *SNRPN*, *PEG3* and *KCNQ1OT1*, but to increased methylation of *H19* [224]. MII oocytes of cloned porcine showed mostly unmethylated profiles of DMR [225]. A recent study on bovine SCNT showed that significant demethylation at the *H19* DMD is attributed to biallelic expression of the imprint, which might lead to decline in the rate of implantation [226]. Moreover, biallelic expression of *H19* in bovine is correlated to hypermethylation of the paternal *H19* differentially methylated domain and locus anomalies cause low SCNT efficiency in cattle [226].

CONTROL OF GENE EXPRESSION DURING GAMETOGENESIS

Gametogenesis and embryogenesis involve epigenetic reprogramming to establish proper epigenetic marks and gene regulation. Generally, epigenetic pattern of the genome first reprograms and reestablishes during gametogenesis. The second round of reprogramming and maintenance happens after fertilization, especially during preimplantation of the embryo [61] (Fig. 2). Gametogenesis in both sexes involves methylation of DMRs, reestablished in a parent-specific manner [60]. Epigenetic reprogramming is mostly characterized during gametogenesis and early embryonic development especially prior to the zygotic implantation [227]. In fact, during gametogenesis (spermatogenesis and oogenesis) the methylation patterns of these genes are erased and reestablished. These modifications are continued after fertilization and during preimplantation specifically within non-imprinted genes [225] (for review see [180]). Gametogenesis involves sex-specific, epigenetic remodeling of male and female germ lines that matures the gametes for fertilization and constitutes proper regulatory processes [180]. Epigenetic reprogramming in sperm begins with DNA demethylation, followed by DNA remethylation and *de novo* methylation to chromatin modification and histone-to-promatine transition [180,228]. Moreover, during spermatogenesis, testis specific linker histones occupy somatic linker variants. Among the histone variants centromere protein A appears to be epigenetically important during spermatogenesis [180]. Spermatozoa have a transcriptionally inactive and highly condensed chromatin structure. During spermatogenesis in rats, paternal-specific imprinted genes are prone to hypomethylation due to estrogen-associated signaling [229]. In the male germ cells, DMR of *IGF2/H19* acquires DNA methylation during spermatogenesis, however, in the female germ cells, the DMR possesses the zinc finger protein CTCF by which the DMR defends against methylation so the allele is able to be expressed [230]. Through fertilization, paternal genome undergoes a series of remodeling events, which are controlled by the activity of the oocyte, and the protamine replaced by oocyte-supplied histone and possessing maternal chromatin related proteins [231].

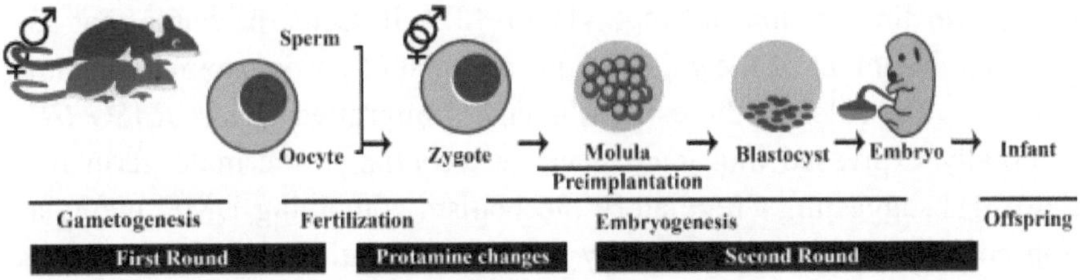

FIGURE 2. Establishment and maintenance of imprinted genes (epigenetic regulation) during mammalian gametogenesis and development. Sex specific establishment of DNA methylation of imprinted genes occurs during gametogenesis. Just after fertilization, protamine changes occur and follow by the second round of reprogramming begins with embryonic preimplantation. After fertilization, active and passive demethylations happen in parental specific manner. de novo methylation happens significantly during both rounds (for review see [61]).

DNA methylation is a sex bias phenomenon. As opposed to the male mouse embryonic germ cells, the female is not that much prone to methylate *IGF2 receptor*, *IDF2* and *H19* [82,232–234]. The same result has been illustrated during the blastocyst stage. It has been shown that in bovine, there is a significant tendency for methylation in the male in comparison to that of the female [81]. Piwi proteins (mili and miwi2) are expressed only in germ line, which are responsible for the establishment of *de novo* DNA methylation in transposons, and it is shown that PiRNAs directs DNA methylation in the male mouse germ cells through which the transposon is silenced [235–237]. In the other words, Piwi/PiRNA complex appears to guide the *de novo* methylation at transposons [235,237] and deactivate transposons within a germline [238]. PiRNAs and siRNAs such as the one located within *AU76*, a pseudogene of *RAN-GAP1*, negatively regulates transposons through their *cis*-acting function in mouse oocytes as well as establishes the methylation of retrotransposons in the male mouse germ line [235,237,239].

Somatic environment of the male/female germ line shows their influence in DNA methylation of imprints [82]. Using sex-reversed mice to evaluate sex-specific methylation pattern *in vivo*, the germ cells are found to be responsible for female/male imprints during oogenesis/spermatogenesis, though sex chromosome constitution shows significant influence on male germ line for imprint methylation [82]. It seems probable that somatic environment of the genital ridge and that of chromosomal constitution have key roles in the establishment of imprinted genes. *RASGRF1*, paternally expressed imprinted gene, is essential in the male germ line [240,241] suggesting a regulatory mechanism containing DMD methylation and the repeat sequences, by which methylation of the germ line is established [79].

EPIGENETIC REGULATION DURING GESTATION

Normal fetal development is dependent on proper development of embryo and placenta. These developments are modulated through epigenetic signals during gestation. Although these molecular signals are controlled by the same epigenetic mechanisms, their regulation is independent of each other and may follow different patterns during embryogenesis in comparison to placental development. During pregnancy, most monoallelic expressed genes carry out in extraembryonic tissues, such as trophoblast and yolk sac, regulate the development and function of placenta [44,201]. In placenta, this regulation seems to be directed by histone modification and ncRNAs through DNA methylation [201]. Embryo-placental development is a complex modulating phenomenon through which imprints undergo necessary maintenance, establishment and/or reestablishment. Placental development is under the control of *IGF2* and its degradation receptor, *IGF2r* [242]. *IGF2*, paternally expressed imprint, codes for embryo-placental growth factors. However, its receptor seems an unorthodox, imprinted gene [243].

Embryogenesis involves global methylation to erase and remethylate the methylation pattern. During early embryogenesis, methylation re-establishment occurs mostly within CpG islands and the imprints regulate in a sex-specific manner based on the new gender [79,244,245]. The demethylation mostly happens during primordial germ cells (PGCs) migration towards the genital ridge [79,246,247]. It is hypothesized that histone replacement and chromatin changes, using DNA repair mechanisms, are in accordance with the epigenetic reprogramming of PGC [61]. Evidences for such associations come from the chromatin modification markers, for instance *H3K9me2/3*, *H3K27me3*, *H3K4me2/3*, *H3K9ac*, *NAP-1* and *HIRA*, during early embryogenesis [248].

Just after fertilization, pre-implantation phase, promatines replace with histones and some level of histone modifications occur. Active demethylation of paternal pronucleus of the zygote starts and follows with passive demethylation during the cleavage states. Re-activation of the inactive X chromosome inactivation is the last significant change of the female embryo during pre-implantation development (for review see [43]). Chromatin modifications during germ line development begin with demethylation of imprinted genes in primordial germ cells, in a sex-dependent manner [246]. Then, during female gametogenesis, this modification proceeds to form primary oocytes, and follows to reestablishment of maternal-specific methylation pattern during growth and maturation of oocyte [123]. In male gametogenesis, a number of modification factors are involved. During spermatogenesis, histones undergo hypoacetylation. Especially, DNMTs are significantly important in the alteration of Leptotene to Pachytene. In this transformation, DNA methylation, histone methylation and histone deacetylation are counterparts [249]. The last modification to produce a mature sperm is the promatine formation (for review see [53]).

EPIGENETIC REGULATION DURING EMBRYOGENESIS

Early embryonic mouse shows high level of DNA methylation and expression of imprinted genes. The epigenetic pattern is maintained in somatic cells but erased in the PGC about 11. 5–12. 5 embryonic day [61]. At this time, the expression of the imprinted genes in PGC is biallelic and reestablishes during prenatal (in a male embryo) and postnatal stages (in a female embryo) [61,246,250]. Mammalian promoters enriched in H3 K27 trimethylation [251] and H3 K4 trimethylation [252] are mostly occupied by polycomb group (PcG). Besides, PcG proteins influence the pluripotency of embryonic stem cell [253–257]. Study on mouse embryos revealed the regulatory mechanism for imprints in which DNA configuration is the key silencing factor. An imprinted gene, *KCNQ1*, is paternally repressed by ncRNA, *KCNQLOT1*, in association with PcG proteins (EZH2 and Rnf2) at *Cdkn1c*, *Cd81* and *Tssc4 cis* genes [167]. An *in vitro* study on the expression of imprinted genes, *H19* and *SNRPN*, in male mouse [258], suggested a mostly intrinsic, sex-specific reestablishment of DNA methylation (after DNA demethylation) in the male germ line. However, there is still a probability that somatic cells at earlier stages may influence DNA methylation [61].

EPIGENETIC REGULATION AND PLACENTAL DEVELOPMENT

In mammals, imprinting has an important role in extraembryonic tissue. Their activation pattern seems to be tissue-specific as they are varied between embryonic imprints and placental imprints, for instance in mouse, placental imprints are mostly paternally repressed [201]. The authors suggested an evolutionary relation between placenta imprints and that of chromosome X repression. These findings propose that independent regulatory mechanisms are active in the embryo and in the placenta [259]. The regulatory mechanism suggested for imprints expression, is DNA methylation through histone modification and ncRNAs [201]. Research on a mouse model postulates that the regulation of the expression of imprints is not very firm in the trophoblast as it is in the embryo [260]. In fact, histone methylation maintains the silencing of the inactive allele of the imprints in mouse extraembryonic tissue, placenta. Further, the absence of histone methyltransferase *G9a* that has aberrant effects on placental imprinted cluster, *Kcnq1*. Retrotransposon-derived *Peg 11* (paternally expressed 11) or *Rtl1* (Retrotransposon-like q) is a paternally expressed imprinted gene responsible for the maintenance of placental development especially in fetal capillary development [261,262]. In placenta, repressive histone modification seems more crucial for the maintenance of imprinted genes [201,260]. Moreover, ethanol-induced growth inhibitory effects on the methylation of paternal allele *H19*, suggests CCCTC-binding factor site as a epigenetic switch in placenta [263]. A recent study on methylation status of placental *PEG10* emphasizes the importance of normal methylation of placental imprints for normal development of SCNT in cloned cattle. The placental *PEG10* shows a similar methylation pattern in the cloned calves, which survived and were healthy, in comparison with normal calves. Further, the cloned calves that died because of developmental failure, showed hypermethylation in *PEG10* [264]. The importance of epigenetic regulation for proper placental development is obvious. Thus, unsurprisingly, SCNT paves the way to have a better understanding of placental development and molecular signals that modulate epigenetic regulation.

TRANSCRIPTIONAL REGULATION BY POLYCOMB PROTEIN

DNA methylation and PcG proteins are two main silencing epigenetic pathways that are in accordance with each other [265]. PcG proteins are epigenetic regulatory proteins combined in numerous protein complexes as well as individual PcG proteins and usually interact with histones. They are classified as polycomb repressive complexes 1 and 2 classes. EED is a PcG, which can modify histone and change chromatin structure. EZH2, a PcG protein, is a regulatory element for the methylation of CpG in which the protein is in direct contact with DNMTs. PcG proteins play a critical role in epigenetic regulation such as in higher organisms X-chromosome inactivation, imprinting regulation and restoring to pluripotent status [266,267]. Their epigenetic role is more in maintaining chromatin structure as well as reestablishment of transcriptional regulation (for review see [268]), especially during differentiation and development [255,269]. PcG are responsible for maintenance of repression of specific developmental genes [180]. CTCF is an essential factor in insulator's function that regulates transcription in mammals [270]. It contains 11 zinc-finger DNA-binding protein [271] with versatile functions [272,273] as well as a transcriptional activator [274] and a repressor [271,275] (see review [276] for more information). Studies show epigenetic activities of CTCF in regulation of imprints [277] and X chromosome inactivation [148,278,279]. However, post-fertilization methylation of the *H19* ICR in a transgenic mouse model shows the necessity of the CTCF binding sites for the maintenance of the imprint pattern after implantation but not during pre-implantation phase [280]. A recent study on *H19* ICR in SCNT bovine embryos reconfirms that significant demethylation of the gene prevents successful implantation of the embryo [226]. These investigators showed that the CTCF binding sites of the paternal allele are mostly unmethylated, and coincided with the expression of the *H19*, though during postimplantation period the methylation pattern and the expression profile of the gene was similar to control. Transcriptional regulation of genes is also associated with chromatin modification enzymes, such as *HDAC1* [265], *G9a* [281], which are associated with DNMTs [282].

CONCLUSION

Creating live and healthy offspring through SCNT technique is only partially explained through epigenetic modifications. Clearly, somatic nuclei need to appropriately reprogram to the pluripotent state from which embryogenesis embarks. Most of the epigenetic modifications are probably mediated by DNA methylation and histone modifications. The epigenetic modifications reviewed here might explain some of the transcriptional regulatory mechanisms in SCNT reprogramming, which could influence embryonic gene expressions and might also affect the placental development during gestation. In contrast to genetic alterations, most epigenetic modifications are reversible, and the modulation of such modifications by reprogramming pluripotent genes in an embryo and placenta increase the amount of successes in animal cloning. In this review, we have highlighted the important and possible epigenetic modifications that probably influence the efficiency of animal cloning. Proper regulation of these could further influence the life span of the cloned livestock via the epigenetic modulation of somatic gene expression. In addition to unraveling the mechanisms that have been described in the past decade, several other mechanisms would require additional careful investigation. As discussed, somatic profiles of DNA methylation, histone modifications and chromatin configuration must be erased and reprogrammed in a precise manner in terms of both timing and location. Embryo-specific marks must be acquired in cloned embryos, similar to its natural counterpart. For proper acquisition epigenetic marks take place during embryogenesis, and this is why understanding of the epigenetic modification through gametogenesis up to fertilization is crucial. Inappropriate modification and reprogramming would affect the embryo-placental development and, consequently, could lead to failure during gestation, abnormalities and syndromes. This review is intended to emphasize the importance of understanding nuclear reprogramming for proper SCNT and the importance of DNA methylation and chromatin modification in nuclear reprogramming. Failures in the reprogramming will influence normal development of embryo and placenta and cause several abnormalities. All efforts to illuminate the complexity of epigenetic reprogramming that produces healthy cloned offsprings are necessary in order to have a better insight into the interaction between genomics and epigenomics.

REFERENCES

1. Bonasio R, Tu S, Reinberg, D. Molecular signals of epigenetic states. *Science* **2010**, 330, 612–616.
2. Mikkelsen TS, Hanna J, Zhang X, Ku M, Wernig M, Schorderet P, Bernstein BE, Jaenisch R, Lander ES, Meissner A. Dissecting direct reprogramming through integrative genomic analysis. *Nature* **2008**, 454, 49–55.
3. Yang, X, Smith, S. L, Tian, X. C, Lewin, H. A, Renard, J. -P, Wakayama, T. Nuclear reprogramming of cloned embryos and its implications for therapeutic cloning. *Nat Genet* **2007**, 39, 295–302.
4. Rideout, W. M, Eggan, K, Jaenisch, R. Nuclear cloning and epigenetic reprogramming of the genome. *Science* **2001**, 293, 1093–1098.
5. Keefer, C. L. Lessons learned from nuclear transfer (cloning). Theriogenology **2008**, 69, 48–54.
6. Yamanaka, S, Blau, H. M. Nuclear reprogramming to a pluripotent state by three approaches. *Nature* **2010**, 465, 704–712.
7. Takahashi, K, Yamanaka, S. Induction of pluripotent stem cells from mouse embryonic and adult fibroblast cultures by defined factors. *Cell* **2006**, 126, 663–676.

8. Pells, S, Di Domenico, A. I, Gallagher, E. J, McWhir, J. Multipotentiality of neuronal cells after spontaneous fusion with embryonic stem cells and nuclear reprogramming in vitro. *Cloning Stem Cells* **2002**, 4, 331–338.
9. Tada, M, Takahama, Y, Abe, K, Nakatsuji, N, Tada, T. Nuclear reprogramming of somatic cells by in vitro hybridization with ES cells. *Curr Biol* **2001**, 11, 1553–1558.
10. Takahashi, K, Tanabe, K, Ohnuki, M, Narita, M, Ichisaka, T, Tomoda, K, Yamanaka, S. Induction of pluripotent stem cells from adult human fibroblasts by defined factors. *Cell* **2007**, 131, 861–872.
11. Yu, J. Y, Vodyanik, M. A, Smuga-Otto, K, Antosiewicz-Bourget, J, Frane, J. L, Tian, S, Nie, J, Jonsdottir, G. A, Ruotti, V, Stewart, R, et al. Induced pluripotent stem cell lines derived from human somatic cells. *Science* **2007**, 318, 1917–1920.
12. Bhutani, N, Brady, J. J, Damian, M, Sacco, A, Corbel, S. Y, Blau, H. M. Reprogramming towards pluripotency requires AID-dependent DNA demethylation. *Nature* **2010**, 463, 1042–1047.
13. Morgan, H. D, Dean, W, Coker, H. A, Reik, W, Petersen-Mahrt, S. K. Activation-induced cytidine deaminase deaminates 5-methylcytosine in DNA and is expressed in pluripotent tissues—Implications for epigenetic reprogramming. *J Biol Chem* **2004**, 279, 52353–52360.
14. Constant, F, Guillomot, M, Heyman, Y, Vignon, X, Laigre, P, Servely, J. L, Renard, J. P, Chavatte-Palmer, P. Large offspring or large placenta syndrome? Morphometric analysis of late gestation bovine placentomes from somatic nuclear transfer pregnancies complicated by hydrallantois. *Biol Reprod* **2006**, 75, 122–130.
15. Tamashiro, K. L. K, Wakayama, T, Blanchard, R. J, Blanchard, D. C, Yanagimachi, R. Postnatal growth and behavioral development of mice cloned from adult cumulus cells. *Biol Reprod* **2000**, 63, 328–334.
16. Hiendleder, S, Wirtz, M, Mund, C, Klempt, M, Reichenbach, H. -D, Stojkovic, M, Weppert, M, Wenigerkind, H, Elmlinger, M, Lyko, F, et al. Tissue-specific effects of in vitro fertilization procedures on genomic cytosine methylation levels in overgrown and normal sized bovine fetuses. *Biol Reprod* **2006**, 75, 17–23.
17. Curchoe, C. L, Zhang, S, Yang, L, Page, R, Tian, X. C. Hypomethylation trends in the intergenic region of the imprinted IGF2 and H19 genes in cloned cattle. *Anim Reprod Sci* **2009**, 116, 213–225.
18. Paoloni-Giacobino, A. Implications of reproductive technologies for birth and developmental outcomes: Imprinting defects and beyond. *Expert Rev Mol Med* **2006**, 8, 1–14.
19. Smith, L, Suzuki, J, Jr, Goff, A, Filion, F, Therrien, J, Murphy, B, Kohan-Ghadr, H, Lefebvre, R, Brisville, A, Buczinski, S. Epigenetic anomalies associated with prenatal survival and neonatal morbidity in cloned calves. *Anim Reprod* **2010**, 7, 197–203.
20. Heyman, Y, Chavatte-Palmer, P, LeBourhis, D, Camous, S, Vignon, X, Renard, J. P. Frequency and occurrence of late-gestation losses from cattle cloned embryos. *Biol Reprod* **2002**, 66, 6–13.
21. Cheng, X, Hashimoto, H, Horton, J. R, Zhang, X. Mechanisms of DNA Methylation, Methyl-CpG Recognition, and Demethylation in Mammals. In *Handbook of Epigenetics*; Trygve, T, Ed, Academic Press: San Diego, CA, USA, 2011; pp. 9–627.

22. Dvir, A, Conaway, J. W, Conaway, R. C. Mechanism of transcription initiation and promoter escape by RNA polymerase II. *Curr Opin Genet Dev* **2001**, 11, 209–214.
23. Sandelin, A, Carninci, P, Lenhard, B, Ponjavic, J, Hayashizaki, Y, Hume, D. A. Mammalian RNA polymerase II core promoters: Insights from genome-wide studies. *Nat Rev Genet* **2007**, 8, 424–436.
24. Tran, K, Gralla, J. D. Control of the timing of promoter escape and RNA catalysis by the transcription factor IIB fingertip. *J Biol Chem* **2008**, 283, 15665–15671.
25. Malik, S, Roeder, R. G. Dynamic regulation of pol II transcription by the mammalian Mediator complex. *Trends Biochem. Sci* **2005**, 30, 256–263.
26. Hoffmann, A, Natoli, G, Ghosh, G. Transcriptional regulation via the NF-κB signaling module. *Oncogene* **2006**, 25, 6706–6716.
27. Carrera, I, Treisman, J. E. Message in a nucleus: Signaling to the transcriptional machinery. *Curr. Opin. Genet. Dev* **2008**, 18, 397–403.
28. Li, B, Carey, M, Workman, J. L. The role of chromatin during transcription. *Cell* **2007**, 128, 707–719.
29. Berger, S. L. The complex language of chromatin regulation during transcription. *Nature* **2007**, 447, 407–412.
30. Taft, R. J, Pang, K. C, Mercer, T. R, Dinger, M, Mattick, J. S. Non-coding RNAs: Regulators of disease. *J. Pathol* **2010**, 220, 126–139.
31. Bonasio, R, Tu, S. J, Reinberg, D. Molecular signals of epigenetic states. *Science* **2010**, 330, 612–616.
32. Chotalia, M, Smallwood, S. A, Ruf, N, Dawson, C, Lucifero, D, Frontera, M, James, K, Dean, W, Kelsey, G. Transcription is required for establishment of germline methylation marks at imprinted genes. *Genes Dev* **2009**, 23, 105–117.
33. Schmitt, S, Prestel, M, Paro, R. Intergenic transcription through a polycomb group response element counteracts silencing. *Genes Dev* **2005**, 19, 697–708.
34. Cedar, H, Bergman, Y. Linking DNA methylation and histone modification: Patterns and paradigms. *Nat. Rev. Genet* **2009**, 10, 295–304.
35. Hanley, B, Dijane, J, Fewtrell, M, Grynberg, A, Hummel, S, Junien, C, Koletzko, B, Lewis, S, Renz, H, Symonds, M, et al. Metabolic imprinting, programming and epigenetics—A review of present priorities and future opportunities. *Br. J. Nutr* **2010**, 104, S1–S25.
36. Hore, T. A, Rapkins, R. W, Graves, J. A. M. Construction and evolution of imprinted loci in mammals. *Trends Genet* **2007**, 23, 440–448.
37. Yen, Z. C, Meyer, I. M, Karalic, S, Brown, C. J. A cross-species comparison of X-chromosome inactivation in Eutheria. *Genomics* **2007**, 90, 453–463.
38. Miranda, T. B, Jones, P. A. DNA methylation: The nuts and bolts of repression. *J. Cell. Physiol* **2007**, 213, 384–390.
39. Lande-Diner, L, Zhang, J, Ben-Porath, I, Amariglio, N, Keshet, I, Hecht, M, Azuara, V, Fisher, A. G, Rechavi, G, Cedar, H. Role of DNA methylation in stable gene repression. *J Biol Chem* **2007**, 282, 12194–12200.
40. Shi, L. J, Wu, J. Epigenetic regulation in mammalian preimplantation embryo development. *Reprod. Biol. Endocrinol* **2009**, 7, doi:10. 1186/1477-7827-7-59.
41. Wang, J. L, Zhang, M, Zhang, Y, Kou, Z. H, Han, Z. M, Chen, D. Y, Sun, Q. Y, Gao, S. R. The histone demethylase JMJD2C is stage-specifically expressed in

preimplantation mouse embryos and is required for embryonic development. *Biol Reprod* **2010**, 82, 105–111.
42. Badr, H, Bongioni, G, Abdoon, A. S. S, Kandil, O, Puglisi, R. Gene expression in the in vitro-produced preimplantation bovine embryos. *Zygote* **2007**, 15, 355–367.
43. Hajkova, P. Epigenetic reprogramming—Taking a lesson from the embryo. *Curr. Opin. Cell Biol* **2010**, 22, 342–350.
44. Coan, P. M, Burton, G. J, Ferguson-Smith, A. C. Imprinted genes in the placenta—A review. *Placenta* **2005**, 26, S10–S20.
45. Fowden, A. L, Coan, P. M, Angiolini, E, Burton, G. J, Constancia, M. Imprinted genes and the epigenetic regulation of placental phenotype. *Prog. Biophys. Mol. Biol* **2011,** 106, 281–288.
46. Ng, H. K, Novakovic, B, Hiendleder, S, Craig, J. M, Roberts, C. T, Saffery, R. Distinct patterns of gene-specific methylation in mammalian placentas: Implications for placental evolution and function. *Placenta* **2010**, 31, 259–268.
47. Schär, P, Fritsch, O. DNA Repair and the Control of DNA Methylation. In Epigenetics and Disease, Gasser, S. M, Li, E, Eds, Springer: Basel, Switzerland, 2011, Volume 67, pp. 51–68.
48. Gronbaek, K, Hother, C, Jones, P. A. Epigenetic changes in cancer. *APMIS* **2007**, 115, 1039–1059.
49. Baumann, C, Daly, C. M, McDonnell, S. M, Viveiros, M. M, de la Fuente, R. Chromatin configuration and epigenetic landscape at the sex chromosome bivalent during equine spermatogenesis. *Chromosoma* **2011**, 120, 227–244.
50. Ho, L, Crabtree, G. R. Chromatin remodelling during development. *Nature* **2010**, 463, 474–484.
51. Lucia, P, Fanti, L, Negri, R, Del Vescovo, V, Fatica, A, Pimpinelli, S. The Heterochromatin Protein 1 positively regulates euchromatic gene expression by RNA binding. Aviable online: http://hdl. handle. net/10101/npre. **2008**. 2687. 1 (accessed on 27 July 2011).
52. Girton, J. R, Johansen, K. M. Chromatin Structure and the Regulation of Gene Expression: The Lessons of PEV in Drosophila. In *Advances in Genetics*, van Veronica, H, Robert, E. H, Eds, Academic Press: San Diego, CA, USA, **2008**, Volume 61, pp. 1–43.
53. Li, E. Chromatin modification and epigenetic reprogramming in mammalian development. *Nat. Rev. Genet* **2002**, 3, 662–673.
54. Simon, J. A, Kingston, R. E. Mechanisms of Polycomb gene silencing: Knowns and unknowns. *Nat. Rev. Mol. Cell Biol* **2009**, 10, 697–708.
55. Girod, P. -A, Nguyen, D. -Q, Calabrese, D, Puttini, S, Grandjean, M, Martinet, D, Regamey, A, Saugy, D, Beckmann, J. S, Bucher, P, et al. Genome-wide prediction of matrix attachment regions that increase gene expression in mammalian cells. *Nat. Methods* **2007**, 4, 747–753.
56. Shiota, K. DNA methylation profiles of CpG islands for cellular differentiation and development in mammals. *Cytogenet. Genome Res* **2004**, 105, 325–334.
57. Kim, J. K, Samaranayake, M, Pradhan, S. Epigenetic mechanisms in mammals. *Cell. Mol. Life Sci* **2009**, 66, 596–612.
58. Vaissiere, T, Sawan, C, Herceg, Z. Epigenetic interplay between histone modifications and DNA methylation in gene silencing. *Mutat. Res* **2008**, 659, 40–48.

59. Hanna, J. H, Saha, K, Jaenisch, R. Pluripotency and cellular reprogramming: Facts, hypotheses, unresolved issues. *Cell* **2010**, 143, 508–525.
60. Reik, W, Dean, W, Walter, J. Epigenetic reprogramming in mammalian development. *Science* **2001**, 293, 1089–1093.
61. Weaver, J. R, Susiarjo, M, Bartolomei, M. S. Imprinting and epigenetic changes in the early embryo. *Mamm. Genome* **2009**, 20, 532–543.
62. Ooi, S. K. T, O'Donnell, A. H, Bestor, T. H. Mammalian cytosine methylation at a glance. *J. Cell Sci* **2009**, 122, 2787–2791.
63. Walsh, C. P, Bestor, T. H. Cytosine methylation and mammalian development. *Genes Dev* **1999**, 13, 26–34.
64. Gopalakrishnan, S, van Emburgh, B. O, Robertson, K. D. DNA methylation in development and human disease. *Mutat. Res* **2008**, 647, 30–38.
65. Chang, H, Zhang, T, Zhang, Z, Bao, R, Fu, C, Wang, Z, Bao, Y, Li, Y, Wu, L, Zheng, X, et al. Tissue-specific distribution of aberrant DNA methylation associated with maternal low-folate status in human neural tube defects. *J. Nutr. Biochem* 2011.
66. Howlett, S. K, Reik, W. Methylation levels of maternal and paternal genomes during preimplantation development. *Development* 1991, 113, 119–127.
67. Monk, M, Boubelik, M, Lehnert, S. Temporal and regional changes in dna methylation in the embryonic, extraembryonic and germ-cell lineages during mouse embryo development. Development 1987, 99, 371–382.
68. Gehring, M, Reik, W, Henikoff, S. DNA demethylation by DNA repair. *Trends Genet* **2009**, 25, 82–90.
69. Mayer, W, Niveleau, A, Walter, J, Fundele, R, Haaf, T. Embryogenesis: Demethylation of the zygotic paternal genome. *Nature* **2000**, 403, 501–502.
70. Oswald, J, Engemann, S, Lane, N, Mayer, W, Olek, A, Fundele, R, Dean, W, Reik, W, Walter, J. Active demethylation of the paternal genome in the mouse zygote. *Curr. Biol* **2000**, 10, 475–478.
71. Doherty, A. S, Mann, M. R. W, Tremblay, K. D, Bartolomei, M. S, Schultz, R. M. Differential effects of culture on imprinted H19 expression in the preimplantation mouse embryo. *Biol Reprod* **2000**, 62, 1526–1535.
72. Mann, M. R. W, Chung, Y. G, Nolen, L. D, Verona, R. I, Latham, K. E, Bartolomei, M. S. Disruption of imprinted gene methylation and expression in cloned preimplantation stage mouse embryos. *Biol Reprod* **2003**, 69, 902–914.
73. Kang, Y. K, Lee, K. K, Han, Y. M. Reprogramming DNA methylation in the preimplantation stage: Peeping with Dolly's eyes. *Curr. Opin. Cell Biol* **2003**, 15, 290–295.
74. Jones, P. A, Takai, D. The role of DNA methylation in mammalian epigenetics. *Science* **2001**, 293, 1068–1070.
75. Ferguson-Smith, A. C, Surani, M. A. Imprinting and the epigenetic asymmetry between parental genomes. *Science* **2001**, 293, 1086–1089.
76. Mayer, W, Niveleau, A, Walter, J, Fundele, R, Haaf, T. Embryogenesis—Demethylation of the zygotic paternal genome. *Nature* **2000**, 403, 501–502.
77. Tremblay, K. D, Saam, J. R, Ingram, R. S, Tilghman, S. M, Bartolomei, M. S. A paternal-specific methylation imprint marks the alleles of the mouse H19 gene. *Nat Genet* 1995, 9, 407–413.

78. Lister, R, Pelizzola, M, Dowen, R. H, Hawkins, R. D, Hon, G, Tonti-Filippini, J, Nery, J. R, Lee, L, Ye, Z, Ngo, Q. M, et al. Human DNA methylomes at base resolution show widespread epigenomic differences. *Nature* **2009**, 462, 315–322.
79. Holmes, R, Soloway, P. D. Regulation of imprinted DNA methylation. *Cytogenet. Genome Res* **2006**, 113, 122–129.
80. Santos, F, Hendrich, B, Reik, W, Dean, W. Dynamic reprogramming of DNA methylation in the early mouse embryo. *Dev. Biol* **2002**, 241, 172–182.
81. Gebert, C, Wrenzycki, C, Herrmann, D, Groger, D, Thiel, J, Reinhardt, R, Lehrach, H, Hajkova, P, Lucas-Hahn, A, Carnwath, J. W, et al. DNA methylation in the IGF2 intragenic DMR is re-established in a sex-specific manner in bovine blastocysts after somatic cloning. *Genomics* **2009**, 94, 63–69.
82. Durcova-Hills, G, Hajkova, P, Sullivan, S, Barton, S, Surani, M. A, McLaren, A. Influence of sex chromosome constitution on the genomic imprinting of germ cells. *Proc. Natl. Acad. Sci.* USA **2006**, 103, 11184–11188.
83. Horsthemke, B. Genomic imprinting and imprinting defects. Med. Genet. **2010**, 22, 385–391.
84. Hou, J, Cui, X. H, Lei, T. H, Liu, L, An, X. R, Chen, Y. F. Aberrant DNA methylation patterns in cultured mouse embryos. *Prog. Nat. Sci.* **2005**, 15, 1079–1083.
85. Beaujean, N, Taylor, J, Gardner, J, Wilmut, I, Meehan, R, Young, L. Effect of limited DNA methylation reprogramming in the normal sheep embryo on somatic cell nuclear transfer. *Biol Reprod* **2004**, 71, 185–193.
86. Wei, Y, Zhu, J, Huan, Y, Liu, Z, Yang, C, Zhang, X, Mu, Y, Xia, P, Liu, Z. Aberrant expression and methylation status of putatively imprinted genes in placenta of cloned piglets. *Cell. Reprogram.* **2010**, 12, 213–222.
87. Bourque, D. K, Avila, L, Penaherrera, M, von Dadelszen, P, Robinson, W. P. Decreased placental methylation at the H19/IGF2 imprinting control region is associated with normotensive intrauterine growth restriction but not preeclampsia. *Placenta* **2010**, 31, 197–202.
88. Balassiano, K, Lima, S, Jenab, M, Overvad, K, Tjonneland, A, Boutron-Ruault, M. C, Clavel-Chapelon, F, Canzian, F, Kaaks, R, Boeing, H, et al. Aberrant DNA methylation of cancer-associated genes in gastric cancer in the european prospective investigation into cancer and nutrition (EPIC-EURGAST). *Cancer Lett.* 2011, 311, 85–95.
89. Chung, J. -H, Lee, H. J, Kim, B. -h, Cho, N. -Y, Kang, G. H. DNA methylation profile during multistage progression of pulmonary adenocarcinomas. *Virchows Arch.* 2011, 459, 201–211.
90. Estecio, M. R. H, Issa, J. -P. J. Dissecting DNA hypermethylation in cancer. *FEBS Lett.* 2011, 585, 2078–2086.
91. Tada, Y, Yokomizo, A, Shiota, M, Tsunoda, T, Plass, C, Naito, S. Aberrant DNA methylation of T-cell leukemia, homeobox 3 modulates cisplatin sensitivity in bladder cancer. *Int. J. Oncol.* 2011, 39, 727–733.
92. Shames, D. S, Minna, J. D, Gazdar, A. F. DNA methylation in health, disease, and cancer. *Curr. Mol. Med.* **2007**, 7, 85–102.
93. Acevedo, L. G, Sanz, A, Jelinek, M. A. Novel DNA binding domain-based assays for detection of methylated and nonmethylated DNA. *Epigenomics* 2011, 3, 93–101.
94. Laird, P. W. Principles and challenges of genome-wide DNA methylation analysis. *Nat. Rev. Genet.* **2010**, 11, 191–203.

95. Harris, R. A, Wang, T, Coarfa, C, Nagarajan, R. P, Hong, C. B, Downey, S. L, Johnson, B. E, Fouse, S. D, Delaney, A, Zhao, Y. J, et al. Comparison of sequencing-based methods to profile DNA methylation and identification of monoallelic epigenetic modifications. *Nat. Biotechnol.* **2010**, 28, 1097–1105.
96. Razin, A, Kantor, B. DNA Methylation in Epigenetic Control of Gene Expression. In *Epigenetics and Chromatin*, Jeanteur, P, Ed, Springer: Berlin, Germany, **2005**, Volume 38, pp. 151–167.
97. Singal, R, Ginder, G. D. DNA methylation. *Blood* **1999**, 93, 4059–4070.
98. Bestor, T, Laudano, A, Mattaliano, R, Ingram, V. Cloning and sequencing of a cDNA-encoding DNA methyltransferase of mouse cells: The carboxyl-terminal domain of the mammalian enzymes is related to bacterial restriction methyltransferases. *J. Mol. Biol.* 1988, 203, 971–983.
99. Chen, T. P, Li, E. Structure and Function of Eukaryotic DNA Methyltransferases. In *Stem Cells in Development and Disease,* Schatten, G. P, Ed, Academic Press: San Diego, CA, USA, **2004**, Volume 60, pp. 55–89.
100. Bestor, T. H. The DNA methyltransferases of mammals. *Hum. Mol. Genet.* **2000**, 9, 2395–2402.
101. Feng, J, Zhou, Y, Campbell, S. L, Le, T, Li, E, Sweatt, J. D, Silva, A. J, Fan, G. P. Dnmtl and Dnmt3a maintain DNA methylation and regulate synaptic function in adult forebrain neurons. *Nat. Neurosci.* **2010**, 13, 423–430.
102. Robert, M. F, Morin, S, Beaulieu, N, Gauthier, F, Chute, I. C, Barsalou, A, MacLeod, A. R. DNMT1 is required to maintain CpG methylation and aberrant gene silencing in human cancer cells. *Nat Genet* **2003**, 33, 61–65.
103. Chen, T, Li, E. Establishment and maintenance of DNA methylation patterns in mammals. *Curr. Top. Microbiol. Immunol.* **2006**, 301, 179–201.
104. Mortusewicz, O, Schermelleh, L, Walter, J, Cardoso, M. C, Leonhardt, H. Recruitment of DNA methyltransferase I to DNA repair sites. *Proc. Natl. Acad. Sci.* USA **2005**, 102, 8905–8909.
105. Grandjean, V, Yaman, R, Cuzin, F, Rassoulzadegan, M. Inheritance of an epigenetic mark: The CpG DNA methyltransferase 1 is required for de novo establishment of a complex pattern of non-CpG methylation. *PLoS One* **2007**, 2, doi:10. 1371/journal. pone. 0001136.
106. Schaefer, M, Lyko, F. Solving the Dnmt2 enigma. *Chromosoma* **2010**, 119, 35–40.
107. Chedin, F. The DNMT3 family of mammalian de novo DNA methyltransferases. In *Modifications of Nuclear DNA and Its Regulatory Proteins*, Cheng, X. D, Blumenthal, R. M, Eds, Academic Press: San Diego, CA, USA, 2011, Volume 101, pp. 255–285.
108. Okano, M, Bell, D. W, Haber, D. A, Li, E. DNA methyltransferases Dnmt3a and Dnmt3b are essential for de novo methylation and mammalian development. *Cell* **1999**, 99, 247–257.
109. Kaneda, M, Okano, M, Hata, K, Sado, T, Tsujimoto, N, Li, E, Sasaki, H. Essential role for de novo DNA methyltransferase Dnmt3a in paternal and maternal imprinting. *Nature* **2004**, 429, 900–903.
110. Bourc'his, D, Xu, G. L, Lin, C. S, Bollman, B, Bestor, T. H. Dnmt3L and the establishment of maternal genomic imprints. *Science* **2001**, 294, 2536–2539.

111. Chedin, F, Lieber, M. R, Hsieh, C. L. The DNA methyltransferase-like protein DNMT3L stimulates de novo methylation by Dnmt3a. *Proc. Natl. Acad. Sci. USA* **2002**, 99, 16916–16921.
112. Gowher, H, Liebert, K, Hermann, A, Xu, G. L, Jeltsch, A. Mechanism of stimulation of catalytic activity of Dnmt3A and Dnmt3B DNA-(cytosine-C5)-methyltransferases by Dnmt3L. *J Biol Chem* **2005**, 280, 13341–13348.
113. Webster, K. E, O'Bryan, M. K, Fletcher, S, Crewther, P. E, Aapola, U, Craig, J, Harrison, D. K, Aung, H, Phutikanit, N, Lyle, R, et al. Meiotic and epigenetic defects in Dnmt3L-knockout mouse spermatogenesis. *Proc. Natl. Acad. Sci.* USA **2005**, 102, 4068–4073.
114. Klose, R. J, Bird, A. P. Genomic DNA methylation: The mark and its mediators. *Trends Biochem.* Sci. **2006**, 31, 89–97.
115. Jurkowska, R. Z, Jurkowski, T. P, Jeltsch, A. Structure and function of mammalian DNA methyltransferases. *Chembiochem* 2011, 12, 206–222.
116. Sakai, Y, Suetake, I, Shinozaki, F, Yamashina, S, Tajima, S. Co-expression of de novo DNA methyltransferases Dnmt3a2 and Dnmt3L in gonocytes of mouse embryos. *Gene Expr. Patterns* **2004**, 5, 231–237. \
117. Suetake, I, Shinozaki, F, Miyagawa, J, Takeshima, H, Tajima, S. DNMT3L stimulates the DNA methylation activity of Dnmt3a and Dnmt3b through a direct interaction. *J Biol Chem* **2004**, 279, 27816–27823.
118. Yamanaka, K. I, Sakatani, M, Kubota, K, Balboula, A. Z, Sawai, K, Takahashi, M. Effects of downregulating DNA methyltransferase 1 transcript by RNA interference on DNA methylation status of the satellite I region and in vitro development of bovine somatic cell nuclear transfer embryos. *J. Reprod. Dev.* **2011,** 57, 393–402.
119. Metivier, R, Gallais, R, Tiffoche, C, Le Peron, C, Jurkowska, R. Z, Carmouche, R. P, Ibberson, D, Barath, P, Demay, F, Reid, G, et al. Cyclical DNA methylation of a transcriptionally active promoter. *Nature* **2008**, 452, 45–50.
120. Ooi, S. L, Henikoff, S. Germline histone dynamics and epigenetics. *Curr. Opin. Cell Biol.* **2007**, 19, 257–265.
121. Gopalakrishnan, S, Van Emburgh, B. O, Shan, J. X, Su, Z, Fields, C. R, Vieweg, J, Hamazaki, T, Schwartz, P. H, Terada, N, Robertson, K. D. A novel DNMT3B splice variant expressed in tumor and pluripotent cells modulates genomic DNA methylation patterns and displays altered DNA binding. *Mol. Cancer Res.* **2009**, 7, 1622–1634.
122. Garzon, R, Liu, S. J, Fabbri, M, Liu, Z. F, Heaphy, C. E. A, Callegari, E, Schwind, S, Pang, J. X, Yu, J. H, Muthusamy, N, et al. MicroRNA-29b induces global DNA hypomethylation and tumor suppressor gene reexpression in acute myeloid leukemia by targeting directly DNMT3A and 3B and indirectly DNMT1. *Blood* **2009**, 113, 6411–6418.
123. Hata, K, Okano, M, Lei, H, Li, E. Dnmt3L cooperates with the Dnmt3 family of de novo DNA methyltransferases to establish maternal imprints in mice. *Development* **2002**, 129, 1983–1993.
124. Feltus, F. A, Lee, E. K, Costello, J. F, Plass, C, Vertino, P. M. Predicting aberrant CpG island methylation. *Proc. Natl. Acad. Sci. USA* **2003**, 100, 12253–12258.

125. Jair, K. W, Bachman, K. E, Suzuki, H, Ting, A. H, Rhee, I, Yen, R. W. C, Baylin, S. B, Schuebel, K. E. De novo CpG island methylation in human cancer cells. *Cancer Res.* **2006**, 66, 682–692.
126. Jia, D, Jurkowska, R. Z, Zhang, X, Jeltsch, A, Cheng, X. D. Structure of Dnmt3a bound to Dnmt3L suggests a model for de novo DNA methylation. *Nature* **2007**, 449, 248–251.
127. Ooi, S. K. T, Qiu, C, Bernstein, E, Li, K. Q, Jia, D, Yang, Z, Erdjument-Bromage, H, Tempst, P, Lin, S. P, Allis, C. D, et al. DNMT3L connects unmethylated lysine 4 of histone H3 to de novo methylation of DNA. *Nature* **2007**, 448, 714–717.
128. Ferguson-Smith, A. C, Greally, J. M. Epigenetics: Perceptive enzymes. *Nature* **2007**, 449, 148–149.
129. Zhang, Y. Y, Rohde, C, Tierling, S, Jurkowski, T. P, Bock, C, Santacruz, D, Ragozin, S, Reinhardt, R, Groth, M, Walter, J, et al. DNA methylation analysis of chromosome 21 gene promoters at single base pair and single allele resolution. *PLoS Genet.* **2009**, 5, doi:10. 1371/ journal. pgen. 1000438.
130. Ooi, S. K. T, Bestor, T. H. The colorful history of active DNA demethylation. *Cell* **2008**, 133, 1145–1148.
131. Hattori, N, Imao, Y, Nishino, K, Ohgane, J, Yagi, S, Tanaka, S, Shiota, K. Epigenetic regulation of Nanog gene in embryonic stem and trophoblast stem cells. *Genes Cells* **2007**, 12, 387–396.
132. Simonsson, S, Gurdon, J. DNA demethylation is necessary for the epigenetic reprogramming of somatic cell nuclei. *Nat. Cell Biol.* **2004**, 6, 984–990.
133. La Salle, S, Mertineit, C, Taketo, T, Moens, P. B, Bestor, T. H, Trasler, J. M. Windows for sex-specific methylation marked by DNA methyltransferase expression profiles in mouse germ cells. *Dev. Biol.* **2004**, 268, 403–415.
134. Lees-Murdock, D. J, Shovlin, T. C, Gardiner, T, De Felici, M, Walsh, C. P. DNA methyltransferase expression in the mouse germ line during periods of de novo methylation. *Dev. Dyn.* **2005**, 232, 992–1002.
135. Davis, T. L, Yang, G. J, McCarrey, J. R, Bartolomei, M. S. The H19 methylation imprint is erased and re-established differentially on the parental alleles during male germ cell development. *Hum. Mol. Genet.* **2000**, 9, 2885–2894.
136. Fedoriw, A. M, Stein, P, Svoboda, P, Schultz, R. M, Bartolomei, M. S. Transgenic RNAi reveals essential function for CTCF in H19 gene imprinting. *Science* **2004**, 303, 238–240.
137. Jelinic, P, Stehle, J. C, Shaw, P. The testis-specific factor CTCFL cooperates with the protein methyltransferase PRMT7 in H19 imprinting control region methylation. *PLoS Biol.* **2006**, 4, e355.
138. Verona, R. I, Mann, M. R. W, Bartolomei, M. S. Genomic imprinting: Intricacies of epigenetic regulation in clusters. *Annu. Rev. Cell Dev. Biol.* **2003**, 19, 237–259.
139. Fitzpatrick, G. V, Soloway, P. D, Higgins, M. J. Regional loss of imprinting and growth deficiency in mice with a targeted deletion of KvDMR1. *Nat Genet* **2002**, 32, 426–431.
140. Mancini-DiNardo, D, Steele, S. J. S, Ingram, R. S, Tilghman, S. M. A differentially methylated region within the gene KCNQ1 functions as an imprinted promoter and silencer. *Hum. Mol. Genet.* **2003**, 12, 283–294.

141. Thorvaldsen, J. L, Duran, K. L, Bartolomei, M. S. Deletion of the H19 differentially methylated domain results in loss of imprinted expression of H19 and IGF2. *Genes Dev* **1998**, 12, 3693–3702.
142. Zhang, Y. J, Qu, L. H. Non-coding RNAs and the acquisition of genomic imprinting in mammals. *Sci. China C Life Sci.* **2009**, 52, 195–204.
143. Peters, J, Robson, J. E. Imprinted noncoding RNAs. *Mamm. Genome* **2008**, 19, 493–502.
144. Latos, P. A, Barlow, D. P. Regulation of imprinted expression by macro non-coding RNAs. *RNA Biol.* **2009**, 6, 100–106.
145. Bourc'his, D, Voinnet, O. A small-RNA perspective on gametogenesis, fertilization, and early zygotic development. *Science* **2010**, 330, 617–622.
146. Moazed, D. Small RNAs in transcriptional gene silencing and genome defence. *Nature* **2009**, 457, 413–420.
147. Rinn, J. L, Kertesz, M, Wang, J. K, Squazzo, S. L, Xu, X, Brugmann, S. A, Goodnough, L. H, Helms, J. A, Farnham, P. J, Segal, E, et al. Functional demarcation of active and silent chromatin domains in human HOX loci by Noncoding RNAs. *Cell* **2007**, 129, 1311–1323.
148. Koerner, M. V, Pauler, F. M, Huang, R, Barlow, D. P. The function of non-coding RNAs in genomic imprinting. *Development* **2009**, 136, 1771–1783.
149. Erhard, F, Zimmer, R. Classification of ncRNAs using position and size information in deep sequencing data. *Bioinformatics* **2010**, 26, i426–i432.
150. Childs, L, Nikoloski, Z, May, P, Walther, D. Identification and classification of ncRNA molecules using graph properties. *Nucleic Acids Res.* **2009**, 37, doi:10. 1093/nar/gkp206.
151. Wutz, A, Gribnau, J. X inactivation Xplained. *Curr. Opin. Genet. Dev.* **2007**, 17, 387–393.
152. Seitz, H, Royo, H, Lin, S. P, Youngson, N, Ferguson-Smith, A. C, Cavaille, J. Imprinted small RNA genes. *Biol. Chem.* **2004**, 385, 905–911.
153. Royo, H, Bortolin, M. L, Seitz, H, Cavaille, J. Small non-coding RNAs and genomic imprinting. Cytogenet. *Genome Res.* **2006**, 113, 99–108.
154. Wang, Y, Medvid, R, Melton, C, Jaenisch, R, Blelloch, R. DGCR8 is essential for microRNA biogenesis and silencing of embryonic stem cell self-renewal. *Nat Genet* **2007**, 39, 380–385.
155. Kanellopoulou, C, Muljo, S. A, Kung, A. L, Ganesan, S, Drapkin, R, Jenuwein, T, Livingston, D. M, Rajewsky, K. Dicer-deficient mouse embryonic stem cells are defective in differentiation and centromeric silencing. *Genes Dev* **2005**, 19, 489–501.
156. Stefani, G, Slack, F. J. Small non-coding RNAs in animal development. *Nat. Rev. Mol. Cell Biol.* **2008**, 9, 219–230.
157. Santoro, F, Barlow, D. P. Developmental control of imprinted expression by macro non-coding RNAs. *Semin. Cell Dev. Biol.* **2011**, 22, 328–335.
158. Martello, G, Zacchigna, L, Inui, M, Montagner, M, Adorno, M, Mamidi, A, Morsut, L, Soligo, S, Tran, U, Dupont, S, et al. MicroRNA control of nodal signalling. *Nature* **2007**, 449, 183–188.
159. Kwon, C, Han, Z, Olson, E. N, Srivastava, D. MicroRNA1 influences cardiac differentiation in Drosophila and regulates notch signaling. *Proc. Natl. Acad. Sci. USA* **2005**, 102, 18986–18991.

160. Deng, Z, Chen, J. -F, Wang, D. -Z. Transgenic overexpression of miR-133a in skeletal muscle. *BMC Musculoskelet. Disord.* **2011,** 12, doi:10. 1186/1471-2474-12-115.
161. Sempere, L. F, Freemantle, S, Pitha-Rowe, I, Moss, E, Dmitrovsky, E, Ambros, V. Expression profiling of mammalian microRNAs uncovers a subset of brain-expressed microRNAs with possible roles in murine and human neuronal differentiation. *Genome Biol.* **2004**, 5, R13–R23.
162. Vo, N. K, Cambronne, X. A, Goodman, R. H. MicroRNA pathways in neural development and plasticity. *Curr. Opin. Neurobiol.* **2010**, 20, 457–465.
163. Hudson, Q. J, Kulinski, T. M, Huetter, S. P, Barlow, D. P. Genomic imprinting mechanisms in embryonic and extraembryonic mouse tissues. *Heredity* **2010**, 105, 45–56.
164. Lewis, A, Mitsuyaj, K, Constancia, M, Reik, W. Tandem repeat hypothesis in imprinting: Deletion of a conserved direct repeat element upstream of H19 has no effect on imprinting in the IGF2-H19 region. *Mol. Cell. Biol.* **2004**, 24, 5650–5656.
165. Umlauf, D, Goto, Y, Cao, R, Cerqueira, F, Wagschal, A, Zhang, Y, Feil, R. Imprinting along the KCNQ1 domain on mouse chromosome 7 involves repressive histone methylation and recruitment of Polycomb group complexes. *Nat Genet* **2004**, 36, 1296–1300.
166. Pandey, R. R, Mondal, T, Mohammad, F, Enroth, S, Redrup, L, Komorowski, J, Nagano, T, Mancini-DiNardo, D, Kanduri, C. Antisense noncoding RNA mediates lineage-specific transcriptional silencing through chromatin-level regulation. *Mol. Cell* **2008**, 32, 232–246.
167. Terranova, R, Yokobayashi, S, Stadler, M. B, Otte, A. P, van Lohuizen, M, Orkin, S. H, Peters, A. H. F. M. Polycomb group proteins EZH2 and Rnf2 direct genomic contraction and imprinted repression in early mouse embryos. *Dev. Cell* **2008**, 15, 668–679.
168. Carmell, M. A, Hannon, G. J. RNase III enzymes and the initiation of gene silencing. *Nat. Struct. Mol. Biol.* **2004**, 11, 214–218.
169. Chu, C. -Y, Rana, T. M. Small RNAs: Regulators and guardians of the genome. *J. Cell. Physiol.* **2007**, 213, 412–419.
170. Filipowicz, W, Bhattacharyya, S. N, Sonenberg, N. Mechanisms of post-transcriptional regulation by microRNAs: Are the answers in sight? *Nat. Rev. Genet.* **2008**, 9, 102–114.
171. Kim, V. N, Han, J, Siomi, M. C. Biogenesis of small RNAs in animals. *Nat. Rev. Mol. Cell Biol.* **2009**, 10, 126–139.
172. Landgraf, P, Rusu, M, Sheridan, R, Sewer, A, Iovino, N, Aravin, A, Pfeffer, S, Rice, A, Kamphorst, A. O, Landthaler, M, et al. A mammalian microRNA expression atlas based on small RNA library sequencing. *Cell* **2007**, 129, 1401–1414.
173. Girard, A, Sachidanandam, R, Hannon, G. J, Carmell, M. A. A germline-specific class of small RNAs binds mammalian Piwi proteins. *Nature* **2006**, 442, 199–202.
174. Klattenhoff, C, Theurkauf, W. Biogenesis and germline functions of piRNAs. *Development* **2008**, 135, 3–9.
175. Szakmary, A, Cox, D. N, Wang, Z, Lin, H. F. Regulatory relationship among piwi, pumilio, and bag-of-marbles in Drosophila germline stem cell self-renewal and differentiation. *Curr. Biol.* **2005**, 15, 171–178.

176. Efroni, S, Duttagupta, R, Cheng, J, Dehghani, H, Hoeppner, D. J, Dash, C, Bazett-Jones, D. P, Le Grice, S, McKay, R. D. G, Buetow, K. H, et al. Global transcription in pluripotent embryonic stem cells. *Cell Stem Cell* **2008**, 2, 437–447.
177. Kimura, H, Tada, M, Nakatsuji, N, Tada, T. Histone code modifications on pluripotential nuclei of reprogrammed somatic cells. *Mol. Cell. Biol.* **2004**, 24, 5710–5720.
178. Kouzarides, T. Chromatin modifications and their function. *Cell* **2007**, 128, 693–705.
179. Kim, J. M, Liu, H. L, Tazaki, M, Nagata, M, Aoki, F. Changes in histone acetylation during mouse oocyte meiosis. *J. Cell Biol.* **2003**, 162, 37–46.
180. Kimmins, S, Sassone-Corsi, P. Chromatin remodelling and epigenetic features of germ cells. *Nature* **2005**, 434, 583–589.
181. Yamanaka, K, Sugimura, S, Wakai, T, Kawahara, M, Sato, E. Acetylation level of histone H3 in early embryonic stages affects subsequent development of miniature pig somatic cell nuclear transfer embryos. *J. Reprod. Dev.* **2009**, 55, 638–644.
182. Zhang, Y. H, Li, J, Villemoes, K, Pedersen, A. M, Purup, S, Vajta, G. An epigenetic modifier results in improved in vitro blastocyst production after somatic cell nuclear transfer. *Cloning Stem Cell*s **2007**, 9, 357–363.
183. Birney, E, Stamatoyannopoulos, J. A, Dutta, A, Guigo, R, Gingeras, T. R, Margulies, E. H, Weng, Z. P, Snyder, M, Dermitzakis, E. T, Thurman, R. E, et al. Identification and analysis of functional elements in 1% of the human genome by the ENCODE pilot project. *Nature* **2007**, 447, 799–816.
184. Tamaru, H, Selker, E. U. A histone H3 methyltransferase controls DNA methylation in Neurospora crassa. *Nature* **2001**, 414, 277–283.
185. Jackson, J. P, Lindroth, A. M, Cao, X. F, Jacobsen, S. E. Control of CpNpG DNA methylation by the KRYPTONITE histone H3 methyltransferase. *Nature* **2002**, 416, 556–560.
186. Fuks, F. DNA methylation and histone modifications: Teaming up to silence genes. *Curr. Opin. Genet. Dev.* **2005**, 15, 490–495.
187. Francis, N. J, Follmer, N. E, Simon, M. D, Aghia, G, Butler, J. D. Polycomb proteins remain bound to chromatin and DNA during DNA replication in vitro. *Cell* **2009**, 137, 110–122.
188. Blobel, G. A, Kadauke, S, Wang, E, Lau, A. W, Zuber, J, Chou, M. M, Vakoc, C. R. A reconfigured pattern of Mll occupancy within mitotic chromatin promotes rapid transcriptional reactivation following mitotic exit. *Mol. Cell* **2009**, 36, 970–983.
189. Hahn, M. A, Wu, X. W, Li, A. X, Hahn, T, Pfeifer, G. P. Relationship between gene body DNA methylation and intragenic H3K9me3 and H3K36me3 chromatin marks. *PLoS One* **2011,** 6, doi:10. 1371/journal. pone. 0018844.
190. Bartke, T, Vermeulen, M, Xhemalce, B, Robson, S. C, Mann, M, Kouzarides, T. Nucleosome-interacting proteins regulated by DNA and histone methylation. *Cell* **2010**, 143, 470–484.
191. Lindroth, A. M, Park, Y. J, McLean, C. M, Dokshin, G. A, Persson, J. M, Herman, H, Pasini, D, Miro, X, Donohoe, M. E, Lee, J. T, et al. Antagonism between DNA and H3K27 methylation at the imprinted RASGRF1 locus. *PLoS Genet.* **2008**, 4, doi:10. 1371/journal. pgen. 1000145.
192. Myant, K, Termanis, A, Sundaram, A. Y. M, Boe, T, Li, C, Merusi, C, Burrage, J, de Las Heras, J. I, Stancheva, I. LSH and G9a/GLP complex are required for developmentally programmed DNA methylation. *Genome Res.* **2011**, 21, 83–94.

193. Fournier, C, Goto, Y. J, Ballestar, E, Delaval, K, Hever, A. M, Esteller, M, Feil, R. Allele-specific histone lysine methylation marks regulatory regions at imprinted mouse genes. *EMBO J.* **2002**, 21, 6560–6570.
194. Rougeulle, C, Navarro, P, Avner, P. Promoter-restricted H3 Lys 4 di-methylation is an epigenetic mark for monoallelic expression. *Hum. Mol. Genet.* **2003**, 12, 3343–3348.
195. Appanah, R, Dickerson, D. R, Goyal, P, Groudine, M, Lorincz, M. C. An unmethylated 3′ promoter-proximal region is required for efficient transcription initiation. *PLoS Genet.* **2007**, 3, 241–253.
196. Weber, M, Hellmann, I, Stadler, M. B, Ramos, L, Paabo, S, Rebhan, M, Schubeler, D. Distribution, silencing potential and evolutionary impact of promoter DNA methylation in the human genome. *Nat Genet* **2007**, 39, 457–466.
197. O'Carroll, D, Erhardt, S, Pagani, M, Barton, S. C, Surani, M. A, Jenuwein, T. The Polycomb-group gene EZH2 is required for early mouse development. *Mol. Cell. Biol.* **2001**, 21, 4330–4336.
198. Tachibana, M, Sugimoto, K, Nozaki, M, Ueda, J, Ohta, T, Ohki, M, Fukuda, M, Takeda, N, Niida, H, Kato, H, et al. G9a histone methyltransferase plays a dominant role in euchromatic histone H3 lysine 9 methylation and is essential for early embryogenesis. *Genes Dev* **2002**, 16, 1779–1791.
199. Lagger, G, O'Carroll, D, Rembold, M, Khier, H, Tischler, J, Weitzer, G, Schuettengruber, B, Hauser, C, Brunmeir, R, Jenuwein, T, et al. Essential function of histone deacetylase 1 in proliferation control and CDK inhibitor repression. *EMBO J.* **2002**, 21, 2672–2681.
200. Tilghman, S. M. The sins of the fathers and mothers: Genomic imprinting in mammalian development. *Cell* **1999**, 96, 185–193.
201. Wagschal, A, Feil, R. Genomic imprinting in the placenta. *Cytogenet. Genome Res.* **2006**, 113, 90–98.
202. Davies, W, Smith, R. J, Kelsey, G, Wilkinson, L. S. Expression patterns of the novel imprinted genes Nap115 and Peg13 and their non-imprinted host genes in the adult mouse brain. *Gene Expr. Patterns* **2004**, 4, 741–747.
203. Smith, R. J, Dean, W, Konfortova, G, Kelsey, G. Identification of novel imprinted genes in a genome-wide screen for maternal methylation. *Genome Res.* **2003**, 13, 558–569.
204. Kagitani, F, Kuroiwa, Y, Wakana, S, Shiroishi, T, Miyoshi, N, Kobayashi, S, Nishida, M, Kohda, T, KanekoIshino, T, Ishino, F. Peg5/Neuronatin is an imprinted gene located on sub-distal chromosome 2 in the mouse. *Nucleic Acids Res.* **1997**, 25, 3428–3432.
205. Kikyo, N, Williamson, C. M, John, R. M, Barton, S. C, Beechey, C. V, Ball, S. T, Cattanach, B. M, Surani, M. A, Peters, J. Genetic and functional analysis of neuronatin in mice with maternal or paternal duplication of distal Chr 2. *Dev. Biol.* **1997**, 190, 66–77.
206. Choi, J. D, Underkoffler, L. A, Wood, A. J, Collins, J. N, Williams, P. T, Golden, J. A, Schuster, E. F, Loomes, K. M, Oakey, R. J. A novel variant of Inpp5f is imprinted in brain, and its expression is correlated with differential methylation of an internal CpG island. *Mol. Cell. Biol.* **2005**, 25, 5514–5522.
207. Peters, J, Beechey, C. Identification and characterisation of imprinted genes in the mouse. *Brief. Funct. Genomics Proteomics* **2004**, 2, 320–333.

208. Branco, M. R, Oda, M, Reik, W. Safeguarding parental identity: Dnmt1 maintains imprints during epigenetic reprogramming in early embryogenesis. *Genes Dev* **2008**, 22, 1567–1571.
209. Ikegami, K, Ohgane, J, Tanaka, S, Yagi, S, Shiota, K. Interplay between DNA methylation, histone modification and chromatin remodeling in stem cells and during development. *Int. J. Dev. Biol.* **2009**, 53, 203–214.
210. Royo, H, Cavaille, J. Non-coding RNAs in imprinted gene clusters. *Biol. Cell* **2008**, 100, 149–166.
211. Wutz, A, Smrzka, O. W, Schweifer, N, Schellander, K, Wagner, E. F, Barlow, D. P. Imprinted expression of the IGF2r gene depends on an intronic CpG island. *Nature* **1997**, 389, 745–749.
212. Birger, Y, Shemer, R, Perk, J, Razin, A. The imprinting box of the mouse IGF2r gene. *Nature* **1999**, 397, 84–88.
213. Kantor, B, Makedonski, K, Green-Finberg, Y, Shemer, R, Razin, A. Control elements within the PWS/AS imprinting box and their function in the imprinting process. *Hum. Mol. Genet.* **2004**, 13, 751–762.
214. Ideraabdullah, F. Y, Abramowitz, L. K, Thorvaldsen, J. L, Krapp, C, Wen, S. C, Engel, N, Bartolomei, M. S. Novel cis-regulatory function in ICR-mediated imprinted repression of H19. *Dev. Biol.* **2011**, 355, 349–357.
215. Edwards, C. A, Ferguson-Smith, A. C. Mechanisms regulating imprinted genes in clusters. *Curr. Opin. Cell Biol.* **2007**, 19, 281–289.
216. Brandeis, M, Kafri, T, Ariel, M, Chaillet, J. R, McCarrey, J, Razin, A, Cedar, H. The ontogeny of allele-specific methylation associated with imprinted genes in the mouse. *EMBO J.* **1993**, 12, 3669–3677.
217. Moore, T, Constancia, M, Zubair, M, Bailleul, B, Feil, R, Sasaki, H, Reik, W. Multiple imprinted sense and antisense transcripts, differential methylation and tandem repeats in a putative imprinting control region upstream of mouse IGF2. *Proc. Natl. Acad. Sci. USA* **1997**, 94, 12509–12514.
218. Feil, R, Walter, J, Allen, N. D, Reik, W. Developmental control of allelic methylation in the imprinted mouse IGF2 and H19 genes. *Development* **1994**, 120, 2933–2943.
219. Kacem, S, Feil, R. Chromatin mechanisms in genomic imprinting. *Mamm. Genome* **2009**, 20, 544–556.
220. Hirasawa, R, Chiba, H, Kaneda, M, Tajima, S, Li, E, Jaenisch, R, Sasaki, H. Maternal and zygotic Dnmt1 are necessary and sufficient for the maintenance of DNA methylation imprints during preimplantation development. *Genes Dev* **2008**, 22, 1607–1616.
221. Yang, L, Chavatte-Palmer, P, Kubota, C, O'Neill, M, Hoagland, T, Renard, J. P, Taneja, M, Yang, X. Z, Tian, X. C. Expression of imprinted genes is aberrant in deceased newborn cloned calves and relatively normal in surviving adult clones. *Mol. Reprod. Dev*. **2005**, 71, 431–438.
222. Zhang, S. Q, Kubota, C, Yang, L, Zhang, Y. Q, Page, R, O'Neill, M, Yang, X. Z, Tian, X. C. Genomic imprinting of H19 in naturally reproduced and cloned cattle. *Biol Reprod* **2004**, 71, 1540–1544.
223. Sato, A, Otsu, E, Negishi, H, Utsunomiya, T, Arima, T. Aberrant DNA methylation of imprinted loci in superovulated oocytes. *Hum. Reprod.* **2007**, 22, 26–35.

224. Market-Velker, B. A, Zhang, L. Y, Magri, L. S, Bonvissuto, A. C, Mann, M. R. W. Dual effects of superovulation: Loss of maternal and paternal imprinted methylation in a dose-dependent manner. *Hum. Mol. Genet.* **2010**, 19, 36–51.
225. Park, C. H, Kim, H. S, Lee, S. G, Lee, C. K. Methylation status of differentially methylated regions at IGF2/H19 locus in porcine gametes and preimplantation embryos. *Genomics* **2009**, 93, 179–186.
226. Suzuki, J, Therrien, J, Filion, F, Lefebvre, R, Goff, A. K, Perecin, F, Meirelles, F. V, Smith, L. C. Loss of methylation at H19 DMD is associated with biallelic expression and reduced development in cattle derived by somatic cell nuclear transfer. *Biol Reprod* 2011, 84, 947–956.
227. Morgan, H. D, Santos, F, Green, K, Dean, W, Reik, W. Epigenetic reprogramming in mammals. *Hum. Mol. Genet.* **2005**, 14, R47–R58.
228. Sassone-Corsi, P. Unique chromatin remodeling and transcriptional regulation in spermatogenesis. *Science* **2002**, 296, 2176–2178.
229. Pathak, S, Kedia-Mokashi, N, Saxena, M, D'Souza, R, Maitra, A, Parte, P, Gill-Sharma, M, Balasinor, N. Effect of tamoxifen treatment on global and insulin-like growth factor 2-H19 locus-specific DNA methylation in rat spermatozoa and its association with embryo loss. *Fertil. Steril.* **2009**, 91, 2253–2263.
230. Delaval, K, Feil, R. Epigenetic regulation of mammalian genomic imprinting. *Curr. Opin. Genet. Dev.* **2004**, 14, 188–195.
231. McLay, D. W, Clarke, H. J. Remodelling the paternal chromatin at fertilization in mammals. *Reproduction* **2003**, 125, 625–633.
232. Labosky, P. A, Barlow, D. P, Hogan, B. L. M. Mouse embryonic germ (EG) cell-lines: Transmission through the germline and differences in the methylation imprint of insulin-like growth-factor 2 receptor (IGF2r) gene compared with embryonic stem (ES) cell-lines. *Development* 1994, 120, 3197–3204.
233. Tada, T, Tada, M, Hilton, K, Barton, S. C, Sado, T, Takagi, N, Surani, M. A. Epigenotype switching of imprintable loci in embryonic germ cells. *Dev. Genes Evol.* **1998**, 207, 551–561.
234. Durcova-Hills, G, Burgoyne, P, McLaren, A. Analysis of sex differences in EGC imprinting. *Dev. Biol.* **2004**, 268, 105–110.
235. Aravin, A. A, Sachidanandam, R, Bourc'his, D, Schaefer, C, Pezic, D, Toth, K. F, Bestor, T, Hannon, G. J. A piRNA pathway primed by individual transposons is linked to de novo DNA methylation in mice. *Mol. Cell* **2008**, 31, 785–799.
236. Carmell, M. A, Girard, A, van de Kant, H. J. G, Bourc'his, D, Bestor, T. H, de Rooij, D. G, Hannon, G. J. MIWI2 is essential for spermatogenesis and repression of transposons in the mouse male germline. *Dev. Cell* **2007**, 12, 503–514.
237. Kuramochi-Miyagawa, S, Watanabe, T, Gotoh, K, Totoki, Y, Toyoda, A, Ikawa, M, Asada, N, Kojima, K, Yamaguchi, Y, Ijiri, T. W, et al. DNA methylation of retrotransposon genes is regulated by Piwi family members MILI and MIWI2 in murine fetal testes. *Genes Dev* **2008**, 22, 908–917.
238. Nicholas, C. R, Chavez, S. L, Baker, V. L, Pera, R. A. R. Instructing an embryonic stem cell-derived oocyte fate: Lessons from endogenous oogenesis. *Endocr. Rev.* **2009**, 30, 264–283.

239. Watanabe, T, Totoki, Y, Toyoda, A, Kaneda, M, Kuramochi-Miyagawa, S, Obata, Y, Chiba, H, Kohara, Y, Kono, T, Nakano, T, et al. Endogenous siRNAs from naturally formed dsRNAs regulate transcripts in mouse oocytes. *Nature* **2008**, 453, 539–543.
240. Yoon, B. J, Herman, H, Sikora, A, Smith, L. T, Plass, C, Soloway, P. D. Regulation of DNA methylation of RASGRF1. *Nat Genet* **2002**, 30, 92–96.
241. Holmes, R, Chang, Y. J, Soloway, P. D. Timing and sequence requirements defined for embryonic maintenance of imprinted DNA methylation at RASGRF1. *Mol. Cell. Biol.* **2006**, 26, 9564–9570.
242. Constancia, M, Hemberger, M, Hughes, J, Dean, W, Ferguson-Smith, A, Fundele, R, Stewart, F, Kelsey, G, Fowden, A, Sibley, C, et al. Placental-specific IGF-II is a major modulator of placental and fetal growth. *Nature* **2002**, 417, 945–948.
243. Kalscheuer, V. M, Mariman, E. C, Schepens, M. T, Rehder, H, Ropers, H. H. The insulin-like growth factor type-2 receptor gene is imprinted in the mouse but not in humans. *Nat Genet* 1993, 5, 74–78.
244. Gardner, D. K, Larman, M. G, Thouas, G. A. Sex-related physiology of the preimplantation embryo. *Mol. Hum. Reprod.* **2010**, 16, 539–547.
245. Biliya, S, Bulla, L. A. Genomic imprinting: The influence of differential methylation in the two sexes. *Exp. Biol. Med.* **2010**, 235, 139–147.
246. Hajkova, P, Erhardt, S, Lane, N, Haaf, T, El-Maarri, O, Reik, W, Walter, J, Surani, M. A. Epigenetic reprogramming in mouse primordial germ cells. *Mech. Dev.* **2002**, 117, 15–23.
247. Yamazaki, Y, Mann, M. R. W, Lee, S. S, Marh, J, McCarrey, J. R, Yanagimachi, R, Bartolomei, M. S. Reprogramming of primordial germ cells begins before migration into the genital ridge, making these cells inadequate donors for reproductive cloning. *Proc. Natl. Acad. Sci. USA* **2003**, 100, 12207–12212.
248. Hajkova, P, Ancelin, K, Waldmann, T, Lacoste, N, Lange, U. C, Cesari, F, Lee, C, Almouzni, G, Schneider, R, Surani, M. A. Chromatin dynamics during epigenetic reprogramming in the mouse germ line. *Nature* **2008**, 452, 877–881.
249. Hazzouri, M, Pivot-Pajot, C, Faure, A. K, Usson, Y, Pelletier, R, Sele, B, Khochbin, S, Rousseaux, S. Regulated hyperacetylation of core histones during mouse spermatogenesis: Involvement of histone-deacetylases. *Eur. J. Cell Biol.* **2000**, 79, 950–960.
250. Szabo, P. E, Mann, J. R. Biallelic expression of imprinted genes in the mouse germline: Implications for erasure, establishment, and mechanisms of genomic imprinting. *Genes Dev* **1995**, 9, 1857–1868.
251. Boyer, L. A, Plath, K, Zeitlinger, J, Brambrink, T, Medeiros, L. A, Lee, T. I, Levine, S. S, Wernig, M, Tajonar, A, Ray, M. K, et al. Polycomb complexes repress developmental regulators in murine embryonic stem cells. *Nature* **2006**, 441, 349–353.
252. Ku, M, Koche, R. P, Rheinbay, E, Mendenhall, E. M, Endoh, M, Mikkelsen, T. S, Presser, A, Nusbaum, C, Xie, X. H, Chi, A. S, et al. Genomewide analysis of PRC1 and PRC2 occupancy identifies two classes of bivalent domains. *PLoS Genet.* **2008**, 4, doi:10. 1371/journal. pgen. 1000242.
253. Kashyap, V, Rezende, N. C, Scotland, K. B, Shaffer, S. M, Persson, J. L, Gudas, L. J, Mongan, N. P. Regulation of stem cell pluripotency and differentiation involves a mutual regulatory circuit of the Nanog, OCT4, and SOX2 pluripotency transcription factors with polycomb repressive complexes and stem cell microRNAs. *Stem Cells Dev.* **2009**, 18, 1093–1108.

254. Atkinson, S, Armstrong, L. Epigenetics in embryonic stem cells: Regulation of pluripotency and differentiation. *Cell Tissue Res.* **2008**, 331, 23–29.
255. Pasini, D, Bracken, A. P, Hansen, J. B, Capillo, M, Helin, K. The polycomb group protein Suz12 is required for embryonic stem cell differentiation. *Mol. Cell. Biol.* **2007**, 27, 3769–3779.
256. Herranz, N, Pasini, D, Diaz, V. M, Franci, C, Gutierrez, A, Dave, N, Escriva, M, Hernandez-Munoz, I, di Croce, L, Helin, K, et al. Polycomb complex 2 is required for E-cadherin repression by the snail1 transcription factor. *Mol. Cell. Biol.* **2008**, 28, 4772–4781.
257. Yuzyuk, T, Fakhouri, T. H. I, Kiefer, J, Mango, S. E. The polycomb complex protein mes-2/E(z) promotes the transition from developmental plasticity to differentiation in *C. elegans* embryos. *Dev. Cell* **2009**, 16, 699–710.
258. Iwahashi, K, Yoshioka, H, Low, E. W, McCarrey, J. R, Yanagimachi, R, Yamazaki, Y. Autonomous regulation of sex-specific developmental programming in mouse fetal germ cells. *Biol Reprod* **2007**, 77, 697–706.
259. Feil, R, Berger, F. Convergent evolution of genomic imprinting in plants and mammals. *Trends Genet.* **2007**, 23, 192–199.
260. Wagschal, A, Sutherland, H. G, Woodfine, K, Henckel, A, Chebli, K, Schulz, R, Oakey, R. J, Bickmore, W. A, Feil, R. G9a histone methyltransferase contributes to imprinting in the mouse placenta. *Mol. Cell. Biol.* **2008**, 28, 1104–1113.
261. Ono, R, Kobayashi, S, Wagatsuma, H, Aisaka, K, Kohda, T, Kaneko-Ishino, T, Ishino, F. A retrotransposon-derived gene, PEG10, is a novel imprinted gene located on human chromosome 7q21. *Genomics* **2001**, 73, 232–237.
262. Sekita, Y, Wagatsuma, H, Nakamura, K, Ono, R, Kagami, M, Wakisaka, N, Hino, T, Suzuki-Migishima, R, Kohda, T, Ogura, A, et al. Role of retrotransposon-derived imprinted gene, Rtl1, in the feto-maternal interface of mouse placenta. *Nat Genet* **2008**, 40, 243–248.
263. Haycock, P. C, Ramsay, M. Exposure of mouse embryos to ethanol during preimplantation development: Effect on dna methylation in the H19 imprinting control region. *Biol Reprod* **2009**, 81, 618–627.
264. Su, J. M, Xu, W. B, Li, Y. Y, Wang, L. J, Wang, Y. S, Zhang, Y. The methylation status of PEG10 in placentas of cloned transgenic calves. *Yi Chuan* **2011**, 33, 533–538.
265. Vire, E, Brenner, C, Deplus, R, Blanchon, L, Fraga, M, Didelot, C, Morey, L, van Eynde, A, Bernard, D, Vanderwinden, J. M, et al. The Polycomb group protein EZH2 directly controls DNA methylation. *Nature* **2006**, 439, 871–874.
266. Pasini, D, Bracken, A. P, Agger, K, Christensen, J, Hansen, K, Cloos, P. A. C, Helin, K. Regulation of Stem *Cell* Differentiation by Histone Methyltransferases and Demethylases. Cold Spring Harb. *Symp. Quant. Biol.* **2008**, 73, 253–263.
267. Schuettengruber, B, Chourrout, D, Vervoort, M, Leblanc, B, Cavalli, G. Genome regulation by polycomb and trithorax proteins. *Cell* **2007**, 128, 735–745.
268. Bantignies, F, Cavalli, G. *Cell*ular memory and dynamic regulation of polycomb group proteins. Curr. Opin. *Cell Biol.* **2006**, 18, 275–283.
269. Ross, P. J, Ragina, N. P, Rodriguez, R. M, Iager, A. E, Siripattarapravat, K, Lopez-Corrales, N, Cibelli, J. B. Polycomb gene expression and histone H3 lysine 27 trimethylation changes during bovine preimplantation development. *Reproduction* **2008**, 136, 777–785.

270. Dunn, K. L, Davie, J. R. The many roles of the transcriptional regulator CTCF. *Biochem. Cell Biol.* **2003**, 81, 161–167.
271. Filippova, G. N, Fagerlie, S, Klenova, E. M, Myers, C, Dehner, Y, Goodwin, G, Neiman, P. E, Collins, S. J, Lobanenkov, V. V. An exceptionally conserved transcriptional repressor, CTCF, employs different combinations of zinc fingers to bind diverged promoter sequences of avian and mammalian c-Myc oncogenes. *Mol. Cell. Biol.* **1996**, 16, 2802–2813.
272. Soshnikova, N, Montavon, T, Leleu, M, Galjart, N, Duboule, D. Functional analysis of CTCF during mammalian limb development. *Dev. Cell* **2010**, 19, 819–830.
273. Essien, K, Vigneau, S, Apreleva, S, Singh, L. N, Bartolomei, M. S, Hannenhalli, S. CTCF binding site classes exhibit distinct evolutionary, genomic, epigenomic and transcriptomic features. *Genome Biol.* **2009**, 10, doi:10. 1186/gb-2009-10-11-r131.
274. Vostrov, A. A, Quitschke, W. W. The zinc finger protein CTCF binds to the APBβ domain of the amyloid β-protein precursor promoter. *J Biol Chem* **1997**, 272, 33353–33359.
275. Cuddapah, S, Jothi, R, Schones, D. E, Roh, T. Y, Cui, K. R, Zhao, K. J. Global analysis of the insulator binding protein CTCF in chromatin barrier regions reveals demarcation of active and repressive domains. *Genome Res.* **2009**, 19, 24–32.
276. Phillips, J. E, Corces, V. G. CTCF: Master weaver of the genome. *Cell* **2009**, 137, 1194–1211.
277. Han, L, Lee, D. H, Szabo, P. E. CTCF is the master organizer of domain-wide allele-specific chromatin at the H19/IGF2 imprinted region. *Mol. Cell. Biol.* **2008**, 28, 1124–1135.
278. Kim, T. H, Abdullaev, Z. K, Smith, A. D, Ching, K. A, Loukinov, D. I, Green, R. D, Zhang, M. Q, Lobanenkov, V. V, Ren, B. Analysis of the vertebrate insulator protein CTCF-binding sites in the human genome. *Cell* **2007**, 128, 1231–1245.
279. Ideraabdullah, F. Y, Vigneau, S, Bartolomei, M. S. Genomic imprinting mechanisms in mammals. *Mutat. Res.* **2008**, 647, 77–85.
280. Matsuzaki, H, Okamura, E, Fukamizu, A, Tanimoto, K. CTCF binding is not the epigenetic mark that establishes post-fertilization methylation imprinting in the transgenic H19 ICR. *Hum. Mol. Genet.* **2010**, 19, 1190–1198.
281. Esteve, P. -O, Chin, H. G, Smallwood, A, Feehery, G. R, Gangisetty, O, Karpf, A. R, Carey, M. F, Pradhan, S. Direct interaction between DNMT1 and G9a coordinates DNA and histone methylation during replication. *Genes Dev* **2006**, 20, 3089–3103.
282. Robertson, A. K, Geiman, T. M, Sankpal, U. T, Hager, G. L, Robertson, K. D. Effects of chromatin structure on the enzymatic and DNA binding functions of DNA methyltransferases DNMT1 and Dnmt3a in vitro. *Biochem. Biophys. Res. Commun.* **2004**, 322, 110–118.

CHAPTER 3

GENOME-WIDE ANALYSIS OF DNA METHYLATION IN HUMAN AMNION

JINSIL KIM, MITCHELL M. PITLICK, PAUL J. CHRISTINE, AMANDA R. SCHAEFER, CESAR SALEME, BELÉN COMAS, VIVIANA COSENTINO, ENRIQUE GADOW, and JEFFREY C. MURRAY

INTRODUCTION

The human amnion is the inner layer of the fetal membranes composed of a monolayer of epithelial cells attached to a basement membrane overlying a collagen-rich stroma [1, 2]. This tissue, which encloses the amniotic fluid, protects the fetus from external mechanical forces and provides an environment that supports fetal movement and growth [3, 4]. The amnion is also a metabolically active tissue involved in the synthesis of various substances with important functions during pregnancy, including prostaglandins and cytokines [1, 5, 6]. It is particularly well known as a major source of prostaglandin E2, a potent molecule mediating cervical ripening and myometrial contraction [7–10], whose levels dramatically increase before and during labor [11, 12].

The amniotic membrane provides most of the tensile strength of the fetal membranes, and alterations in its integrity can lead to undesirable pregnancy outcomes such as preterm premature rupture of membranes (PPROMs) [1, 13], which complicates 3% of all pregnancies and is responsible for approximately one-third of all preterm births (PTBs) [14]. Given the important role of the amnion in the maintenance of pregnancy and parturition, investigation into molecular events occurring in this tissue may contribute to a better understanding of physiological and pathological processes involved in pregnancy.

Considering that the amniotic fluid is in a constantly changing state, it may be critical that the amnion properly responds to environmental cues from the amniotic fluid to accommodate the dynamic needs of the fetus, which could be mediated through epigenetic processes. A previous study by Wang et al. [15] has shown that matrix metalloproteinase 1 (MMP1), whose genetic variation is associated with susceptibility to PPROM [16], is regulated at the epigenetic level, specifically by DNA methylation, and that MMP1 promoter methylation status correlates with its expression in the amnion and association with PPROM. This finding suggests that the amnion represents an intriguing source of tissue for studying epigenetic events of potential physiological and pathological relevance.

In this study, we performed genome-wide methylation profiling of human term and preterm amnion in order to explore the possible importance of DNA methylation in physiologic labor as well as the etiology of PTB. In addition, independent of the genome-wide methylation study, we carried out methylation analysis of the promoter region of the oxytocin receptor (OXTR) gene whose role in human parturition is well established [17]. Given that OXTR expression in the amnion increases in association with the onset of labor [18] and that its aberrant methylation in other tissue types has been implicated in autism [19], a disorder that has been associated with PTB [20, 21], we sought to investigate if DNA methylation could represent one mechanism regulating OXTR gene function in the contexts of normal parturition and prematurity.

MATERIALS AND METHODS

PLACENTAL TISSUE COLLECTION AND PREPARATION

Fresh human placentas were collected in 2009 and 2010 at the University of Iowa Hospitals and Clinics in IA, USA and Instituto de Maternidad y Ginecología Nuestra Señora de las Mercedes in Tucumán, Argentina with signed informed consent and an institutional review board approval. We examined 121 placentas from three groups of patients undergoing: term cesarean delivery without labor (term no labor (TNL) group, n = 18), normal term vaginal delivery (term labor (TL) group, n = 40), and spontaneous preterm (<37 weeks of gestation) delivery (preterm labor (PTL) group, n = 63). Gestational age (GA) was determined using the first day of the last menstrual period as well as by ultrasound examination and was confirmed by assessment at birth. Each placenta was dissected into fetal (amnion, chorion) and maternal (decidua basalis) components within an hour of delivery. The amnion and chorion obtained from the extraplacental membranes (reflected membranes) were separated by blunt dissection under sterile conditions. Decidual tissue samples were macroscopically isolated from the surface of the basal plate of the placenta. After being cut into small pieces, the dissected tissues were placed in RNA later solution (Applied Biosystems, Carlsbad, CA, USA) and stored per manufacturer's recommendations until used. A subset of these samples was selected for genome-wide methylation analysis on the basis of their informativity in relation to our previous gene expression profiling study (unpublished). Additional samples used for validation experiments were selected primarily based on the quality of DNA or RNA extracted from the tissue samples.

DNA PREPARATION AND METHYLATION STANDARDS

Genomic DNA was extracted from placental tissue samples using the DNeasy Blood & Tissue Kit (QIAGEN, Valencia, CA, USA) following the manufacturer's protocol. The quality of the extracted DNA was evaluated by agarose gel electrophoresis. 500ng of DNA was bisulfite-converted using the EZ DNA Methylation Kit (Zymo Research, Irvine, CA, USA) according to the manufacturer's instructions, and used in subsequent experiments. Universal Methylated Human DNA Standard (Zymo Research), which is enzymatically methylated in vitro at all cytosines in CpG dinucleotides, was used as a positive control in the Illumina Infinium methylation assay. We also used Human Methylated and Non-methylated DNA Standards (Zymo Research) as positive and negative controls for methylation-specific PCR. Both of the standards are purified from DNMT1 and DNMT3b double-knockout HCT116 cells, but the methylated standard is enzymatically methylated at all cytosines in CpG dinucleotides.

GENOME-WIDE DNA METHYLATION ANALYSIS

ILLUMINA INFINIUM METHYLATION ASSAY

DNA methylation profiling was performed by the W.M. Keck Biotechnology Resource Laboratory at Yale University, using the Illumina Infinium HumanMethylation27 BeadChip (Illumina, San Diego, CA, USA). Details of the design and general properties of this platform have been previously described [22]. A total of 24 samples were assayed on two BeadChips (12 samples per chip) following the standard protocol provided by Illumina. The samples examined included 9 individual and 1 pooled amnion samples each from the TNL and TL groups, one pooled amnion sample from the PTL group obtained by combining 6 individual samples, and 3 controls (methylated DNA control treated with M.SssI methyltransferase (New England Biolabs, Ipswich, MA, USA), Universal Methylated Human DNA Standard (Zymo Research), and bisulfite-untreated control). These samples were selected from among patients who had participated in our previous gene expression profiling study (unpublished), performed independently of the current work. Based on this previous study, which showed heterogeneous global gene expression patterns among PTL samples, we only included one pooled PTL sample to assess a group DNA methylation average. The samples were arranged randomly on each chip and were processed in a blinded fashion. Table 1 summarizes the clinical characteristics of the three groups of samples studied.

TABLE 1. Clinical characteristics of the three subject groups studied by genome-wide DNA methylation profiling.

Parameter	TNL (n = 9)1	TL (n = 9)1	PTL (n = 6)2
Gestational age (weeks)3	39.1 ± 0.8	38.8 ± 0.8	33.5 ± 2.6
Race			
White	4	3	3
Black	0	0	1
Other	5	6	2
Maternal age at delivery (years)3	29.7 ± 5.5 (Range 22–38)	27 ± 4.4 (Range 20–33)	28.7 ± 3.5 (Range 25–33)
Antibiotics during pregnancy or labor			
Yes	6	1	5
No	3	7	0

Table 1 continued...

Unknown	0	1	1
Birth weight (grams)3	3508.3 ± 267.2	3354.1 ± 348.6	2102.2 ± 724.6
Infant gender			
Female	6	4	2
Male	3	5	4

1Examined both individually and as a pooled sample.

2Examined as a pooled sample.

3Data are presented as mean ± standard deviation (SD).

Abbreviations: TNL: term no labor; TL: term labor; PTL: preterm labor.

QUALITY CONTROL AND STATISTICAL ANALYSIS

Data analysis was conducted on a fee-for-service basis by the W.M. Keck Biostatistics Resource at Yale University with GenomeStudio Methylation Module v1.0 (Illumina). We evaluated the quality of the data based on the signals of assay built-in control probes (staining, hybridization, target removal, extension, bisulfite conversion, methylation signal specificity, background determination, and overall assay performance) and three experimental controls (two positive methylated controls and one non-bisulfite-converted control), and confirmed the reliability of our data. Principal component analysis (PCA) demonstrated that there is no significant batch effect among the three groups of samples examined. The methylation status of each interrogated CpG site was determined employing the β-value (defined as the fraction of methylation, calculation details described in a previous study [23]) method. An average β-value (AVG_Beta) for each CpG locus ranging from 0 (unmethylated) to 1 (completely methylated) was extracted utilizing the GenomeStudio software and used in further analyses. For the determination of differential methylation between two given groups, we used the Illumina custom error model. This model assumes a normal distribution of the methylation value (β) among replicates corresponding to a set of biological conditions (TNL, TL, and PTL). We prioritized differentially methylated CpG sites by difference score (DiffScore). DiffScore, which takes into account background noise and sample variability [24], was calculated using the following formula: DiffScore = 10sgn (βcondition − βreference)log 10P, where βcondition = βTL/βPTL,

βreference = βTNL or βcondition = βTNL/βPTL, βreference = βTL. The resulting differentially methylated CpG sites were annotated with respect to their nearest gene based on the information provided by Illumina. A more detailed description of the Illumina custom error model and the DiffScore has been provided previously [25].

FUNCTIONAL ENRICHMENT ANALYSIS

Differentially methylated genes (DMGs) with a DiffScore of >20 (equivalent to P-value of <0.01) were evaluated for functional enrichment using predefined gene sets from the Molecular Signatures Database (MSigDB) [26]. We searched for significantly enriched gene sets by computing overlaps between the lists of DMGs and the CP collection (canonical pathways, 880 gene sets) or the C5 collection (GO gene sets, 1454 gene sets) in the MSigDB. Gene sets with a P-value (based on the hypergeometric distribution) less than 0.05 were considered significant.

BISULFITE SEQUENCING (BS)

To validate methylation differences revealed by the genome-wide methylation assay, we performed bisulfite sequencing on urocortin (UCN), a gene identified as differentially methylated between the TL and PTL groups, and OXTR, a gene whose methylation status has recently been shown to be important in the pathogenesis of autism [19]. We investigated the methylation status of OXTR, given its significant role in parturition [17] and its labor-associated expression pattern in the amnion [18], which makes it a potential candidate gene for PTB. There are two OXTR CpG sites targeted by the Illumina Infinium BeadChip assay, both of which were not identified as being differentially methylated. However, because there is currently no evidence supporting the biological importance of the regions containing the two sites, we focused our BS analysis on CpG sites of known biological significance that are located in a different region of the OXTR gene. Primers for UCN were designed to cover the CpG site identified as being differentially methylated by genome-wide methylation profiling, using the default parameters of MethPrimer [27]. PCR amplification using the primer pair results in a 278bp product that spans part of the promoter, exon 1, and part of intron 1 of UCN (-439 to -162 relative to translation start site (TSS)) containing 16 CpG sites. For OXTR, we used the same primers and PCR conditions as those used in the previous study [19]. PCR amplification using the primer set results in a 358bp product that spans the OXTR promoter (-1195 to -838 relative to TSS) containing 22 CpG sites that has been associated with tissue-specific OXTR expression [28] and the development of autism [19]. The regions examined in both genes were located within CpG islands. We carried out our analysis using the same samples assayed on the BeadChips and eight additional independent PTL samples (TNL, n = 9; TL, n = 9; PTL, n = 14). Bisulfite-converted DNA was PCR-amplified using ZymoTaq DNA polymerase (Zymo Research). The resulting PCR products were run on an agarose gel and cloned into the pGEM-T Easy vector (Promega, Madison, WI, USA). Individual clones were isolated, amplified following standard protocols, and purified using the PureLink Quick Plasmid Miniprep Kit (Invitrogen, Carlsbad, CA, USA) per manufacturer's instructions. Ten clones per sample, on average, were isolated and sequenced at the University of Iowa DNA facility. Percentage methylation was determined for each CpG site similarly as done in previous work [19]. Statistical analysis was conducted using SigmaPlot 11.0 (Systat Software, San Jose, CA, USA). Significance of differential methylation (DM) was assessed using thet-test (two-tailed), Mann-Whitney (M-W) rank sum test (two-sided), one-way ANOVA, or Kruskal-Wallis (K-W) one-way ANOVA by ranks, as indicated in the text and/or figure legends. Post hoc analysis following ANOVA was performed using either the Holm-Sidak or Dunn's Method. A $P < 0.05$ was considered significant.

ENCYCLOPEDIA OF DNA ELEMENTS (ENCODE) CHIP-SEQ DATA

We examined the potential functional significance of the region of the OXTR gene containing CpG sites with statistically different DNA methylation status (CpGs-959 and -1084) using the ChIP-seq data from the ENCODE project available in the University of California Santa Cruz

(UCSC) genome browser [29, 30]. We specifically used the suppressor of zeste 12 homolog (Drosophila) (SUZ12) and Pol2 ChIP-seq data generated by the laboratories of Michael Snyder at Stanford University and Vishy Iyer at the University of Texas Austin. The ChIP-Seq data were obtained using human cells (NT2-D1 for the SUZ12 data; GM18526, 18951, 19099, 19193, and ProgFib for the Pol2 data).

METHYLATION-SPECIFIC PCR (MSP)

Validation of DM was additionally carried out using methylation-specific PCR (MSP). Two pairs of primers (unmethylated and methylated) for each of the lysophosphatidic acid receptor 5 (LPAR5), paternally expressed 10 (PEG10), and solute carrier family 30 member 3 (SLC30A3) genes were designed using the MSP-specific default parameters of the MethPrimer program [27]. Bisulfite-converted DNA extracted from amnion tissues (TNL, n = 9; TL, n = 9; PTL, n = 14) was PCR-amplified using Biolase DNA polymerase (Bioline, Taunton, MA, USA). The resulting PCR products were visualized on a 2% agarose gel. Human Methylated and Non-methylated DNA Standards from Zymo Research were used as positive and negative controls.

RNA EXTRACTION AND REAL-TIME QRT-PCR

Total RNA was extracted from amnion (TNL, n = 14; TL, n = 34; PTL, n = 59) and decidua (TNL, n = 12; TL, n = 16; PTL, n = 31) tissues using TRIzol reagent (Invitrogen) according to the manufacturer's protocol. The quality of extracted RNA was checked using the Agilent 2100 Bioanalyzer (Agilent Technologies, Santa Clara, CA, USA). Reverse transcription was carried out with the High Capacity cDNA Reverse Transcription Kit (Applied Biosystems), using random hexamers as primers following the manufacturer's instructions. Real-time qRT-PCR was performed using synthesized cDNA as a template, gene-specific primers (UCN and OXTR) and Power SYBR Green PCR Master Mix (Applied Biosystems). The reactions (including no-template controls) were run in triplicate on the

7900HT Fast Real-Time PCR System (Applied Biosystems) using ACTB (beta actin) [31] as an endogenous reference. Data were analyzed with the SDS 2.4 software (Applied Biosystems), employing the comparative CT method [32]. Absence of nonspecific amplification was confirmed by dissociation curve analysis. Samples with a value that falls outside ±2 standard deviations of the group mean were defined as outliers and removed from the study. Statistical analysis was performed similarly as described above in the bisulfite sequencing section. Data were presented as mean ± standard error of the mean (SEM).

RESULTS

GENOME-WIDE PATTERNS OF DNA METHYLATION AND DIFFERENTIALLY METHYLATED CPG LOCI BETWEEN TERM (NON-LABORED AND LABORED) AND PRETERM AMNION TISSUES

To investigate the possible involvement of epigenetic mechanisms in the physiology of normal labor and the pathogenesis of PTB, we examined the genome-wide methylation profiles of the amnion obtained following term (TNL and TL, n = 9 for each) and preterm (PTL, n = 6) deliveries using the Illumina Infinium BeadChip platform. The overall levels of DNA methylation in the experimental samples were low with third quartile AVG_Beta values between 0.4 and 0.55. Principal component analysis (PCA) placed the pooled TL and PTL samples close to each other and very distant from the pooled TNL sample (Figure 1), which indicates that the genome-wide methylation patterns in amnion tissues from the two spontaneous labor groups (regardless of gestational age (GA) at delivery) are more similar to each other than to those observed in non-labor tissues.

We also performed gene/locus level analysis of differential methylation (DM), searching for methylation changes associated with labor and/or PTB at specific CpG sites. Using the Illumina custom error model algorithm, we identified 65 CpG sites in 64 and 61 autosomal genes each that are differentially methylated between the TNL and TL groups and the TL and PTL groups, respectively with a DiffScore of >30 (equivalent to

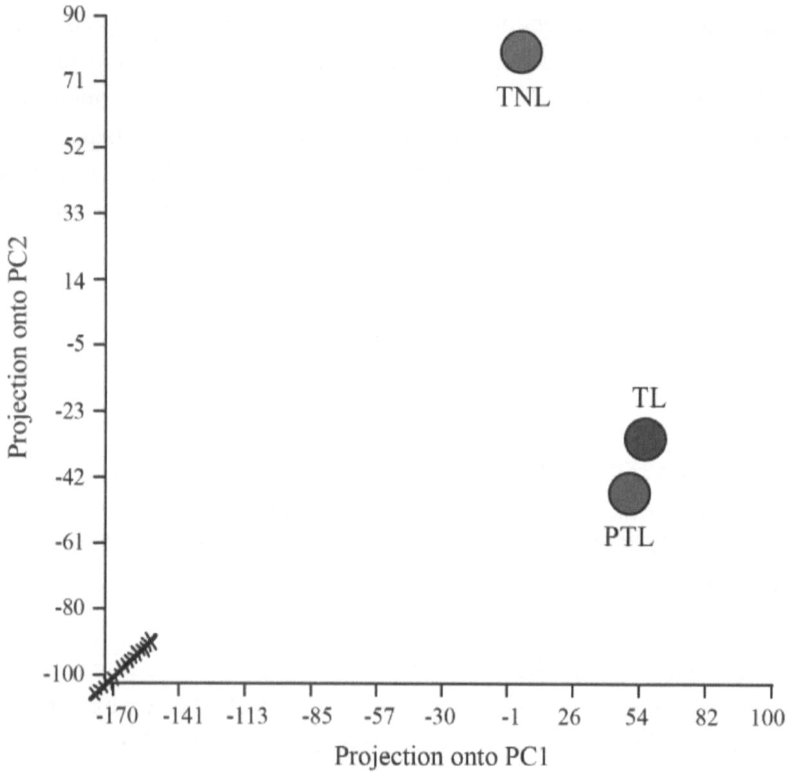

FIGURE 1. Principal component analysis (PCA) plot of DNA methylation profiles in term (non-labored and labored) and preterm amnion. Each colored dot represents a pooled DNA sample from term no labor (TNL), term labor (TL), or preterm labor (PTL) group. Note that the TNL sample is placed distantly from the TL or PTL samples, indicating that the TNL group displays distinctly different methylation patterns compared to the other two groups.

P-value of <0.001). Listed in Table 2 are the 15 most highly differentially methylated genes (DMGs). It was noted that among the genes with differentially methylated sites, although very few, were those belonging to special classes of genes, including noncoding RNAs and imprinted genes (such as Down syndrome critical region gene 10 (DSCR10), FBXL19 antisense RNA 1 (FBXL19-AS1), and paternally expressed 10 (PEG10) as shown in Table 2), many of which have regulatory functions in diverse biological processes.

TABLE 2. List of top 15 differentially methylated autosomal genes in amnion tissues from term (TNL, TL) and preterm (PTL) deliveries ranked by statistical significance (1).

TNL versus TL2			TL versus PTL3		
Gene	Locus	CpG island4	Gene	Locus	CpG island4
IL32	16p13.3	No	TOB1	17q21	No
EDARADD	1q42.3	No	PNPLA3	22q13.31	No
STK19	6p21.3	No	ZNF671	19q13.43	No
EXTL1	1p36.1	No	DAB2IP 7	9q33.1–q33.3	No
HLA-DQB2	6p21	No	MFNG	22q12	No
MFSD3	8q24.3	Yes	UCN	2p23–p21	Yes
RAB31	18p11.3	Yes	EXOC3L2	19q13.32	Yes
PNPLA3	22q13.31	No	SLC44A2	19p13.1	Yes
GRHPR	9q12	Yes	FBXL19-AS1 6	16p11.2	Yes
MPHOSPH10	2p13.3	No	DLGAP5	14q22.3	Yes
PEG10 5	7q21	No	SLC30A3	2p23.3	Yes
DSCR10 6	21q22.13	No	CHFR	12q24.33	No
SRRD	22q12.1	Yes	C11orf1	11q23.1	No
POLI	18q21.1	Yes	SLC24A4	14q32.12	No
OSTalpha	3q29	No	PI4KB	1q21	No

1Statistical significance was determined based on P-values calculated from DiffScores. All genes listed here have a DiffScore >40 (corresponding to P-value of <0.0001).

2Genes most highly methylated in the TL group compared to the TNL group.

3Genes most highly methylated in the PTL group compared to the TL group.

4Defined by the CpG island track in the UCSC Genome Browser.

5An imprinted gene.

6Non-protein coding genes.

7A gene identified as having three non-island CpG sites with a DiffScore >40.

Abbreviations: TNL: term no labor; TL: term labor; PTL: preterm labor.

FUNCTIONAL ENRICHMENT ANALYSIS

To determine the biological significance of DMGs, functional annotation analysis was performed. Our approach involved examining the extent of overlap between our lists of DMGs and predefined annotated gene sets from the MSigDB [26] (see Section 2 for further details). For this analysis, we used gene lists with a less stringent P-value cutoff of <0.01

(corresponding to a DiffScore of >20), given the small number of DMGs (n = 65) with a P-value below 0.001. We found that 7 gene sets were significantly overrepresented ($P < 0.05$) in the list of 110 DMGs between the TNL and TL groups. The seven enriched gene sets included cation transport, ion channel activity, and those shown in Table 3, most of which are highly relevant to molecular processes involved in physiologic labor. Among the 186 DMGs between the TL and PTL groups, 17 gene sets were overrepresented. Many of the enriched gene sets were found to be associated with the regulation of cell behavior and extracellular matrix-cell interactions, including focal adhesion, cell junction, cell-substrate adherens junction, and integrin binding (Table 3).

BISULFITE SEQUENCING (BS) ANALYSIS OF DIFFERENTIAL METHYLATION

To validate DM detected by genome-wide methylation profiling, we performed BS analysis on UCN, a gene identified as being overmethylated in the PTL group compared with the TL group with a DiffScore >50 (Table 2). We performed the same analysis on one additional gene named oxytocin receptor (OXTR) whose mRNA and protein expression has been shown to be markedly upregulated in association with labor in primary human amnion epithelial cells [18]. Previous studies have demonstrated that the methylation status of the promoter region of this gene is associated with tissue-specific OXTR expression [28] and the development of autism [19], a disorder linked to PTB [20, 21, 33, 34]. These findings intrigued us to investigate whether DNA methylation could represent one mechanism regulating the labor-associated activity of OXTR in the amnion. We selected the two genes (UCN and OXTR), given their crucial role in normal labor and parturition, which makes them potential candidate genes for PTB. Details on the regions amplified, samples used in the BS experiments, and statistical tests performed for the analysis of the sequencing results are given in Section 2 and Figure 2.

All 16 CpG dinucleotides interrogated in the UCN gene showed some degree of methylation with the ones at positions -361, -335, and -319, being more highly methylated (22.9–55.7%, Table 4) compared with those at other

TABLE 3. Gene sets overrepresented among differentially methylated genes in amnion tissues from term (TNL, TL) and preterm (PTL) deliveries (1).

TNL versus TL	
Gene set[2]	**P-value[3]**
HEART_DEVELOPMENT	0.012
POSITIVE_REGULATION_OF_CYTOKINE_PRODUCTION	0.017
GATED_CHANNEL_ACTIVITY	0.024
REGULATION_OF_HEART_CONTRACTION	0.041
REGULATION_OF_CYTOKINE_PRODUCTION	0.044
TL versus PTL	
NEGATIVE_REGULATION_OF_TRANSFERASE_ACTIVITY	0.007
ADHERENS_JUNCTION[4]	0.011
HEPARIN_BINDING	0.014
FOCAL_ADHESION_FORMATION	0.02
FOCAL_ADHESION	0.024

[1]Presented are the top 5 most significantly enriched gene sets from C5 collection (GO gene sets).

[2]Defined in the Molecular Signatures Database (MSigDB).

[3]The cutoff for statistical significance was P = 0.05.

[4]Also identified as being enriched (P = 0.049) in the analysis performed with the CP collection.

Abbreviations: TNL: term no labor; TL: term labor; PTL: preterm labor.

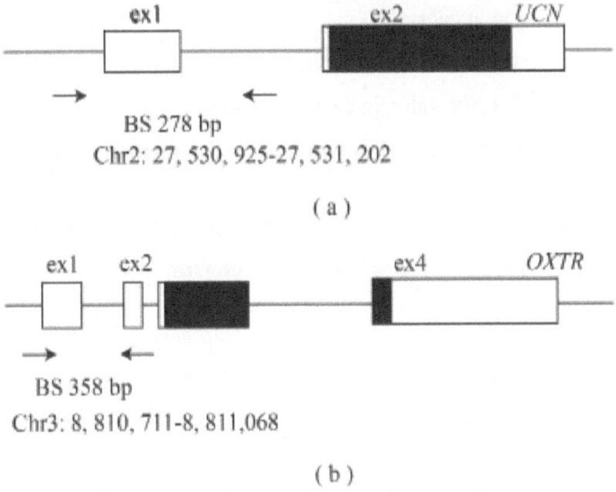

FIGURE 2. Schematic representation of CpG island regions of UCN (a) and OXTR (b) analyzed by bisulfite sequencing (BS). Black horizontal arrows denote BS PCR primer binding sites. Solid box: coding region; open box: untranslated region. The expected PCR product sizes and positions of the primer binding sites (chromosome and base count, NCBI Build GRCh37/hg19) are indicated.

positions (1.1–17.1%). All except two CpG sites were overmethylated in the PTL samples compared to the TL samples, showing the expected direction of DM. However, the differences were not statistically significant.

TABLE 4. UCN and OXTR promoter methylation status in the amnion and decidua from term (TNL, TL) and preterm (PTL) deliveries (1).

				UCN				
Site2		**Amnion**				**Decidua**		
	TL		PTL	P-value	TL		PTL	P-value
-1903	2.2%		8.6%	0.39	6.7%		14.8%	0.07
-2793	2.2%		11.4%	0.07	15.6%		12.5%	0.97
-3194	24.4%		22.9%	0.83	19.9%		30.8%	0.08
-3354	23.3%		27.1%	0.56	23.2%		25.8%	0.75
-3614	48.9%		55.7%	0.33	32.1%		36.4%	0.48
				OXTR				
Site2		Amnion				Decidua		
	TNL	TL	PTL	P-value	TNL	TL	PTL	P-value
-860	24.4%	22.2%	24.3%	0.96	20%	15.6%	24.6%	0.19
-901	37.8%	30%	45.7%	0.17	45.6%	38.9%	46.9%	0.66
-924	56.7%	62.2%	68.6%	0.29	60%	52.2%	67.2%	0.09
-9345	50%	41.1%	59.3%	0.22	46.7%	42.2%	57.1%	0.02
-9595	43.3%	24.4%	27.1%	0.014	33.3%	33.3%	30.6%	0.91
-10845	4.4%	4.4%	5.7%	0.97	10%	0%	8.7%	0.008

1Presents average % methylation at each CpG site.

2Nucleotide positions relative to translation start site.

3CpG sites in UCN with the lowest P-value in each tissue type.

4CpG sites methylated at higher levels in both tissues than the average methylation level of all sites examined.

5CpG sites with statistically significant (P < 0.05) differential methylation in either tissue.

Abbreviations: TNL: term no labor; TL: term labor; PTL: preterm labor.

For OXTR, since we had no priori data on the methylation status of the 22 CpG sites in amnion tissue, all three groups of samples (TNL, TL, and PTL) were examined. Consistent with the finding of Gregory et al. [19], 5 CpG sites at positions -959, -934, -924, -901, and -860 showed the highest levels (22.2–68.6%, Table 4) and variation in methylation, whereas very little or no methylation (0–5.7%) was observed at the other sites. We found that one (CpG-959) of the five sites was significantly differentially methylated among the three groups tested (one-way ANOVA, $P = 0.014$, Table 4). Pairwise comparisons (Holm-Sidak test) revealed significant differences between the TNL and TL groups ($P = 0.017$) and the TNL and PTL groups ($P = 0.025$) and borderline significant difference between the TL and PTL groups ($P = 0.050$), demonstrating more distinct differences in methylation at this site between non-labor and labor tissues than between term and preterm tissues.

To determine if the observed DM also occurs in other parts of the placenta where the genes are known to be expressed [17, 35], we extended our study to decidua tissues from the same groups of individuals. The decidua, which is of maternal origin, unlike the amnion of fetal origin [4], was selected, given that the function of OXTR in parturition has been well demonstrated in maternal tissue [17], and therefore, the examination of the decidua, along with the amnion, may allow us to compare the methylation state of the OXTR gene and possibly its importance in both fetal and maternal tissues.

The overall methylation patterns observed in the decidua were similar to those identified in the amnion. However, unlike in the amnion tissues, the methylation levels not at CpG-959, but at different sites (CpGs-934 and -1084), were found to be statistically significantly different ($P = 0.02$, 0.008, resp., K-W one-way ANOVA by ranks) among the three groups of the decidua tissues (Table 4). The CpG-1084 site, interestingly, was completely unmethylated in the TL group, whereas it was methylated to some small degree in the other two groups (TNL, 10%; PTL, 8.7%) (Table 4). Significant differences between the TL and TNL or PTL groups were confirmed by Dunn's post hoc test ($P < 0.05$). In the case of CpG-934, the difference was significant only between the TL and PTL groups. Taken together, it appears that there exist compartment-specific OXTR methylation patterns in the placenta.

ANALYSIS OF UCN AND OXTR GENE EXPRESSION IN THE AMNION AND DECIDUA

To evaluate the functional significance of the methylation status of the two genes, we performed gene expression analysis using qRT-PCR on an extended set of amnion and decidua tissues ($n = 107$, 59, resp.) from the three groups.

Although the DM of UCN was not validated by BS, we observed a statistically significant 2.3-fold increase in its transcript levels in the PTL amnion samples compared to the TNL and TL samples ($P < 0.001$, K-W one-way ANOVA by ranks, Figure 3). There was also a statistically significant, but less

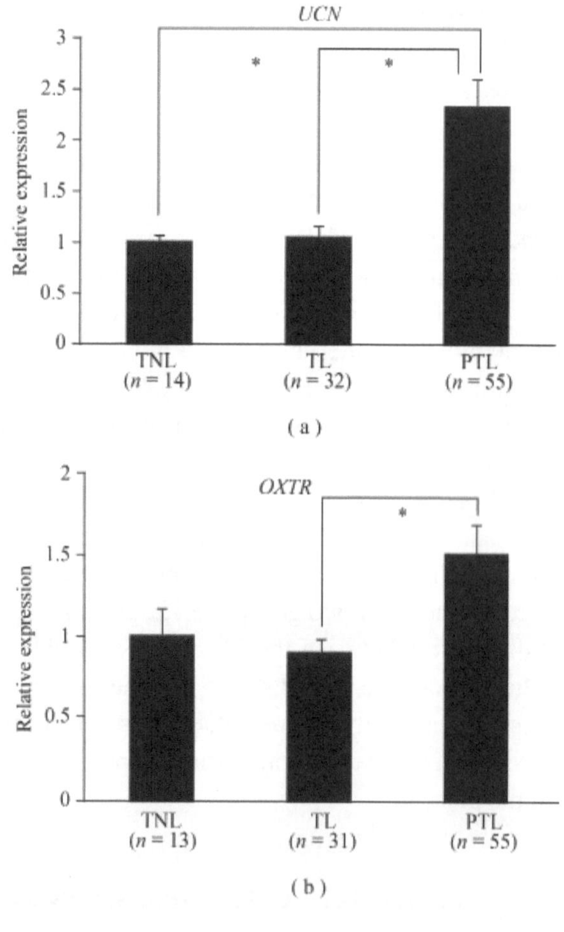

FIGURE 3. UCN and OXTR mRNA expression levels in term (non-labored and labored) and preterm amnion. Expression levels were normalized to that of beta-actin (ACTB). Experiments were performed in triplicate. Data presented are mean ± standard error of the mean (SEM). Asterisks represent statistically significant differences ($P < 0.05$, K-W one-way ANOVA by ranks followed by Dunn's post hoc test) between specified groups.

than twofold increase in OXTR mRNA levels in the PTL amnion samples compared with the TL samples ($P < 0.05$, K-W one-way ANOVA by ranks, Dunn's post hoc test, Figure 3). The results were not replicated in the decidua samples for either gene. These findings suggest that the upregulation of UCN is specific to the amnion from spontaneous preterm deliveries, and that the DM observed in OXTR may not correlate with OXTR expression given that methylation generally plays a role in gene silencing.

METHYLATION-SPECIFIC PCR (MSP) ANALYSIS OF DIFFERENTIAL METHYLATION

As an alternative approach to validate DNA methylation differences captured by our genome-wide methylation study, we carried out methylation-specific PCR (MSP) for selected 3 DMGs between the TNL and TL groups (PEG10) and between the TL and PTL groups (LPAR5 and SLC30A3). Analysis of the same set of amnion samples used in BS revealed no intergroup differences in PEG10 and LPAR5 methylation (data not shown). However, the methylation status of SLC30A3 was in good agreement with our genome-wide methylation data with methylated MSP products present in 10 out of 14 (71%) PTL samples and none of the TL samples (Figure 4).

DISCUSSION

The present study investigated if there exist unique genome-wide methylation signatures that distinguish among term (non-labored and labored) and preterm amnion tissues. Our methylation profiling revealed a higher degree of similarity between the methylation patterns in the TL and PTL pooled samples than those observed in the TNL pooled sample, suggesting the potential role of methylation in the regulation of labor, independent from GA. We identified a relatively small number of DMGs between the TNL and TL groups and the TL and PTL groups (65 genes each) at the $P < 0.001$ significance level. This observation may be attributed to the small sample size and the sample-to-sample variability related to GA. Gene set enrichment analysis of those genes revealed significant overrepresentation

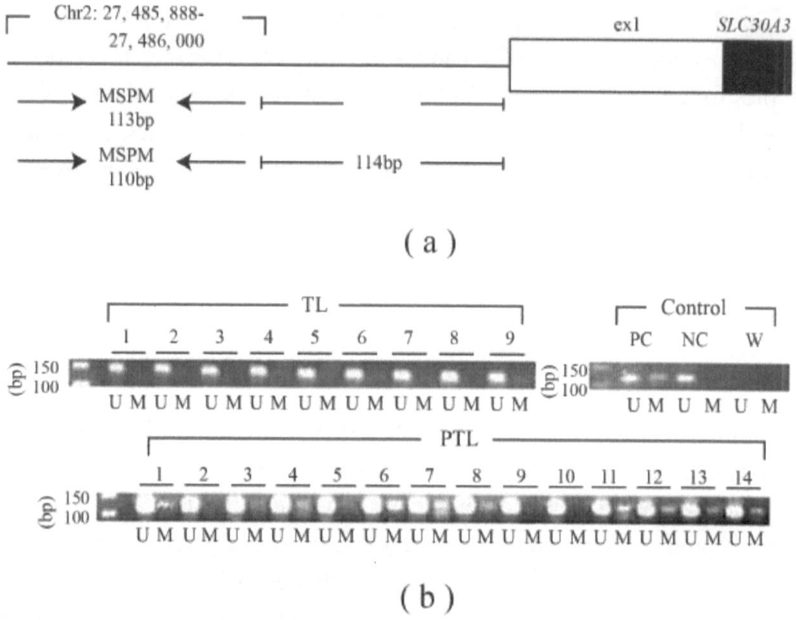

FIGURE 4. Methylation-specific PCR (MSP) analysis of SLC30A3. (a) Schematic representation of MSP primer binding sites. Black horizontal arrows: methylated-specific primer (MSPM) binding sites; gray horizontal arrows: unmethylated-specific primer (MSPU) binding sites. The expected PCR product sizes and positions of the primer annealing sites (chromosome and base count, NCBI Build GRCh37/hg19) are indicated. Solid box: coding region; open box: untranslated region. (b) Agarose gel electrophoresis

of pathways that appear to be functionally relevant (Table 3). The enrichment of pathways related to ion transport, ion channel activity, and cytokine production among the DMGs between the TNL and TL groups reflects biochemical and molecular events associated with the onset of labor, which, along with hormonal factors, help to initiate parturition. These results are at least partially in line with previous gene expression profiling studies reporting labor-associated cytokine-related gene signatures in human amniotic [36] and chorioamniotic [37] membranes. The overrepresentation of heart-(development and contraction) related gene sets may be explained by the presence of myofibroblasts in the connective tissue of the amnion [38], which have contractile ability [39], and hence are involved in heart rhythm regulation [40] and, possibly, prevention of excessive distension of the amniotic membrane [38]. The DMGs between the TL and PTL groups were enriched in gene sets involved in cell adhesion, cell-cell and cell-extracellular matrix interactions, which have crucial roles in the modulation of cellular behavior and tissue maintenance and organization [41]. This observation confirms the importance of intact fetal membranes as a critical factor in the maintenance of pregnancy. Another overrepresented gene set was the negative regulation of transferase activity. Given the versatile roles of transferases, differential methylation of this group of genes (including HEXIM1, SFN, CBLC, and DUSP2) may influence a wide range of cellular processes in a way that interferes with timely onset of labor and parturition. Among these genes, DUSP2 has previously been documented as being significantly upregulated following interleukin-1β (IL-1β) stimulation in myometrial cells [42], suggesting its potential role in the mediation of uterine contractions. It would be intriguing to examine how the activity of DUSP2 in the amnion may contribute to the process of parturition.

Our study at the individual gene level using BS revealed three CpG sites (CpGs-934, -959, and -1084) in OXTR that exhibit significant DM among the three groups of amnion and decidua tissues, which are of fetal and maternal origin, respectively [4]. Subsequent gene expression analysis demonstrated no correlation between gene expression and methylation and therefore, the functional significance of the observed DM remains undetermined. Previous work showed that site-specific methylation can result in transcriptional alterations through its effects on the interaction of transcription factors (TFs) with its cognate DNA sequence [43]. Currently, there are no known TF binding sites around CpG-959, which was previously identified as significantly hypermethylated in peripheral blood mononuclear cells from autistic patients compared with those from control patients [19]. However, Gregory et al. [19] have indicated that CpG-934, whose differential methylation has also been associated with autism, falls within predicted binding domains for v-rel reticuloendotheliosis viral oncogene homolog (avian) (c-Rel), zinc fingers and homeoboxes 2 (ZHX2), and lectin, galactoside-binding, soluble, 4 (LGALS4). Using ENCODE ChIP-seq data available in the UCSC genome browser, we also found that CpG-1084 falls within putative binding sites for SUZ12 and Pol2 (see

Section 2 for more details), which warrants future studies to dissect the impact of the methylation status at this specific dinucleotide on the interactions between these TFs and their binding sites.

Despite the lack of any significant difference in UCN methylation levels between the TL and PTL groups, our observation of a significant, more than 2-fold increase in UCN mRNA levels in the PTL amnion tissues compared with the term tissues suggests a potential role of this gene in the etiology of PTB, which encodes an endogenous ligand for corticotropin releasing hormone receptor (CRHR) that mediates the action of CRH, one of the major endocrine factors in parturition [44]. Given that there are several putative binding sites for TFs (such as C7EBP, GATA, and MyoD) [45] upstream of the region examined in this study, it would be intriguing to investigate whether the methylation status of CpG dinucleotides encompassing those sites correlates with the observed gene expression patterns. It would also be worthwhile to examine if mechanisms other than methylation underlie the transcriptional regulation of UCN in the amnion.

Our MSP analysis identified another gene (SLC30A3) that might play a role in pathogenic processes of PTB. This gene, also known as ZNT3, encodes a zinc transporter responsible for zinc efflux from the cytoplasm to extracellular spaces or intracellular organelles [46]. Given the differential expression of SLC30A3 in relation to dietary zinc and/or glucose supply in mouse placenta [47] and beta cells [48], it is postulated that its dysregulated expression due to aberrant methylation in human amnion may influence nutritional homeostasis during pregnancy, ultimately, leading to PTB.

Our work was limited by the small sample size and the lack of control for gender-specific methylation differences [49, 50]. Another major limitation is that the PTL tissues were examined as a pooled sample, not individually. Previous studies have demonstrated that pooled DNA samples can be used to provide a reliable estimate of average group methylation when analyzed using high-throughput techniques such as MALDI-TOF mass spectrometry [51, 52]. Therefore, a DNA pooling approach using such systems could be employed in future studies for large-scale assessment of methylation variations in maternal and fetal tissues. Very recently, it has been shown that neonatal DNA exhibits a considerable degree of GA-associated variability in DNA methylation patterns [53]. Given this finding, a precisely stratified analysis based on GA may allow a more accurate characterization of DNA methylation profiles associated with term and preterm pregnancies.

CONCLUSION

This work provides preliminary evidence that DNA methylation changes may play at least a partial role in physiologic labor and the etiology of PTB, and suggests that DNA methylation profiles, together with other types of biological data, hold a promise for the identification of genes involved in normal parturition and preterm birth.

REFERENCES

1. Benirschke K. *Pathology of the Human Placenta*. New York, NY, USA, Springer, **2012**.
2. Niknejad H, Peirovi H, Jorjani M, Ahmadiani A, Ghanavi J, Seifalian AM. Properties of the amniotic membrane for potential use in tissue engineering. *European Cells and Materials* **2008**, 15, 88–99.
3. Toda A, Okabe M, Yoshida T, Nikaido T. The potential of amniotic membrane/amnion-derived cells for regeneration of various tissues. *Journal of Pharmacological Sciences* **2007**, 105(3), 215–228.
4. Carlson BM. *Human Embryology and Developmental Biology*. Philadelphia, PA, USA, Mosby/Elsevier, **2009**.
5. Keelan JA, Sato T, Mitchell MD. Regulation of interleukin (IL)-6 and IL-8 production in an amnion-derived cell line by cytokines, growth factors, glucocorticoids, and phorbol esters. *American Journal of Reproductive Immunology* **1997**, 38(4), 272–278.
6. Keelan JA, Sato T, Mitchell MD. Interleukin (IL)-6 and IL-8 production by human amnion, regulation by cytokines, growth factors, glucocorticoids, phorbol esters, and bacterial lipopolysaccharide. *Biology of Reproduction* **1997**, 57(6), 1438–1444.
7. Ackerman WE, Summerfield TLS, Vandre DD, Robinson JM, Kniss DA. Nuclear factor-kappa B regulates inducible prostaglandin E synthase expression in human amnion mesenchymal cells. *Biology of Reproduction* **2008**, 78(1), 68–76.
8. Bernstein P, Leyland N, Gurland P, Gare D. Cervical ripening and labor induction with prostaglandin E2 gel, a placebo-controlled study. *American Journal of Obstetrics and Gynecology* **1987**, 156(2), 336–340.
9. Rayburn WF. Prostaglandin E2 gel for cervical ripening and induction of labor, a critical analysis. *American Journal of Obstetrics and Gynecology* 1989, 160(3), 529–534.
10. Gibb W. The role of prostaglandins in human parturition. *Annals of Medicine* **1998**, 30(3), 235–241.

11. Haluska GJ, Kaler CA, Cook MJ, Novy MJ. Prostaglandin production during spontaneous labor and after treatment with RU486 in pregnant rhesus macaques. *Biology of Reproduction* **1994**, 51(4), 760–765.
12. Schellenberg JC, Kirkby W. Production of prostaglandin F(2α) and E2 in explants of intrauterine tissues of guinea pigs during late pregnancy and labor. *Prostaglandins* **1997**, 54(3), 625–638.
13. Oyen ML, Calvin SE, Landers DV. Premature rupture of the fetal membranes, is the amnion the major determinant? *American Journal of Obstetrics and Gynecology* **2006**, 195(2), 510–515.
14. Mercer BM. Preterm premature rupture of the membranes. *Obstetrics and Gynecology* **2003**, 101(1), 178–193.
15. Wang H, Ogawa M, Wood JR, et al. Genetic and epigenetic mechanisms combine to control MMP1 expression and its association with preterm premature rupture of membranes. *Human Molecular Genetics* **2008**, 17(8), 1087–1096.
16. Fujimoto T, Parry S, Urbanek M, et al. A single nucleotide polymorphism in the matrix metalloproteinase-1 (MMP-1) promoter influences amnion cell MMP-1 expression and risk for preterm premature rupture of the fetal membranes. *The Journal of Biological Chemistry* **2002**, 277(8), 6296–6302.
17. Gimpl G, Fahrenholz F. The oxytocin receptor system, structure, function, and regulation. *Physiological Reviews* **2001**, 81(2), 629–683.
18. Terzidou V, Blanks AM, Kim SH, Thornton S, Bennett PR. Labor and inflammation increase the expression of oxytocin receptor in human amnion. *Biology of Reproduction* **2011**, 84(3), 546–552.
19. Gregory SG, Connelly JJ, Towers AJ, et al. Genomic and epigenetic evidence for oxytocin receptor deficiency in autism. *BMC Medicine* **2009**, 7, article 62
20. Larsson HJ, Eaton WW, Madsen KM, et al. Risk factors for autism, perinatal factors, parental psychiatric history, and socioeconomic status. *American Journal of Epidemiology* **2005**, 161(10), 916–925.
21. Limperopoulos C, Bassan H, Sullivan NR, et al. Positive screening for autism in ex-preterm infants, prevalence and risk factors. *Pediatrics* **2008**, 121(4), 758–765.
22. Bibikova M, Le J, Barnes B, et al. Genome-wide DNA methylation profiling using Infinium assay. *Epigenomics* **2009**, 1(1), 177–200.
23. Chowdhury S, Erickson SW, MacLeod SL, et al. Maternal genome-wide DNA methylation patterns and congenital heart defects. *PLoS One* **2011**, 6(1)e16506
24. Chudin E, Kruglyak S, Baker SC, Oeser S, Barker D, McDaniel TK. A model of technical variation of microarray signals. *Journal of Computational Biology* **2006**, 13(4), 996–1003.
25. Naumova OY, Lee M, Koposov R, Szyf M, Dozier M, Grigorenko EL. Differential patterns of whole-genome DNA methylation in institutionalized children and children raised by their biological parents. *Development and Psychopathology* **2012**, 24(1), 143–155.
26. Subramanian A, Tamayo P, Mootha VK, et al. Gene set enrichment analysis, a knowledge-based approach for interpreting genome-wide expression profiles. *Proceedings of the National Academy of Sciences of the United States of America* **2005**, 102(43), 15545–15550.

27. Li LC, Dahiya R. MethPrimer, designing primers for methylation PCRs. *Bioinformatics* **2002**, 18(11), 1427–1431.
28. Kusui C, Kimura T, Ogita K, et al. DNA methylation of the human oxytocin receptor gene promoter regulates tissue-specific gene suppression. *Biochemical and Biophysical Research Communications* **2001**, 289(3), 681–686.
29. Birney E, Stamatoyannopoulos JA, Dutta A, et al. Identification and analysis of functional elements in 1% of the human genome by the ENCODE pilot project. *Nature* **2007**, 447(7146), 799–816.
30. Rosenbloom KR, Dreszer TR, Pheasant M, et al. ENCODE whole-genome data in the UCSC genome browser. *Nucleic Acids Research* **2010**, 38(1), D620–D625.gkp961
31. Ahn K, Huh JW, Park SJ, et al. Selection of internal reference genes for SYBR green qRT-PCR studies of rhesus monkey (Macaca mulatta) tissues. *BMC Molecular Biology* **2008**, 9, article 78
32. Livak KJ, Schmittgen TD. Analysis of relative gene expression data using real-time quantitative PCR and the 2-ΔΔCT method. *Methods* **2001**, 25(4), 402–408.
33. Johnson S, Hollis C, Kochhar P, Hennessy E, Wolke D, Marlow N. Autism spectrum disorders in extremely preterm children. *Journal of Pediatrics* **2010**, 156(4), 525.e2–531.e2.
34. Johnson S, Marlow N. Preterm birth and childhood psychiatric disorders. *Pediatric Research* **2011**, 69(5), 11R–18R.
35. Gu Q, Clifton VL, Schwartz J, Madsen G, Sha JY, Smith R. Characterization of urocortin in human pregnancy. *Chinese Medical Journal* **2001**, 114(6), 618–622.
36. Han YM, Romero R, Kim JS, et al. Region-specific gene expression profiling, novel evidence for biological heterogeneity of the human amnion. *Biology of Reproduction* **2008**, 79(5), 954–961.
37. Haddad R, Tromp G, Kuivaniemi H, et al. Human spontaneous labor without histologic chorioamnionitis is characterized by an acute inflammation gene expression signature. *American Journal of Obstetrics and Gynecology* **2006**, 195(2), 394.e12–405.e12.
38. Wang T, Schneider J. Myofibroblasten im bindegewebe des menschlichen amnions. *Zeitschrift für Geburtshilfe und Perinatologie* **1982**, 186, 164–169.
39. Tomasek JJ, Gabbiani G, Hinz B, Chaponnier C, Brown RA. Myofibroblasts and mechano, regulation of connective tissue remodelling. *Nature Reviews Molecular Cell Biology* **2002**, 3(5), 349–363.
40. Rohr S. Myofibroblasts in diseased hearts, new players in cardiac arrhythmias? *Heart Rhythm* **2009**, 6(6), 848–856.
41. DuFort CC, Paszek MJ, Weaver VM. Balancing forces, architectural control of mechanotransduction. *Nature Reviews Molecular Cell Biology* **2011**, 12(5), 308–319.
42. Chevillard G, Derjuga A, Devost D, Zingg HH, Blank V. Identification of interleukin-1β regulated genes in uterine smooth muscle cells. *Reproduction* **2007**, 134(6), 811–822.
43. Bélanger AS, Tojcic J, Harvey M, Guillemette C. Regulation of UGT1A1 and HNF1 transcription factor gene expression by DNA methylation in colon cancer cells. *BMC Molecular Biology* **2010**, 11, article 9
44. Florio P, Vale W, Petraglia F. Urocortins in human reproduction. *Peptides* **2004**, 25(10), 1751–1757.

45. Zhao L, Donaldson CJ, Smith GW, Vale WW. The structures of the mouse and human urocortin genes (Ucn and UCN). *Genomics* 1998, 50(1), 23–33.
46. Cousins RJ, Liuzzi JP, Lichten LA. Mammalian zinc transport, trafficking, and signals. *The Journal of Biological Chemistry* **2006**, 281(34), 24085–24089.
47. Helston RM, Phillips SR, McKay JA, Jackson KA, Mathers JC, Ford D. Zinc transporters in the mouse placenta show a coordinated regulatory response to changes in dietary zinc intake. *Placenta* **2007**, 28(5-6), 437–444.
48. Smidt K, Jessen N, Petersen AB, et al. SLC30A3 responds to glucose- and zinc variations in β-cells and is critical for insulin production and in vivo glucose-metabolism during β-cell stress. *PLoS One* **2009**, 4(5)e5684
49. Yuen RKC, Peñaherrera MS, von Dadelszen P, McFadden DE, Robinson WP. DNA methylation profiling of human placentas reveals promoter hypomethylation of multiple genes in early-onset preeclampsia. *European Journal of Human Genetics* **2010**, 18(9), 1006–1012.
50. Cotton AM, Avila L, Penaherrera MS, Affleck JG, Robinson WP, Brown CJ. Inactive X chromosome-specific reduction in placental DNA methylation. *Human Molecular Genetics* **2009**, 18(19), 3544–3552.
51. Docherty SJ, Davis OSP, Haworth CMA, Plomin R, Mill J. DNA methylation profiling using bisulfite-based epityping of pooled genomic DNA. *Methods* **2010**, 52(3), 255–258.
52. Docherty SJ, Davis OS, Haworth CM, Plomin R, Mill J. Bisulfite-based epityping on pooled genomic DNA provides an accurate estimate of average group DNA methylation. *Epigenetics Chromatin* **2009**, 2, article 3
53. Schroeder JW, Conneely KN, Cubells JC, et al. Neonatal DNA methylation patterns associate with gestational age. *Epigenetics* **2011**, 6(12), 1498–1504.

CHAPTER 4

ASSESSING CAUSAL RELATIONSHIPS IN GENOMICS: FROM BRADFORD-HILL CRITERIA TO COMPLEX GENE-ENVIRONMENT INTERACTIONS AND DIRECTED ACYCLIC GRAPHS

SARA GENELETTI, VALENTINA GALLO, MIQUEL PORTA, MUIN J. KHOURY, and PAOLO VINEIS

INTRODUCTION

Observational studies of human health and disease (basic, clinical and epidemiological) are vulnerable to methodological problems -such as selection bias and confounding- that make causal inferences problematic. Gene-disease associations are no exception, as they are commonly investigated using observational designs. However, as compared to studies of environmental exposures, in genetic studies it is less likely that selection of subjects (e.g., cases and controls in a case-control study) is affected by genetic variants. Confounding is also less likely, with the exception of linkage disequilibrium (i.e., the attribution of a genetic effect to a specific gene rather than to an adjacent one) and population stratification (when cases and controls are drawn from different ethnic populations). There is in fact some empirical evidence suggesting that gene-disease associations are less prone to confounding (e.g., by socio-economic status) than associations between genes and environmental and lifestyle variables [1]. There are some well-known methodological challenges in interpreting the causal significance of gene-disease associations; they include epistasis, linkage disequilibrium, and gene-environment interactions (GEI) [2].

A rich body of knowledge exists in medicine and epidemiology on assessment of causal relationships involving personal and environmental causes of disease; it includes seminal causal criteria developed by Austin Bradford Hill and more recently applied directed acyclic graphs (DAGs). Perhaps unsurprisingly, such knowledge has seldom been applied to assess causal relationships in clinical genetics and genomics, even when studies aimed at making inferences relevant for human health. Conversely, incorporating genetic causal knowledge into clinical and epidemiological causal reasoning is still a largely unexplored task.

APPLYING CAUSAL GUIDELINES TO GENETIC STUDIES

For several decades, guidelines to assess causality have been a powerful tool in clinical and epidemiological research, as well as in the professional practice of medicine and epidemiology outside academia [4-7]. Causal guidelines usually include a series of criteria that help assess which observed associations are potentially causal. They were introduced initially by Bradford-Hill in the debate about the role of smoking in the aetiology of lung cancer; given the issue, they were meant for observational studies only, but many of the criteria can be applied to clinical trials and other experimental studies as well [8]. Although Hill did not have genetic epidemiology in mind at the time, today his criteria remain relevant to causal assessment in this field and, as we will show, to many areas of human genetics as well.

Hill's approach is based on nine criteria: 1) Strength of association; 2) Consistency; 3) Specificity of association; 4) Temporality; 5) Biological gradient (dose-response relationship); 6) Biological plausibility; 7) Coherence; 8) Experimental evidence (e. g. reproducibility in animal models); and 9) Analogy. Statistical significance was not listed but discussed separately by Hill [8].

One major criticism leveled at Hill's approach is that it considers one causal factor at a time and is not intended to tackle complex relationships and interactions, such as those encountered in modern molecular medicine and genomics, which deal with chains of mediators and not only directly acting exposures. However, even complex situations can often be decomposed into simpler constituents, and in such case Hill's criteria can be applied fruitfully. This is a main motivation behind the present work.

In 2006, a Human Genome Epidemiology Network (HuGENet) workshop in Venice was devoted to the development of standardized criteria for the assessment of the credibility of cumulative evidence on gene-disease associations. This led to synopses on various topics in genetic epidemiology; e.g., on DNA repair [9], and on Parkinson's disease [10]. Briefly, according to the Venice guidelines [2] each gene-disease association is graded on the basis of the amount of evidence, replication, and protection from bias. These guidelines contributed to modifying the approach to genetic inferences using Hill's criteria that we adopt here.

Main theoretical issues underlying the application of Hill's criteria in genetics and genomics are shown in Appendix 1 [11-29]; below we will show how these criteria can be applied to an example of gene-environment interaction. Interactions here are defined as "the interdependent operation of two or more causes to produce, prevent, or control an effect" [2].

In summary, Hill's causal criteria and related logical tools that have long been applied fruitfully to clinical and epidemiological research may also be applied productively to research in genetics. However, genetic research has fundamental differences from clinical and epidemiological research. For example, in genetics confounding can be the consequence of events that may not be directly addressed at the other levels, including haplotype blocks, allelic heterogeneity, overdominance, and epistasis [15]. Selection bias is more easily measurable in genomic studies, because we have the null hypothesis represented by Hardy-Weinberg equilibrium (HWE); i.e., we expect independent assortment of alleles in the population, whereas a similar reasoning cannot be applied to daily life exposures. Hardy-Weinberg equilibrium is based on assumptions of population genetics related to the lack of selection, inbreeding, migration; departure from HWE can thus point towards the possibility of gross bias (such as genotyping errors or selection bias).

Explicit guidelines for causal assessment are more popular in clinical and epidemiological research than in genetics [3,30]. The reasons for that have seldom been addressed. They are probably related to the different nature of the objects, factors, mechanisms and processes that we study at each level. However, genetic guidelines on causality do exist and, in fact, have interesting similarities with Hill's criteria: (a) linkage to a particular region of the human genome (LOD>3); (b) one or more independent mutations that are concordant with disease status in affected families (specificity, strength of association); (c) defects that lead to macrochanges in the protein (specificity, coherence); (d) putative mutations that are not present in a sample from a control population (specificity); or (e) presence of some other line of biological evidence (including expression, knockout data, etc.) [15]. Criteria (a), (b) and (c) refer to background knowledge. But it is in particular criterion (e) that supports the causal association by conferring coherence with previous knowledge [3,15].

DIRECTED ACYCLIC GRAPHS AS TOOLS TO CLARIFY ASSOCIATIONS AND COMPLEX CAUSAL RELATIONSHIPS

Directed acyclic graphs (DAGs) have a long tradition in science. They are a rigorous way of visualising complex systems, clarifying ideas, complementing the formulation of hypotheses, and guiding quantitative analyses. There has been much debate on the exact nature and roles of DAGs in the biomedical literature. The most widespread approach in the health sciences is the causal DAG approach promoted by Greenland, Robins, Hernán and colleagues [31-33], and the equivalent mathematical framework of counterfactuals [34]. In causal DAG approaches, the directed edges in a DAG represent causal relationships. Whilst the causal DAG framework is appealing and intuitive, we wish to draw attention to an alternative approach to causal inference, the Decision Theoretic Framework (DTF), which is based on a formal treatment of conditional independences (a non-graphical version of the 'd-separation criteria') [35]. Appendix 2 provides additional details on statistics and assumptions underlying the DTF [36-38]. This approach has recently become increasingly popular in epidemiology, in particular to assess the role of genes as instrumental variables for causal inference [39]. DTF retains the advantages of the causal DAG approach but overcomes some of its limitations. In particular, as DTF uses DAGs to describe the relationships between variables, it retains the capacity of DAGs to clearly and formally visualise complex systems. In contrast to the causal DAG approach where all directed edges are assumed to represent causal relationships, DTF takes a more conservative view where the edges represent statistical associations (and the lack of edges represents independence). Causality in DTF is viewed as external knowledge that can be added to the DAGs and allows some of the edges to be interpreted as causal. There are three reasons for this conservative viewpoint. The first is that it entails fewer assumptions about the existence and direction of causal relationships between variables. The second is that it is not necessary to include all possible causes or covariates in a DAG, only the variables of interest, making DTF more flexible than the causal DAG approach. The third is that when we perform a statistical analysis of observational data, we obtain measures of association (not causation) between variables. We explain this concept in more detail below.

A main problem when making causal inferences in clinical and epidemiological research is that most data are observational. This is also true for a substantial part of basic biomedical research. It is certainly an issue in human genetics, where there is usually no randomization (except in circumstances where Mendelian randomization can be applied [1,3,39,40]), and knowledge of the genetic pathways is tenuous or incomplete. In such circumstances we must be careful to distinguish causal relationships from associations resulting from unobserved biases or chance.

DAGs can still be used to make causal inference, but the causal element is an external assumption that needs to be explicitly incorporated into the DAG rather than implicit in the direction of an edge. We use a DAG to visualise complex associations, but when we only have observational data at our disposal, we must find other ways to assess a) whether a particular association is causal and not due to confounding or other bias, and b) what the direction of this association is.

The problem of inferring causality from observational data in the presence of unobserved confounding is simply described in the DAGs in Figure 1. In the DAG on the left hand side X is the putative cause—e.g., a particular environmental exposure such as urban pollution—Y is the disease outcome under investigation, and U a set of confounders, many of which will typically be unobserved. Epidemiologists are interested in the existence, direction and strength of the X-Y association and whether this can be considered causal. (They are not necessarily interested in whether the other relationships in the DAG are causal). However, they are often unable to capture all this information from observational studies due to the presence of unobserved confounders U. Even when there is no direct association—i.e., there is no edge between X and Y as in the DAG on the right hand side of the Figure 1—the presence of U (this time as a common parent) will result in a statistical association between the two. Again, the question is, how do we distinguish a causal association from a statistical association when only observational data are available?

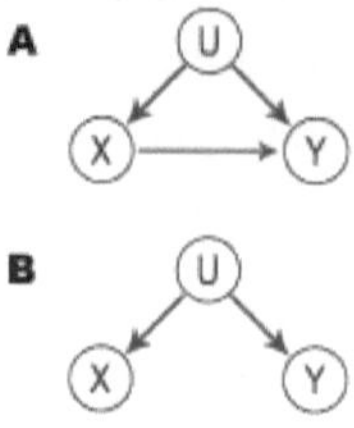

FIGURE 1. DAG demonstrating the ideas of confounding. A: U is an unobserved confounder for the association between X and Y and X is a cause of Y. B: U is an unobserved confounder for the association between X and Y but X is not a cause of Y. From purely observational data these two situations cannot be separated.

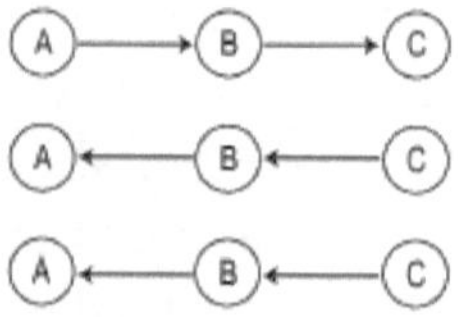

FIGURE 2. Three DAGs exhibiting the same conditional independence but with different causal interpretations.

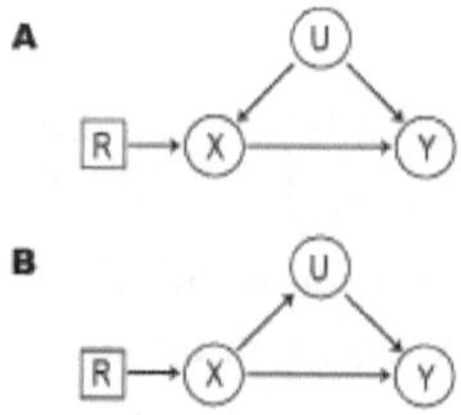

FIGURE 3. DAG with a randomisation node R. R indicates whether X is randomised or allowed to arise naturally. A: U is a confounder. B: U is a mediator. Randomisation allows us to distinguish between these situations.

One way to answer this question is by incorporating prior knowledge in Hill's scheme (or similar criteria) with DAGs to determine which edges can be considered causal. This is the approach we propose in this chapter and that we describe in detail below. Another way of introducing causality is by adding so-called intervention or randomisation variables to a DAG and to the corresponding probability statements. A more detailed description of such variables is given in Appendix 2. As a thorough explanation is beyond the scope of this paper we refer the interested reader to Dawid [41], Didelez [42], Geneletti [43], and Lauritzen [44].

For the remainder of this chapter, the DAGs we use can be viewed as heuristic tools to understand gene-environment relationships.

PARKINSON'S DISEASE: PESTICIDES, AND GENE-ENVIRONMENT INTERACTIONS

In order to illustrate our methods, we present a case study based on Parkinson's disease. First we present a short description of the disease and a summary literature review of its genetic component; we focus in particular on a recently identified genetic form. Second, we use graphical methods to propose and assess hypotheses on how the risk factors might interact. Third, we apply Hill's criteria to each of the hypothesised associations to assess causality in light of the available evidence.

Parkinson's disease is the most common neurodegenerative disorder after Alzheimer's disease, affecting 16-19 new individuals per 100,000 persons each year in developed countries [45]. Characterized by bradykinesia, resting tremor, rigidity and postural instability, it is also one of the most common late-life movement disorders. The pathological characteristic of the disease is a selective loss of pigmented neurons, most prominently in the substantia nigra (one of the brain basal ganglia) accompanied by a characteristic α-synuclein-positive inclusion bodies in neurons (Lewy bodies) [45]. While the causes of Parkinson's disease remain unknown, significant progress is being made in elucidating genetic and environmental risk factors and the neurodegenerative process underlying the disease. Appendix 3 summaries the key evidences to date on environmental and genetic risk factors for Parkinson's disease [46-49].

A deletion of the DJ-1 gene in a Dutch family and a mutation conferring a functionally inactive form in an Italian family associated with early onset PD were first observed in 2001 [50], and confirmed in 2003 [51] (as is convention, we use italics to indicate the gene and non-italics to indicate the protein; thus, DJ-1 means the gene, and DJ-1 means the protein). DJ-1 is involved in many cell processes including oncogenic transformation, gene expression and chaperon activity, and it mediates oxidative stress responses [52]. A recent meta-analysis of the association between pesticides and Parkinson's disease [53] concludes that the epidemiologic evidence suggests a fairly consistent association between exposure to pesticides and risk of developing Parkinson's disease. In particular, among the herbicides, paraquat has been found to be most strongly associated with the risk of the disease (with odds ratios ranging from 1.25 to 3.22). Toxicological evidence suggests that both paraquat and rotenone exert a neurotoxic action that might play a role in the etiopathogenic process of Parkinson's disease. Moreover, clinical symptoms of Parkinson's disease have been reproduced in rats by chronic administration of paraquat [54]. Evidence from animal experiments shows that knockout models of *Drosophila melanogaster* (fruit fly) lacking DJ-1 function, display a marked and selective sensitivity to the environmental oxidative insults exerted by both paraquat and rotenone [54]; this suggests that there is an interaction between these toxicants and the DJ-1 genotype [3]. On the basis of these data, it is sensible to hypothesise an interaction between DJ-1, exposure to some pesticides, and risk of Parkinson's disease in humans as well. Using Hill's criteria we can say that the hypothesis has biological plausibility; also, testing the hypothesis entails testing Hill's criterion of analogy (i.e., testing that there are analogous causal mechanisms in certain animal models and in humans). To test the hypothesis, further investigation is needed in order to estimate the effect of the interaction between DJ-1 and exposure to specific pesticides in humans on the risk of developing Parkinson's disease. We can construct a logic framework displaying (a) the association of paraquat (P) with Parkinson's disease (Y); (b) the association of DJ-1 with Parkinson's diseases; and (c) the interaction of DJ-1 with exposure to paraquat. We can also assume the existence of confounding between the exposure to paraquat and the disease outcome (Cp), and between DJ-1 and disease outcome (Cd) (Figure 4). First we are going to propose a graphical method to untangle the relationship between these two risk factors and Parkinson's disease; in a second step we will evaluate the associations from a more strictly causal point of view.

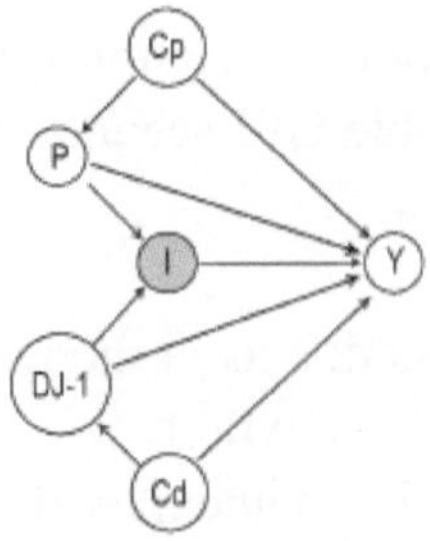

FIGURE 4. DAG showing all possible one-way relationships for gene-environment interactions based on the observed variables.

CASE STUDY: THE DJ-1 GENE, EXPOSURE TO PARAQUAT AND RISK OF PARKINSON'S DISEASE

The process we describe in this section has two components. The first uses DAGs as a visual tool to explore a range of possible interaction scenarios. The second uses DAGs as a formal tool to describe the formal dependence among the variables in the problem. These two components go hand in hand, as intuition about the problem will generally guide the first whilst the second will reflect information in the observed data as well as considerations about what is biologically plausible. In a second instance, which is beyond the scope of this chapter, the interaction quantitative effects can be estimated. How the latter step is done will depend both on the nature of the data available and crucially on the model for interaction. We assume an additive interaction model for simplicity; however, the DAGs work equally well with a multiplicative model as they describe associations rather than their exact mathematical nature.

We consider first the case study of gene-environment interactions (GEI) involving risk of Parkinson's disease, the DJ-1 gene and exposure to paraquat described above. To do this we use simplified versions of models proposed by Khoury et al. [55] and Ottman [56]. Subsequently, we consider fruit fly experiments where the associations between Parkinson's, DJ-1 and paraquat have been ascertained, and we present this as the ideal situation to make causal inference. The approach we are proposing can be also used to tackle a range of other complex problems.

In order to look at possible GEI scenarios we need to introduce some simple notation:

gene: DJ-1 = d* variant (deletion as in the Dutch families or inactivity as in the Italian families); DJ-1 = d wild type
pesticides: P = p* exposed; P = p unexposed
disease: Y = 1 with Parkinson's disease; Y = 0 without Parkinson's disease

The crux of this approach is the introduction of an interaction variable I. It is determined by the values of the genetic and environmental exposure variables. In simple terms, it acts like a switch and is turned "on" when the parents (a parent P of another variable X has an edge pointing into X, and X is a child of P) take on some values, and "off" when the parents have other values. In the current context this is typically the presence of the genetic exposure (i.e., the genetic variant) and/or the environmental exposure that leads to an increase in disease risk, which turns the interaction "on". Thus, in addition to the above variables, we also define:

interaction: I = 1 ("on") if there is an interaction and I = 0 ("off") if there isn't.

The exact nature of the interaction depends on the contexts sketched below.

For the sake of simplicity, we assume that I is a deterministic variable. What we mean by this is that unlike the other variables in the problem, I is not random. Once the value of its parents is known, then so is the value of I. This might be considered unduly restrictive if there are other potential parents in the interaction, which are suspected but unobserved. It is possible in these cases to view I as a random variable, where its variability is associated with that of the unobserved interactant. However, in this chapter we focus on the simplest case and thus we make the following assumption:

1. DJ-1 and P are the only parents of the interaction variable I. Another assumption that is generally plausible, provided that the exposure

does not modify the genetic structure (e.g., the exposure does not cause somatic mutations) is that:

2. There is no a priori association between the gene and the external exposure; this is represented by the absence of a directed edge between DJ-1 and P in the DAGs below.

Generally, this is a plausible assumption provided that the exposure does not modify the genetic structure [57]. In this specific example, this assumption is likely to be true. However, with other environmental exposures this assumption does not hold. For example, the association of some lifestyle factors with genotypes predisposing (or causing) Parkinson's disease is possible as the dopaminergic system is involved in rewarding mechanisms and it is hypothesized to influence some seeking behaviours and addiction (i.e., smoking or alcohol drinking) [58].

The idea of I as a variable to represent interaction is similar to the sufficient component cause (SCC) variables in VanderWeele and Robins [59]. We feel however that our approach presents a few advantages over the SCC framework. As we do not need to incorporate all the sufficient causes (we are not using a causal DAG), the structure of our DAGs is less cumbersome. Also, although for the sake of simplicity we have defined I in terms of binary exposures, we can easily extend it if we are considering multi-valued or continuous exposures. The DAG in Figure 4 shows a complex situation we can imagine, given assumptions 1 and 2, in which there is confounding between both the exposure to paraquat and the disease outcome (Cp) as well as confounding between DJ-1 and the disease (Cd), and no other variables are postulated. Confounding between both exposure to paraquat and the disease might be due, for example, to the fact that people exposed to paraquat may also be more likely to smoke, a factor that is negatively associated with the risk of Parkinson's disease [60]. Confounding between DJ-1 and the disease might be due to the involvement of the dopamine-mediated rewarding system [58]. Any observational study—any study of these issues in humans—is unlikely to observe all potential confounders. Nevertheless, just to simplify our model, we also assume that:

3. There are no further confounders between either the gene and the outcome or the exposure and the outcome. This is represented by the

absence of additional variables and corresponding directed edges in the DAGs below.

Now we turn our attention to looking at the case by evaluating the plausibility of a few different GEI scenarios. As mentioned above, these are loosely based on Khoury et al. [55]. For each of the models that we consider below, we present a more formal description in Appendix 4.

MODEL I

Both exposure and genotype are required to increase risk as in Figure 5. Here, if I is "on" then there is an association between the disease and the genetic exposure and the environmental exposure to pesticides when both are present. If on the other hand I is "off" then there is no association -in other words, Parkinson's is only associated with DJ-1 and paraquat exposure through the interaction itself. This is an extreme form of interaction that is unlikely to occur in the pathogenesis of common diseases. Does this model describe the relationship between DJ-1, exposure to pesticides and Parkinson's disease? For this to be the case, all the Dutch and Italian families with the variant DJ-1 and Parkinson's would also have to have been exposed to pesticides. Further, the incidence of Parkinson's amongst the families with the gene variant would have to be the same on average as that of those without the gene variant (if unexposed to pesticides). Similarly, those exposed to pesticides would have to have the same incidence as those not exposed to the pesticides without the DJ-1 variant. This is clearly not the case.

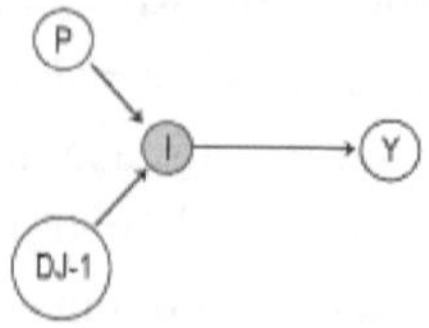

FIGURE 5. Both DJ-1 gene and pesticide exposure need to be present to activate the interaction.

MODEL II

The exposure to pesticides increases the risk of disease but the presence of the gene variant alone does not increase the risk of disease, although the variant further increases the risk of disease in the exposed population (Figure 6). In this model, I is switched on and off by P. When P = p* (exposure to pesticides) I = 1, indicating that the interaction is switched "on" and the presence of the variant in DJ-1 and Parkinson's is influential. When P = p then I = 0 and whether DJ-1 is the variant or wild-type form makes no difference to the outcome Y. It is possible that in some cases exposure to P is protective; i.e., I would take the opposite value of P in a binary situation. In more complex situations, the effect of P might be such that only certain values of P result in interactions and in these cases the values of I and P would not be the same. In this instance, we have that Y depends directly on exposure P; however, Y depends on DJ-1 only through the interaction and the exposure when this is present -i. e. when P = p*.

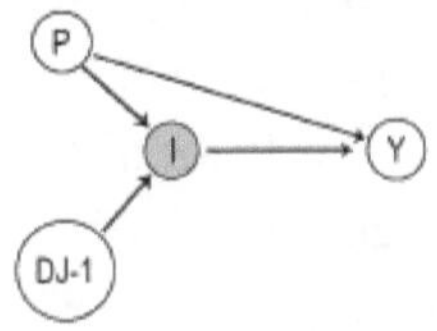

FIGURE 6. Pesticide has an effect but DJ-1 only has an effect if pesticide exposure is present.

This model is also not a plausible description of the relationship between the three variables based on the evidence at hand, as it would mean that all the families with the variant and Parkinson's would have to also have been exposed to pesticides.

MODEL III

Exposure to pesticides exacerbates the effect of the gene variant but has no effect on persons with the normal genotype. In this model, I is switched on and off by DJ-1. The model does not provide either a plausible explanation of the available evidence (Figure 7).

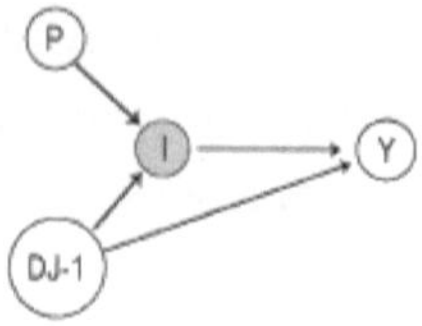

FIGURE 7. DJ-1 has an effect but pesticide only has an effect if the gene mutation is present.

MODEL IV

The environmental exposure and the gene variant both have some effect of their own but together they further modify the effect of the other. Here I is a function of both P and DJ-1 and is defined as follows: I is "on" if and only if both P and DJ-1 are "on" otherwise I is "off". Here there are also direct associations between P and Y and DJ-1 and Y other than through I; this indicates that there are effects of P on Y irrespective of DJ-1, and effects of DJ-1 on Y irrespective of P. From the data we cannot distinguish between DAGs A and B in Figure 8.

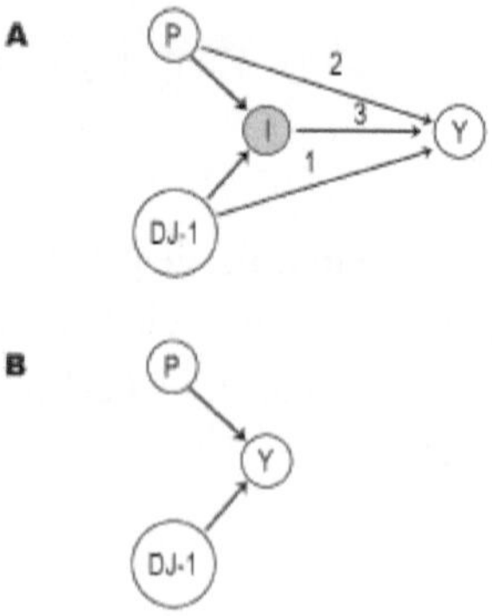

FIGURE 8. Both DJ-1 and the pesticide have an effect and there is a possible interaction in A but not in.

A core issue with these models is that I is essentially unobservable in humans living under normal conditions; these biological interactions can only be tested in animal experiments. Thus, in humans we cannot disentangle the two DAGs above apart without further information (VanderWeele and Robins [61] provide some tests to determine which individuals present Y only when the interaction I is "on" provided there is no unmeasured confounding). In order to be able to fully tell them apart, an experiment can be conducted or the relative risks can be compared (see Appendix 1).

In light of the evidence on Parkinson's disease, we have to favour one of the two models IV above the other three, as it would appear that both the genetic and the environmental exposure have separate (independent) effects on the risk of Parkinson's. However, from the data on humans we cannot distinguish between the two "type IV" models until we run a study to determine the presence of an interaction. In the case of the Drosophila experiments (see section below) the interaction model on the left-hand side provides a better explanation, as flies with the mutation that have been exposed demonstrate further sensitivity to exposure to pesticides than those who do not have the mutation.

The example we have shown exemplifies, we think, a common situation concerning the interaction between metabolic genes and environmental exposures (e. g. arylamines and NAT2, PAH and GSTM1 and many others) but has the peculiarity that experiments in Drosophila have been done (see below).

EXPERIMENTAL EVIDENCE: THE CASE OF THE DROSOPHILA

The DAGs above alone cannot be directly used for causal inference unless additional assumptions are made or experiments conducted. The reason is the limited information on potential confounders (and intermediate variables, etc.) that can influence the relationship between the three observed variables. For the sake of making the DAGs clear, we have assumed that there are no confounders; however this is unlikely to be the case in practice as Parkinson's is a multifactorial disease. The method we have proposed can however be extended to include confounders and intermediate variables.

In the case of Drosophila the situation is simpler. Meulener et al. [49] show that both exposure to pesticides and the mutation of DJ-1 may be associated with increased risk of neural degeneration. Further, the combination of the two has also been demonstrated to aggravate the condition, as the flies, which had the DJ-1 gene knocked out exhibited a ten-fold increase in sensitivity to paraquat (which would indicate a supra-multiplicative interaction).

As in this case both the genetic make-up and the exposure status of the flies have been intervened upon under controlled conditions, we can make causal inference based on this data by introducing randomisation variables into our DAG. The DAG in Figure 9 is an augmented DAG [38] that includes randomisation variables Rp and Rd. These tell us whether P or DJ-1 are being randomised or not and allow us to make inferences about interventions and, hence, causality using DAGs. For a more detailed discussion see Appendix 2.

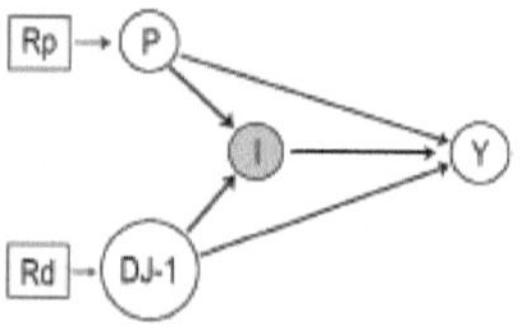

FIGURE 9. DAG representing the fruit-fly experiment where interventions were performed both on the genetic make-up and the pesticide exposure. The interaction can therefore be identified.

The DAG in Figure 9 implied that for the Drosophila at least we can state that exposure to pesticides causes an increased risk of neural damage, as does the presence of the mutated DJ-1 gene. Also as the combined presence of the mutation and paraquat further increases the risk of neural damage, we can ascertain the presence of an interaction. It should be noted that DAGs do not specify or constrain the model of statistical interaction, which can follow either an additive or a multiplicative null hypothesis model.

In the case of humans, we cannot assume such randomisation variables exist (except in Mendelian randomisation which, however, applies to gene variants only, and not to exposure); thus, we cannot expand the DAG in Figure 6. On the other hand, etiologic factors and clinical phenotypes are usually more diverse in human diseases than in animal models; inferences to human diseases from relatively simple animal experiments have well known limitations. An avenue for progress lies in integrating DAGs with the inductive reasoning implicit in Hill's guidelines.

APPLICATION OF CAUSAL GUIDELINES TO DJ-1 AND EXPOSURE TO PARAQUAT FOR PARKINSON'S DISEASE

Following the DAG approach, we established the relationship between genes and some environment exposures in promoting Parkinson's disease, and we proposed different interaction models between DJ-1, pesticides and Parkinson's disease. In order to apply Hill's causal guidelines to the DAGs we are going to work with (Figure 6A), we need to label each of the edges. Throughout the rest of this section we use the following labels:

- The edge between DJ-1 and Parkinson's disease is referred to as [edge 1],
- The edge between exposure to pesticides and Parkinson's disease is referred to as [edge 2],
- The interaction between DJ-1 and the exposure to pesticides in causing Parkinson's disease is called [edge 3].

Hill's guidelines are discussed in a slightly different order than in the original version and statistical significance is omitted because it refers to the contingent evaluation of each study and does not require a specific discussion in relation to genomics.

STRENGTH OF ASSOCIATION

DJ-1 has been seen to be lacking in Dutch families with Parkinson's disease, and to be functionally inactive because of a point mutation in the Italian families studied by Bonifati and cols [51]. The deletion showed complete cosegregation with the disease allele in the Dutch family [51]; also in the Italian family the homozygous mutation showed complete cosegregation with the disease haplotype, and absence from large numbers of control chromosomes [62]. Although the function of the DJ-1 protein is unknown, these data suggest a strong association between the DJ-1 gene and the occurrence of Parkinson's disease in certain families [edge 1]. To establish the strength of the association between specific environmental factors and a disease is far more complicated, mainly due to the quality of exposure assessment, the latency period, and body concentrations during the lifecourse. A meta-analysis of the association of pesticides and Parkinson's disease points out that both pesticide exposure in general and selective exposure to paraquat seem to be associated with Parkinson's disease, with odds ratios ranging from 1. 25 (95% C. I. : 0. 34 - 4. 36) to 3. 22 (95% C. I. : 2. 41 to 4. 31) [53] [edge 2]. With respect to the interaction parameter, there is as yet no epidemiological study that has tested whether there is an interaction between DJ-1 and pesticides; thus neither the existence nor the strength of such an association are known. However, knockout models of *Drosophila melanogaster* (fruit fly) lacking DJ-1 function, display a marked and selective sensitivity to the environmental oxidative insults exerted by both paraquat and rotenone [49], suggesting an interaction between these toxicants and the DJ-1 genotype [edge 3] in animal models and, consequently, that in humans the interaction between the chemicals and DJ-1 is biologically plausible (as can be seen, Hill's criteria often "interact", i.e., they are often related to each other, as in this paragraph the strength of association is related to the biological plausibility).

CONSISTENCY OF THE ASSOCIATION

After the first variants described, different variants of the DJ-1 gene associated with the same Parkinson's disease phenotype have been found in patients of Ashkenazi Jewish and Afro-Caribbean origins [63,64] [edge 1]. The association of paraquat and rotenone with Parkinson's disease is more consistent in animals (in which these two toxicants are often used to produce animal models of the disease) [54] than in humans. In environmental epidemiological studies in humans, the association has been found substantially consistent across studies, although some associations did not reach statistical significance, mainly due to limited sample size. In a study in Taiwan, where paraquat is routinely used in rice fields, a strong association between paraquat exposure and Parkinson's disease was found; the hazard increased by more than six times in subjects exposed for more than 20 years [64]. A dose-response curve with length of exposure was also observed in plantation workers in Hawaii [65], and British Columbia [66]. In a population-based case-control study in Calgary, occupational herbicide use was the only significant predictor of Parkinson's disease in multivariable analysis [67]. However, in another population-based case-control study in Washington, the odds ratio of 1.67 did not reach statistical significance (95% CI: 0. 22-12. 76) [68] [edge 2]. There is yet no evidence from human studies to confirm the consistency of GEIs in the causation of Parkinson's disease [edge 3]. Furthermore, genes other than DJ-1 may be involved in the etiopathogenic process, and so may be exposures other than pesticides, and other GEIs. Since environmental conditions vary substantially across the globe, and the role of one gene, one exposure or one GEI is often dependent on other genes, exposures and GEIs, lack of consistency is to be expected in studies conducted in different settings, and in particular when studies focus only on a few GEIs and overlook other interactions.

SPECIFICITY OF THE ASSOCIATION

The specificity of the association between DJ-1 gene mutations and Parkinson's disease [edge 1] will be clearer once the data on the pathological features of the DJ-1 patients will be available (see Appendix 3). Chronic systemic exposure to rotenone has been demonstrated to cause highly selective nigrostriatal dopaminergic degeneration associated with characteristic movement disorders in rats [54] [edge 2]. Similarly, paraquat caused a significant loss of nigral dopaminergic neurons in mice compared to controls [69] [edge 2]. Once an appropriate epidemiological study is set up aimed at studying GEIs in this context, results from the pathological analysis of the sample subjects will help to answer important questions regarding the aetiological pathway of the disease [edge 3].

TEMPORALITY

This criterion does not apply directly to genotype, as it is determined at conception and it remains constant over time (see Appendix 1) [edge 1]. However, temporality is crucial if we go beyond genetic effects and consider epigenetic mechanisms; e.g., gene regulation by environmental factors [14,16-18]. This problem goes beyond the present contribution, but is worth mentioning. Concerning pesticides, temporality might be a concern given that all studies on GEI in Parkinson's disease are case-control studies, which are particularly prone to selection bias, disease progression bias, and so-called "reverse causality" [3,70,71]. In this case, while it is unlikely that suffering from Parkinson's disease would have influenced past exposure to pesticides or their metabolism, it could have influenced recall. The observed dose-response relationship, with 20 years of exposure required [53], favours the existence of a true association, and is compatible with disease characteristics of neurodegeneration, making the temporality pattern suggestive of a causal role [edge 2].

BIOLOGICAL GRADIENT

This criterion does not apply since we are dealing with a recessive model of inheritance. Nonetheless, a co-dominant model should not be completely ruled out as a careful neurological evaluation of heterozygote subjects might point out some sub-clinical changes [edge 1]. A dose-response relationship between toxicant exposure and neural loss in animal experiments has been observed [72]. In addition, several studies observed a positive correlation with duration of exposure to, and high dose of, herbicides and insecticides in humans [53] [edge 2].

BIOLOGICAL PLAUSIBILITY

Biological plausibility of the DJ-1 mutation awaits the discovery and characterisation of the encoded protein [edge 1]; the capability of some toxicants to induce a progressive cellular loss in the substantia nigra and to be responsible for a progressive clinical syndrome with an intervening latent period has been hypothesized [54] [edge 2]. It is, therefore, plausible that these two factors may interact during the course of life producing Parkinson's symptoms in genetically susceptible individuals [edge 3].

COHERENCE WITH PREVIOUS KNOWLEDGE

Confirmation of the presence of different mutations on the same DJ-1 gene in families with other background origins but manifesting the same symptoms supports the involvement of the gene in the disease [63,64] [edge 1]. A role of herbicides in neurodegeneration has also been studied with generally confirmatory results [edge 2].

All these considerations taken together suggest that there may be a potential interaction between exposure to certain pesticides and the DJ-1 mutation in the risk of developing Parkinson's disease. However, as no studies on humans have yet been specifically conducted to investigate this issue, we can use the evidence only as a reason to further explore this interaction, perhaps by conducting a more targeted study. As mentioned, it is likely that other factors (both genetic and environmental) also contribute to the final development of the disease.

In the example above we have shown that the DAG approach can be complemented by the use of Hill's guidelines when no experimental evidence can be brought to bear on a particular gene-environment interaction.

CONCLUSIONS

While medical and epidemiologic evidence is routinely assessed to determine the causal nature of relationships involving personal and environmental causes of disease, genetic associations have so far not undergone similar scrutiny. However, like epidemiologic studies, genetic studies are also commonly based on observational studies, and may thus be affected by similar weaknesses. As the contribution of genetics to the understanding of disease etiology becomes more important, causal assessment of genetic and genomic evidence will become a key issue [73].

We have explored two complementary ways to tackle causality in gene-environment interactions. The application of causal guidelines to genetics is not straightforward, and it becomes very complex, in particular, if one wants to study gene-environment interactions, as we have illustrated with Parkinson's disease. Hill's criteria were developed to examine one factor at a time and have seldom been applied to evaluate the causal nature of complex relationships involving several exposures. On the other hand, graphical approaches like DAGs are effective in making potential causal networks explicit, but are insufficient to establish the strength of evidence (e.g., edges cannot be interpreted as causal without some kind of additional external support). This seems to be a general problem of causal networks, not only gene-environment interactions.

The graphical approach is useful in particular for clarifying complex causal pathways. We have applied it to a simple example where the inner workings (i.e., the detailed biological mechanisms in animal models) of the interaction are not completely known. The approach we propose uses the statistically formal representation of DAG models. This is in contrast to Weinberg's paper [74] which, although invaluable in highlighting the pros and cons of DAG models, does not actually use DAGs, but heuristic diagrams not dissimilar to those proposed by Ottman [56] and, over 35 years ago, Susser [75]. In the approach advocated by VanderWeele and Robins [76], DAGs are considered implicitly causal. We feel that this can be overly confident when the bases for inference are observational studies, which is generally the case in human genetic studies. Thus, we propose a more conservative approach that involves assessing the causal properties of each individual relationship.

A final caveat to interpreting DAGs involving genes as causal is whether genetic variants can be considered causes of diseases [30]; in a strict sense this issue is unresolved. It is generally accepted that the causal nature of a relationship can be assured when interventions (such as those performed in experiments) take place. This is because controlled interventions usually (and more easily) guarantee that the association investigated is not confounded (but this is not an absolute rule). VanderWeele and Robins [61,76] assume that genes can be considered causes of diseases, without discussing the implications or bringing additional information such as Hill's criteria into play; we believe that this is a strong assumption: knowledge on the mechanisms that govern the subclinical development and clinical course of complex diseases is rather limited.

In summary, we believe that the DAG and causal criteria-based approaches can complement one another, as one helps to assess the strength of evidence, while the other disentangles—in a visual but also formal way—the role played by genes, environmental exposures, and their interactions. The method we suggest can easily be extended to more complex situations and in particular to the understanding of gene-gene associations and interaction. The problems we raise are likely to become more relevant as genome-wide association studies provide new candidate genes for a variety of diseases, Mendelian randomization is used to assess exposure-disease associations, and gene-environment interactions are further investigated in genetics and epigenetics.

REFERENCES

1. Lawlor DA, Harbord RM, Sterne JA, Timpson N, Davey SG: Mendelian randomization: using genes as instruments for making causal inferences in epidemiology. *Stat Med* **2008**, 27:1133-1163.
2. Ioannidis JP, Boffetta P, Little J, O'Brien TR, Uitterlinden AG, Vineis P, Balding DJ, Chokkalingam A, Dolan SM, Flanders WD, Higgins JP, McCarthy MI, McDermott DH, Page GP, Rebbeck TR, Seminara D, Khoury MJ: Assessment of cumulative evidence on genetic associations: interim guidelines. *Int J Epidemiol* **2008**, 37:120-132.
3. Porta M, ed: *A Dictionary of Epidemiology*. 5th edition. New York: Oxford University Press; **2008**. p. 34-37, 65-66, 82-84, 100-103, 116, 129-130, 152-154, 237-238
4. Greenland S, ed: *Evolution of Epidemiologic Ideas. Annotated Readings on Concepts and Methods*. Chestnut Hill, MA: Epidemiology Resources; 1987.
5. Morabia A: *A History of Epidemiologic Methods and Concepts*. Basel: Birkhäuser / Springer; **2004**.
6. Fletcher RH, Fletcher SW: *Clinical Epidemiology -the Essentials*. 4th edition. Philadelphia: Lippincott Williams & Wilkins; **2005**.
7. Haynes RB, Sackett DL, Guyatt GH, Tugwell P: *Clinical epidemiology. How to do clinical practice research.* 3rd edition. Philadelphia: Lippincott, Williams & Wilkins; **2006**.
8. Hill AB: The environment and disease: association or causation? *Proc R Soc Med* **1965**, 58:295-300.
9. Vineis P, Manuguerra M, Kavvoura FK, Guarrera S, Allione A, Rosa F, Di Gregorio A, Polidoro S, Saletta F, Ioannidis JP, Matullo G: A field synopsis on low-penetrance variants in DNA repair genes and cancer susceptibility. *J Natl Cancer Inst* **2009**, 101:24-36.

10. 1Maraganore DM, de Andrade M, Elbaz A, Farrer MJ, Ioannidis JP, Krüger R, Rocca WA, Schneider NK, Lesnick TG, Lincoln SJ, Hulihan MM, Aasly JO, Ashizawa T, Chartier-Harlin MC, Checkoway H, Ferrarese C, Hadjigeorgiou G, Hattori N, Kawakami H, Lambert JC, Lynch T, Mellick GD, Papapetropoulos S, Parsian A, Quattrone A, Riess O, Tan EK, Van Broeckhoven C: Genetic Epidemiology of Parkinson's Disease (GEO-PD) Consortium. Collaborative analysis of alpha-synuclein gene promoter variability and Parkinson disease. *JAMA* **2006**, 296:661-670.
11. Ioannidis JP, Trikalinos TA, Khoury MJ: Implications of small effect sizes of individual genetic variants on the design and interpretation of genetic association studies of complex diseases. *Am J Epidemiol* **2006**, 164:609-614.
12. Yang Q, Khoury MJ, Friedman J, Little J, Flanders WD: How many genes underlie the occurrence of common complex diseases in the population? Int J Epidemiol **2005**, 34:1129-1137.
13. Ioannidis JP, Trikalinos TA: Early extreme contradictory estimates may appear in published research: the Proteus phenomenon in molecular genetics research and randomized trials. *J Clin Epidemiol* **2005**, 58:543-549.
14. Lee DH, Jacobs DR Jr, Porta M: Hypothesis: a unifying mechanism for nutrition and chemicals as lifelong modulators of DNA hypomethylation. *Environ Health Perspect* **2009**, 117:1799-1802.
15. Glazier AM, Nadeau JH, Aitman TJ: Finding genes that underlie complex traits. *Science* **2002**, 298:2345-2349.
16. Jirtle RL, Skinner MK: Environmental epigenomics and disease susceptibility. *Nat Rev Genet* **2007**, 8:253-62.
17. Feinberg AP: Phenotypic plasticity and the epigenetics of human disease. *Nature* **2007**, 447:433-440.
18. Edwards TM, Myers JP: Environmental exposures and gene regulation in disease etiology. *Environ Health Perspect* **2007**, 115:1264-1270.
19. Bellivier F, Henry C, Szöke A, Schürhoff F, Nosten-Bertrand M, Feingold J, Launay JM, Leboyer M, Laplanche JL: Serotonin transporter gene polymorphisms in patients with unipolar or bipolar depression. *Neurosci Lett* 1998, 255:143-146.
20. Kim YI: 5,10-Methylenetetrahydrofolate reductase polymorphisms and pharmacogenetics: a new role of single nucleotide polymorphisms in the folate metabolic pathway in human health and disease. *Nutr Rev* **2005**, 63:398-407.
21. Neasham D, Gallo V, Guarrera S, Dunning A, Overvad K, Tjonneland A, Clavel-Chapelon F, Linseisen JP, Malaveille C, Ferrari P, Boeing H, Benetou V, Trichopoulou A, Palli D, Crosignani P, Tumino R, Panico S, Bueno de Mesquita HB, Peeters PH, van Gib CH, Lund E, Gonzalez CA, Martinez C, Dorronsoro M, Barricarte A, Navarro C, Quiros JR, Berglund G, Jarvholm B, Khaw KT, et al.: Double-strand break DNA repair genotype predictive of later mortality and cancer incidence in a cohort of non-smokers. *DNA Repair* **2008**.
22. Marcus PM, Vineis P, Rothman N: NAT2 slow acetylation and bladder cancer risk: a meta-analysis of 22 case-control studies conducted in the general population. *Pharmacogenetics* **2000**, 10:115-122.
23. Vineis P, McMichael A: Interplay between heterocyclic amines in cooked meat and metabolic phenotype in the etiology of colon cancer. *Cancer Causes Control* **1996**, 7:479-486.

24. Vineis P, Pirastu R: Aromatic amines and cancer. *Cancer Causes Control* **1997**, 8:346-355.
25. Ochs-Balcom HM, Wiesner G, Elston RC: A meta-analysis of the association of N-acetyltransferase 2 gene (NAT2) variants with breast cancer. *Am J Epidemiol* **2007**, 166:246-254.
26. Borlak J, Reamon-Buettner SM: N-acetyltransferase 2 (NAT2) gene polymorphisms in colon and lung cancer patients. *BMC Med Genet* **2006**, 7:58.
27. Postmenopausal estrogen use and heart disease *N Engl J Med* 1986, 315:131-136.
28. Vineis P, Anttila S, Benhamou S, Spinola M, Hirvonen A, Kiyohara C, Garte SJ, Puntoni R, Rannug A, Strange RC, Taioli E: Evidence of gene gene interactions in lung carcinogenesis in a large pooled analysis. *Carcinogenesis* **2007**, 28:1902-1905.
29. Nicholson SJ, Witherden AS, Hafezparast M, Martin JE, Fisher EM: Mice, the motor system, and human motor neuron pathology. *Mamm Genome* **2000**, 11:1041-1052.
30. Porta M, Álvarez-Dardet C: How is causal inference practised in the biological sciences? *J Epidemiol Community Healt* **2000**, 54:559-560.
31. Greenland S, Pearl J, Robins JM: Causal diagrams for epidemiologic research. *Epidemiology* **1999**, 10:37-48.
32. Hernán MA, Robins JM: Instruments for causal inference: an epidemiologist's dream? *Epidemiolog* **2006**, 17:360-372.
33. Hernán MA, Robins JM: Causal Inference. New York: Chapman & Hall/CRC; **2010**.
34. Pearl J: *Causality: Models, Reasoning, and Inference*. Cambridge, U. K.: Cambridge University Press; **2009**.
35. Dawid AP: Conditional independence in statistical theory. With discussion. *J Roy Statist Soc B* **1979**, 41:1-31.
36. Cartwright N: *Nature's capacities and their measurement*. New York: Oxford University Press, **1994**.
37. Davey SG, Ebrahim S: 'Mendelian randomization': can genetic epidemiology contribute to understanding environmental determinants of disease? *Int J Epidemiol* **2003**, 32:1-22.
38. Dawid AP: Influence diagrams for causal modelling and inference. *Intern Statist Rev* **2002**, 70:161-189.
39. Didelez V, Sheehan N: Mendelian randomization as an instrumental variable approach to causal inference. *Stat Methods Med Res* **2007**, 16:309-330.
40. Chen L, Davey SG, Harbord RM, Lewis SJ: Alcohol intake and blood pressure: a systematic review implementing a Mendelian randomization approach. *PLoS Med* **2008**, 5:e52.
41. Dawid AP: Causal inference without counterfactuals. *J Am Statist Ass* **2000**, 95:407-448.
42. Didelez V, Sheenan N: Mendelian randomisation: why epidemiology needs a formal language for causality. In *Causality and probability in the sciences*. College Publications London, London; **2007**.
43. Geneletti S: Identifying direct and indirect effects in a non-counterfactual framework. *J Roy Stat Soc B* **2007**, 69:199-215.
44. Lauritzen S: *Graphical models*. Oxford, UK: Oxford University Press, **1996**.
45. Nelson LM, Tanner CM, Van Den Eeden SK, McGuire V: *Neuroepidemiology*. Oxford, UK: Oxford University Press; **2004**.

46. Kuehn BM: Scientists probe role of genes, environment in Parkinson disease. *JAMA* **2006**, 295:1883-1885.
47. Farrer MJ: Genetics of Parkinson disease: paradigm shifts and future prospects. *Nat Rev Genet* **2006**, 7:306-318.
48. Clements CM, McNally RS, Conti BJ, Mak TW, Ting JP: DJ-1, a cancer- and Parkinson's disease-associated protein, stabilizes the antioxidant transcriptional master regulator Nrf2. *Proc Natl Acad Sci USA* **2006**, 103:15091-15096.
49. Meulener M, Whitworth AJ, Armstrong-Gold CE, Rizzu P, Heutink P, Wes PD, Pallanck LJ, Bonini NM: Drosophila DJ-1 mutants are selectively sensitive to environmental toxins associated with Parkinson's disease. *Curr Biol* **2005**, 15:1572-1577.
50. van Duijn CM, Dekker MC, Bonifati V, Galjaard RJ, Houwing-Duistermaat JJ, Snijders PJ, Testers L, Breedveld GJ, Horstink M, Sandkuijl LA, van Swieten JC, Oostra BA, Heutink P: Park7, a novel locus for autosomal recessive early-onset parkinsonism, on chromosome 1p36. *Am J Hum Genet* **2001**, 69:629-634.
51. Bonifati V, Rizzu P, van Baren MJ, Schaap O, Breedveld GJ, Krieger E, Dekker MC, Squitieri F, Ibanez P, Joosse M, van Dongen JW, Vanacore N, van Swieten JC, Brice A, Meco G, van Duijn CM, Oostra BA, Heutink P: Mutations in the DJ-1 gene associated with autosomal recessive early-onset Parkinsonism. *Science* **2003**, 299:256-259.
52. Bossy-Wetzel E, Schwarzenbacher R, Lipton SA: Molecular pathways to neurodegeneration. *Nat Med* **2004**, 10(Suppl):S2-S9.
53. Brown TP, Rumsby PC, Capleton AC, Rushton L, Levy LS: Pesticides and Parkinson's disease-is there a link? *Environ Health Perspect* **2006**, 114:156-164.
54. Betarbet R, Sherer TB, MacKenzie G, Garcia-Osuna M, Panov AV, Greenamyre JT: Chronic systemic pesticide exposure reproduces features of Parkinson's disease. *Nat Neurosci* **2000**, 3:1301-1306.
55. Khoury MJ, Adams MJ Jr, Flanders WD: An epidemiologic approach to ecogenetics. *Am J Hum Genet* **1988**, 42:89-95.
56. Ottman R: An epidemiologic approach to gene-environment interaction. *Genet Epidemiol* **1990**, 7:177-185.
57. Davey Smith G, Lawlor DA, Harbord R, Timpson N, Day I, Ebrahim S: Clustered environments and randomized genes: a fundamental distinction between conventional and genetic epidemiology. *PLoS Med* **2007**, 4:e352.
58. Alcaro A, Huber R, Panksepp J: Behavioral functions of the mesolimbic dopaminergic system: an affective neuroethological perspective. *Brain Res Rev* **2007**, 56:283-321.
59. VanderWeele TJ, Robins JM: Directed acyclic graphs, sufficient causes, and the properties of conditioning on a common effect. *Am J Epidemiol* **2007**, 166:1096-1104.
60. Quik M: Smoking, nicotine and Parkinson's disease. *Trends Neurosci* **2004**, 27:561-568.
61. VanderWeele TJ, Robins JM: The identification of synergism in the sufficient-component-cause framework. *Epidemiology* **2007**, 18:329-339.
62. Bonifati V, Rizzu P, Squitieri F, Krieger E, Vanacore N, van Swieten JC, Brice A, van Duijn CM, Oostra B, Meco G, Heutink P: DJ-1 (PARK7), a novel gene for autosomal recessive, early onset Parkinsonism. *Neurol Sci* **2003**, 24:159-160.
63. Hague S, Rogaeva E, Hernandez D, Gulick C, Singleton A, Hanson M, Johnson J, Weiser R, Gallardo M, Ravina B, Gwinn-Hardy K, Crawley A, St George-Hyslop PH,

Lang AE, Heutink P, Bonifati V, Hardy J, Singleton A: Early-onset Parkinson's disease caused by a compound heterozygous DJ-1 mutation. *Ann Neurol* **2003**, 54:271-274.

64. Liou HH, Tsai MC, Chen CJ, Jeng JS, Chang YC, Chen SY, Chen RC: Environmental risk factors and Parkinson's disease: a case-control study in Taiwan. *Neurology* 1997, 48:1583-1588.
65. Petrovitch H, Ross GW, Abbott RD, Sanderson WT, Sharp DS, Tanner CM, Masaki KH, Blanchette PL, Popper JS, Foley D, Launer L, White LR: Plantation work and risk of Parkinson disease in a population-based longitudinal study. *Arch Neurol* **2002**, 59:1787-1792.
66. Hertzman C, Wiens M, Bowering D, Snow B, Calne D: Parkinson's disease: a case-control study of occupational and environmental risk factors. Am J Ind Med 1990, 17:349-355.
67. Semchuk KM, Love EJ, Lee RG: Parkinson's disease and exposure to agricultural work and pesticide chemicals. *Neurology* 1992, 42:1328-1335.
68. Firestone JA, Smith-Weller T, Franklin G, Swanson P, Longstreth WT Jr, Checkoway H: Pesticides and risk of Parkinson disease: a population-based case-control study. *Arch Neurol* **2005**, 62:91-95.
69. Corasaniti MT, Bagetta G, Rodino P, Gratteri S, Nistico G: Neurotoxic effects induced by intracerebral and systemic injection of paraquat in rats. *Hum Exp Toxicol* **1992**, 11:535-539.
70. Bertrand KA, Spiegelman D, Aster JC, Altshul LM, Korrick SA, Rodig SJ, Zhang SM, Kurth T, Laden F: Plasma organochlorine levels and risk of non-Hodgkin lymphoma in a cohort of men. *Epidemiology* **2010**, 21:172-180.
71. Porta M, Pumarega J, López T, Jariod M, Marco E, Grimalt JO: Influence of tumor stage, symptoms and time of blood draw on serum concentrations of organochlorine compounds in exocrine pancreatic cancer. *Cancer Causes Contro* **2009**, 20:1893-1906.
72. McCormack AL, Thiruchelvam M, Manning-Bog AB, Thiffault C, Langston JW, Cory-Slechta DA, Di Monte DA: Environmental risk factors and Parkinson's disease: selective degeneration of nigral dopaminergic neurons caused by the herbicide paraquat. *Neurobiol Dis* **2002**, 10:119-127.
73. Wacholder S, Chatterjee N, Caporaso N: Intermediacy and gene-environment interaction: the example of CHRNA5-A3 region, smoking, nicotine dependence, and lung cancer. *J Natl Cancer Inst* **2008**, 100:1488-1491.
74. Weinberg CR: Can DAGs clarify effect modification? *Epidemiology* **2007**, 18:569-572.
75. Susser M: *Causal thinking in the health sciences*. New York: Oxford University Press; 1973.
76. VanderWeele TJ, Robins JM: Four types of effect modification: a classification based on directed acyclic graphs. *Epidemiology* 2007, 18:561-568.

CHAPTER 5

THE BIOLOGY OF LYSINE ACETYLATION INTEGRATES TRANSCRIPTIONAL PROGRAMMING AND METABOLISM

JIGNESHKUMAR PATEL, RAVI R. PATHAK, and SHIRAZ MUJTABA

INTRODUCTION

DNA methylation and lysine modifications comprise major epigenetic processes on chromatin, which alter nucleosomal architecture leading to gene activation or repression [1-3]. Dynamic post-translational modifications (PTMs) occurring in the proximity of a gene promoter are one of the hallmarks of epigenetic regulation of gene expression [4]. Although an individual lysine residue may undergo mutually exclusive multiple PTMs, including acetylation, methylation, neddylation, ubiquitination and sumoylation, multiple lysines of a single protein can undergo diverse modifications [5,6]. Functionally, these site-specific PTMs, which are established during transcriptional programming, impart flexibility to regulate cellular processes in response to diverse physiological and external stimuli. PTMs impact functional capabilities of a protein, thus validating the notion that biological complexities are not restricted only by the number of genes [7]. To elucidate the functional consequences of a single PTM or combinatorial PTMs occurring on chromatin, the histone code hypothesis proposes to integrate the gene regulatory ability of a site-specific histone modification within its biological context [8,9]. In quintessence, a site-specific PTM serves as a mark to recruit a chromatin-associated protein complex(es) that participates in controlling gene activity, thereby, regulating cell fate decisions [10]. For instance, within chromatin, depending on the site and degree of the modification, lysine methylation can cause either gene activation or repression; lysine acetylation on histones is associated with chromatin relaxation contributing to gene activation; and the biochemical outcome of lysine ubiquitination or sumoylation is dynamic turnover of proteins. In addition, although the role of methylation in modulating non-histone proteins, including transcription factor activity, is only beginning to be understood, acetylation of transcription factors can affect their DNA-binding ability, stability, nuclear translocation and capacity to activate target genes [7,11].

Accumulating studies focusing on model systems of viral infection and the DNA-damage response have supported the role for lysine acetylation in enhancing molecular interactions between transcription factors and the transcriptional machinery on a gene promoter, leading to modulation of a specific downstream target [3,12-14]. Mechanistically, addition of an acetyl group to a lysine residue alters the positive charge of the ε-amino group, thereby impacting electrostatic properties that prevent hydrogen bonding and generating a circumferential hydrophobic milieu. Subsequently, this alteration of charge could facilitate acetylation-directed molecular interactions. Historically, almost four decades ago, acetylation of histones was first speculated to be involved in gene transcription. However, it was not until 1996 that one of the first lysine acetyltransferase (KAT), HAT-A from *Tetrahymena*, was cloned and characterized [15]. Very recently, combinatorial approaches with high-affinity acetyl-lysine antibodies, mass spectrometry (MS) and stable-isotope amino-acid labeling (SILAC) techniques detected almost 2000 acetylated proteins in the cell [16,17]. Further, the functional implications of each of these PTMs will have to be determined; one of the major tasks will be to distinguish a dynamic acetyl mark(s) specific for a pathway from a set of pre-existing global marks. Studies demonstrate that lysine acetylation can initiate molecular interplay leading to at least one of the two biochemical outcomes: 1) recruit co-activator complexes via conserved modular domains such as bromodomains; 2) engage co-repressor complexes through lysine deacetylases (KDACs) [18,19]. Published studies have utilized trichostatin A (TSA) or other KDAC inhibitors to highlight the biochemical significance of acetylation [20,21]. Long-term therapeutic aspirations stem from the pharmacological inhibition of KDACs that provides clinical benefits in models of human disease. Histone deacetylation reverts the electrostatic characteristics of chromatin in a manner that favors gene repression. Interestingly, a recent genome-wide chromatin immunoprecipitation analysis revealed preferential association of KDACs with active genes, suggesting that KDACs do not simply turn off genes, but rather function to fine-tune gene expression levels [22].

Cellular-wide proteomic analyses on protein acetylation revealed a large number of acetylated proteins, mostly enzymes involved in intermediary metabolism in the cytoplasm as well as the mitochondrion [16,23,24]. These findings support a larger role of acetylation extending beyond the nucleus mainly toward the regulation of metabolic enzymes by at least two mechanisms: 1) acetylation-mediated modulation of metabolic enzymatic activity; and 2) influencing their protein stability [17,25,26]. Given the frequent occurrence of metabolic dysregulation in human diseases, including diabetes, obesity and cancer, acetylation could play a pivotal role in the progression of these diseases. Particularly in cancers, it is well known that the transcriptional functions of the tumor suppressor p53 are affected by alterations in tumor-cell metabolism [27,28].

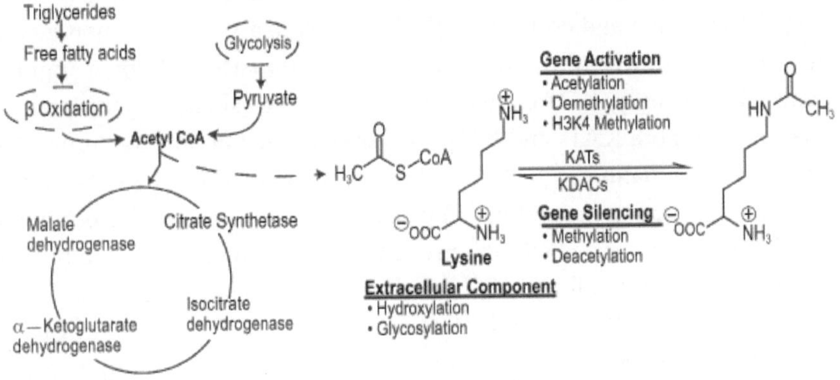

FIGURE 1. A lysine residue targeted by co-factors and enzymes mediating epigenetic events that regulate cellular processes.

THE VERSATILE AND CONSERVED NATURE OF LYSINE ACETYLATION

The versatile nature of the amino acid lysine is exhibited not only by its ability to undergo a wide range of epigenetic modifications implicated in chromatin signaling networks but also by its indispensable structural role in extracellular matrices. The ε-amino group participates in hydrogen bonding and acts as a general base in catalysis. This unusual chemical plasticity within a lysine residue eliminates steric hindrance to allow histone-modifying enzymes that are central to transcriptional regulation to perform acetylation and methylation as well as subsequent deacetylation and demethylation (Figure 1).

Lysine acetylation was initially identified in histones, so KATs and KDACs were referred to as histone acetyltransferases (HATs) and deacetylases (HDACs), respectively. There are three major groups of KATs: Gcn5-related N-acetyltransferases (GNATs); E1A-associated protein of 300 kDa (p300; KAT3A) and CBP (KAT3B); and MYST proteins [10,29]. Known KDACs are divided into classes I, II and IV and the sirtuin family (also known as class III KDACs). In humans, there are KDAC1, -2, -3, and -8 (class I); KDAC4, -5, -6, -7, -9, and -10 (class II); and KDAC11 (class IV)[30]. There are seven members of the sirtuin family in humans (SIRT1-7) [22,31]. Wang and colleagues [22] recently analyzed the genome-wide localization of KDACs and their KAT counterparts in human immune cells. Surprisingly, KDACs were not recruited to silenced gene promoters. Instead, both KATs and KDACs were enriched on inactive promoters that had methylation of histone H3 at lysine 4 (H3K4me) and were also enriched on active promoters. The occurrence of KDACs on promoters imply deacetylation, which will prevent RNA polymerase II from binding to genes that are standing by to be activated but should not yet be switched on. For instance, KDACs might also contribute to the removal of undesired basal acetylation. Collectively, these results indicate a major role for KDACs in the maintenance of gene activation.

Several studies have described acetylated proteins from mouse liver, human leukemia cells, and more recently from human liver cells [17,23,25]. Out of the 1047 acetylated proteins from human liver, 135 overlapped with 195 acetylated proteins from mouse liver. However, only 240 acetylated

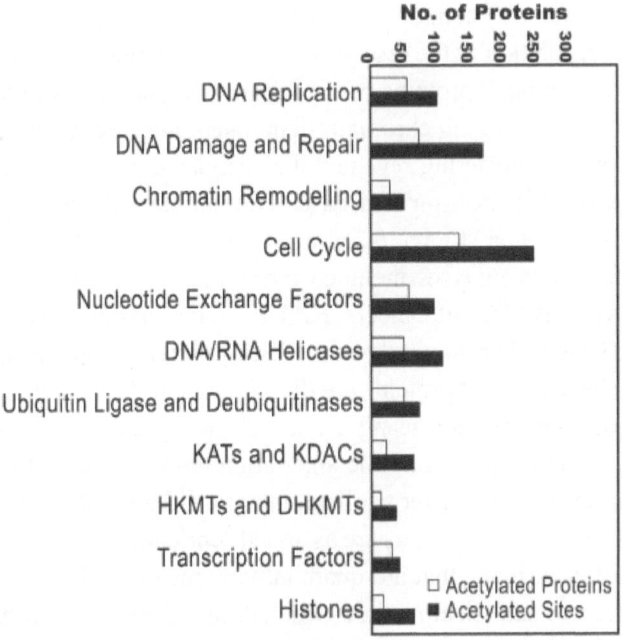

FIGURE 2. Graphic and qualitative representation of the functional distribution of acetylated proteins in a human cancer cell line.

proteins were common between the human liver and leukemia cells, suggesting that differential profiles of acetylated-proteins could be physiologically relevant and also cell-type-dependent [16]. In leukemic cells, using high-resolution MS, 3600 lysine acetylation sites were identified on 1750 proteins [16]. Our analysis of the supplementary data from that study using the functional annotation clustering tool DAVID 6. 7 showed that the lysine-acetylated proteins can be categorized into more than 500 functional clusters, thereby extending our knowledge of the cellular events that are regulated by acetylation [32,33]. This functional annotation clustering tool identifies related genes or proteins by measuring the similarity of their global annotation profiles based on the hypothesis that if two genes have similar annotation profiles, they should be functionally related. Using this rationale, the method identifies broader gene groups whose members share major biological features. Based on the output generated by this tool the acetylated proteins were determined to be involved in the regulation of numerous processes such as mRNA processing, proteolysis, GTP binding, stress responses, regulation of cell death, immune system development, neuron development and differentiation, and regulation of the protein kinase cascade. Interestingly, more than 500 acetylated unique proteins with multiple acetylation sites were categorized as being involved in chromatin-templated processes (Figure 2). Functional annotation clustering revealed the acetylated proteins to be involved in regulation of Parkinson's disease, Huntington's disease, Alzheimer's disease and glycogen storage disease. An important functional cluster that emerged from our analyses included more than 50 acetylated proteins involved in various types of cancers. KAT3B, retinoblastoma and tumor suppressor p53 figured prominently in the list of proteins implicated in human diseases. The acetylated proteins are distributed into 526 functional clusters generated by the software using a heuristic fuzzy clustering concept that measures relationships among the annotation terms based on the degree of their co-association to cluster similar annotations into functional annotation groups. The annotation clusters are assigned "enrichment scores" in decreasing order of occurrence that are quantitatively measured by some common and well-known statistical methods, including Fisher's exact test, binomial probability and hypergeometric distribution.

Lysine acetylation is a prevalent modification in enzymes that catalyze intermediary metabolism, and our analyses extended the scope of this regulation [17]. Lysine acetylated proteins are involved in the metabolism of carbohydrates, lipids, nucleotides, amino acids, secondary metabolites, and xenobiotics. Acetylation also regulates the relative activities of key enzymes controlling the course of glycolysis versus gluconeogenesis, and the branching between the citrate cycle and glyoxylate bypass. This modulation within metabolic pathways is directed by a KAT and KDAC pair whose expression levels are synchronized according to growth conditions. Reversible acetylation of metabolic enzymes ensures rapid cellular responses to environmental changes through prompt sensing of cellular energy status and flexibly altering reaction rates.

Until very recently, lysine acetylation was known only in eukaryotic cellular processes, although its existence in prokaryotes was predicted. Substantiating this idea, very recently, it was shown that reversible lysine acetylation regulates acetyl-coenzyme A synthetase activity in *Salmonella enterica* [25]. Acetylation of metabolic enzymes that depend on a carbon source indicates that acetylation may mediate adaptation to various carbon sources in *S. enterica*, which has only one major bacterial protein acetyltransferase, Pat, and one nicotinamide adenine dinucleotide (NAD+)-dependent deacetylase, CobB. To determine whether and how lysine acetylation globally regulates metabolism in prokaryotes, Zhao *et al.* determined the overall acetylation status of *S. enterica*proteins under fermentable glucose-based glycolysis and under oxidative citrate-based gluconeogenesis [17]. Moreover, those authors demonstrated that key metabolic enzymes of *S. enterica* were acetylated in response to different carbon sources concomitantly with changes in cell growth and metabolic flux.

In addition to the epigenetic modifications on lysines that occur in chromatin, collagen contains hydroxylysine, which is derived from lysine by lysyl hydroxylase. Furthermore, allysine is a derivative of lysine produced by the action of lysyl oxidase in the extracellular matrix and is essential in crosslink formation and stability of collagen and elastin. Similarly, *O*-glycosylation of lysine residues in the endoplasmic reticulum or Golgi apparatus is used to mark certain proteins for secretion from the cell. Interestingly, lysine is metabolized in mammals to give acetyl-CoA, via an initial transamination with α-ketoglutarate, which is then utilized as a substrate by KATs. Bacterial degradation of lysine yields cadaverine by decarboxylation. Although histidine and arginine are also basic amino acids, they are not subjected to PTM as is lysine. Taken together, these findings signify that mechanisms regulating metabolism may be evolutionarily conserved from bacteria to mammals. Furthermore, characterization of acetylation-mediated regulatory mechanisms in bacteria would offer new perspectives in advancing our understanding of many hitherto unknown biological processes. In the remainder of this chapter, we concentrate on a few important proteins that require acetylation to execute their functions and which have been widely investigated but still remain a subject of intense of biochemical research.

THE ACETYLATION-DIRECTED TRANSCRIPTIONAL PROGRAM: ACETYLATION ENGENDERED CHROMATIN MILIEU ON GENE REGULATION

The earliest explanation of histone acetylation was in the physicochemical context that nucleosomes and chromatin impose a barrier to transcription. Subsequently, it became apparent that lysine acetylation neutralizes the positive charges on histone tails (Figure 3), relaxing their electrostatic grip on DNA to cause nucleosomal remodeling that exposes transcription factor binding sites [34]. Furthermore, because acetylated lysine moieties on histone tails could serve as recruitment sites for bromodomain-containing cofactors or reversal of charge by KDACs, this not only suggested an additional mechanism for KAT-directed gene activation [35], but also established that acetylation, like phosphorylation, creates a new scaffold to recruit proteins to the nucleosome. Notably, in charge-neutralization models, acetylation of multiple lysine residues—that is, hyperacetylation—in a single histone tail should produce a stronger effect than mono-acetylation. By contrast, in bromodomain-recruitment models, in which adjacent amino acids determine specificity, a single lysine residue on a histone tail is paramount, and possible hyperacetylation of the entire tail may not be expected to contribute further to recruitment [36]. It is also possible that for a specific lysine residue, both modalities of acetylation may be physiologically relevant and apply in different circumstances, as suggested from *in vitro* studies of H4-K16 acetylation [37,38]. Moreover, adding to the complexities of acetylation, it cannot be ruled out that acetylated moieties may also recruit KDACs to regulate tightly and temporally transcriptional activation, as mentioned above.

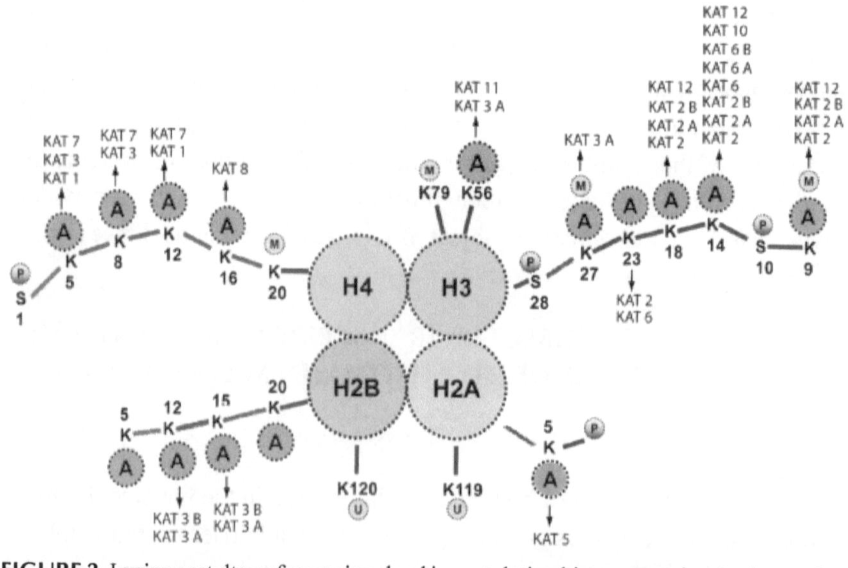

FIGURE 3. Lysine acetyltransferases involved in acetylating histone proteins. In chromatin, A denotes acetylation; M, methylation; P, phosphorylation and U, ubiquitination.

THE PUZZLING CROSSTALK ON CHROMATIN BETWEEN POST-TRANSLATIONALLY MODIFIED SITES

ACETYLATION VERSUS METHYLATION

A growing body of evidence suggests that independent of their proximity, co-existing histone modifications can have synergistic or antagonistic effects on gene expression. This also highlights that epigenetic marks are not deposited or recognized in isolation but comprise a complex and interrelated collection of modifications at adjacent residues on a given nucleosome of a gene promoter. The correlation between different histone modifications is particularly clear for acetylation of histone H3 (Figure 3) and methylation of histone H3 at lysine 4. This is consistent with the observed co-localization of these marks, which show correlated distribution patterns both on a chromosome-wide scale during X inactivation and over the coding regions of individual genes[39,40]. These correlations may arise due to physical links between histone-modifying enzymes such that they are co-recruited to the same loci. Both KMT2A/MLL1, a lysine methyltransferase (KMT) that can generate H3K4me marks [41], and Chd1, the chromatin remodeler that is subsequently recruited by this methyl mark, associate with KAT activities [42,43], whereas the LSD1 complex that removes some of these methyl marks contains the lysine deacetylases KDAC1 and KDAC2 [44]. However, the interaction could also arise due to the mechanism of action of these enzymes. For example, the SET domain of KMT2A has a preference for acetylated substrates [41].

ACETYLATION VERSUS PHOSPHORYLATION

Multiple cellular processes are associated with histone phosphorylation: DNA damage induces phosphorylation on serine 139 of H2A (H2AS139p) [10,45]; transcription, upon mitogenic stimulation, on H3S10p [10]; mitosis on H3S10p and H3S28p; apoptosis, depending on the stimulus used, on H4S1p, H3S10p, H2BS32p, and H2AS32p [9,10]. Serum stimulation induces the PIM1 kinase to phosphorylate pre-acetylated histone H3 at the FOSL1 enhancer [46]. The adaptor protein 14-3-3 binds the phosphorylated nucleosome and recruits KAT8/MOF, which triggers acetylation of H4K16 [47]. Histone crosstalk generates the nucleosomal recognition code composed of H3K9ac/H3S10p/H4K16ac that determines the nucleosome platform for binding of the bromodomain protein BRD4 [48,49]. Recruitment of the positive transcription elongation factor b (P-TEFb) via BRD4 induces the release of the promoter-proximal paused RNA polymerase II and increases its processivity. Thus, the single phosphorylation on H3S10 at the FOSL1 enhancer triggers a cascade of events, which activates transcriptional elongation. Increasing evidence also show that several types of modifications are linked and, in particular, one modification may influence the presence of a nearby modification [9,10,50]. This has been demonstrated for H3K14ac and H3S10p on the histone H3 tail, as well as, for H3S10p and H3K9me on the same tail [51]. Whereas the first pair of modifications has been coupled to activation of gene expression, increasing evidence indicates that H3K9me results in decreased H3S10p and is thereby responsible for silencing.

IMPACT OF ACETYLATION ON THE FUNCTIONING OF TRANSCRIPTION FACTORS

The tumor suppressor protein p53 functions as a transcription factor to orchestrate a transcriptional program that controls many target genes during a wide variety of stress responses[52,53]. After sensing a genetic aberration (such as DNA damage), the sequence-specific DNA-binding ability of p53 enables it to directly participate in controlling target gene transcription to alter cellular responses. In addition to being a DNA-binding transcription factor, p53, following stress, undergoes extensive PTMs to enhance its function as a transcription factor in controlling cellular decisions that could culminate in cell-cycle arrest, senescence or apoptosis [52,53] (Figure 4). Interestingly, p53 has a short half-life; however, depending on the nature of the stress, the cell type and the profile of PTMs rendered, p53 will promptly execute a transcriptional program beneficial to the cell [54]. In addition, through protein-protein interactions, p53 can bind to and recruit general transcription proteins, TAFs (TATA-binding protein-associated factors), to induce transcription of target genes [55-57]. Recent experiments have shown that p53 can also engage with KATs, including KAT3B, KAT3A, KAT5/Tip60 and KAT2B/PCAF to the promoter region of genes [58-60].

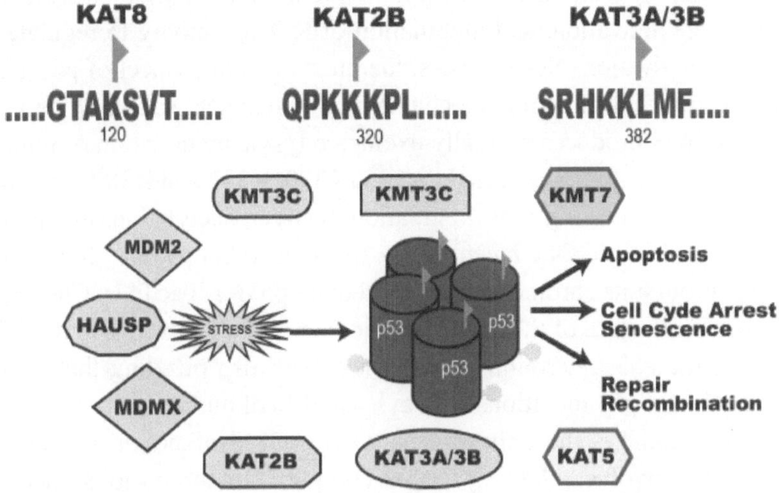

FIGURE 4. Lysine acetyltransferases responsible for mediation of the stress response by p53. The green flag stands for acetylation and the yellow ball-stick represents phosphorylation of p53 during stress.

In 1997, Gu and Roeder showed that acetylation of p53 on its C-terminal lysines by KAT3A/3B is crucial for p53 activation during DNA damage [59,60]. Subsequently, the biochemical significance of p53 acetylation was established in cancer cell lines under various genotoxic stresses and oncogenic Ras activation that lead to the interaction of acetylated p53 with KAT3A and PML [61-63]. In parallel, it was proposed that p53, upon activation, undergoes a wave of phosphorylation on its N-terminus that precludes its degradation by MDM2 and concomitantly brings in KAT3A/3B to acetylate the C-terminal end of p53 [64,65]. At this stage, KAT3A/3B-catalyzed p53 acetylation was implicated in enhancing p53's DNA-binding ability, nuclear localization and co-activator recruitment functions [11]. Later, KAT2B was also shown to acetylate lysine 320 of human p53 and lysine 317 of mouse p53 [62]. Recent identification of the p53K120ac site in the p53 DNA-binding region supports a direct role for acetylation in p53-DNA interactions. Lysine 120 on p53 is acetylated by KAT5 and/or KAT8/MOF [58,66]. Taken together, the impact of acetylation on p53 function can be attributed to the inhibition of nonspecific DNA-binding, recruitment of bromodomain-containing co-activators for target-gene activation, and modulation of KDAC activity to regulate target-gene activation. Nevertheless, the most puzzling aspect of p53 PTMs remains to be clarified: the mechanism by which acetylation of the C-terminus of p53 produces mutually exclusive lysine methylation or ubiquitination. Recent studies revealed that p53K370, K372 and K382 can also be methylated, indicating cross-regulation between acetylation, methylation and ubiquitination. One recent study speculates that p53K372me recruits KAT5 through its chromodomain to mediate p53K120ac [67]. Clearly, the biochemical nature of the multi-layered and mutually exclusive modifications of the p53 C-terminus is complex. It is also puzzling that, despite association of p53 mutations with at least 50% of human cancers, Li Fraumeni syndrome is the only disease where p53 dysfunction is known to be directly involved. Adding further complexity, homozygous mice with seven lysines on p53 mutated to arginine are viable and apparently phenotypically normal [68]. Similarly, in mice with six K→R mutations in p53, expression of the protein was unaffected [69].

Since the discovery of p53 acetylation 15 years ago, numerous studies have revealed quite unexpected complexity. However, these studies also provide valuable lessons for investigating acetylation of other proteins. One realization is that, like histone acetylation, p53 acetylation does not act alone but forms an integral part of an intricate, multisite modification program. One of the strongest pieces of evidence to support the idea that the PTMs of p53 are relevant to the p53 regulatory mechanism is the fact that KDAC inhibitors have been shown to simultaneously increase the levels of acetylated p53 and induce apoptosis or senescence in cancerous and normal cells[70]. Although the PTMs of p53 are certainly important, our ability to properly identify which ones are relevant under what conditions remains rudimentary.

In addition to p53, transcription factors of the nuclear factor kappa-light-chain-enhancer of activated B cells (NF-B) family are essential regulators of the inflammatory and immune responses [71]. Acetylation of

RelA/p65 by KAT3A/3B probably is associated with transcriptional activation [72,73]. Although multiple acetylation sites on p65 protein have been reported (e.g., lysine 310, 314, 315), it's the acetylation of lysine 310 that has been observed to enhance transcriptional activity without altering binding to DNA or IκB [74-76]. Acetylation of lysine 310 is blocked either in the absence of serine 276 phosphorylation or by overexpression of catalytically inactive PKAc [73]. Thus, it is speculated that phosphorylation of serine 276 on p65 triggers recruitment of KAT3A/3B that next acetylates lysine 310 on p65. It is further proposed that, in addition to p65 phosphorylation, IKKα also promotes acetylation through direct phosphorylation of N-CoR/SMRT, which displaces KDAC3 from the SMRT corepressor complex [77]. IKKα is found associated with the κB sites of NF-κB-responsive genes and stimulus-induced phosphorylation of H3 serine 10 [74,78].

In addition to cellular transcription factors, viral proteins interact to manipulate the function host's nuclear factors. It is established that control of the immune network by a human pathogenic virus starts with cooptation of the host's transcription and replication machineries. Interestingly, KATs also acetylate viral proteins; and subsequent molecular events occurring post-acetylation of viral proteins aid in control of the host's transcriptional machinery [12,79]. The best-known example is HIV transactivator protein Tat, which undergoes acetylation on lysines 28 and 50 to promote rapid replication of the HIV proviral genome [79,80]. Acetylation-mediated interactions between the cellular transcription machinery and viral proteins offer new therapeutic avenues, especially since anti-HIV drugs targeted against HIV proteins have been reported to cause drug resistance [80]. Taken together, acetylation of chromatin and transcription factors is a widespread phenomenon that not only facilitates gene regulation but also participates in numerous other cellular processes.

NEXUS LINKING DYSREGULATION OF METABOLISM AND P53 TRANSCRIPTION FUNCTIONS IN CANCERS

The availability of proper nutrients directly supports the synthesis of biological macromolecules that promote the growth and survival of cells and the organism. In contrast, starvation could limit cellular growth in order to sustain self-survival by using energy primarily from the breakdown of macromolecules rather than by synthesis. Clearly, metabolic pathways are tightly regulated to produce energy to allow efficient cellular growth and survival programs [81]. Evidently, tumor cells depend on metabolic changes for their continued growth and survival, and these alterations enhance the uptake of glucose and glutamine by cancer cells [82]. Therefore, components of metabolic pathways could provide new opportunities to explore potential therapeutic targets in the treatment of malignant disease. In most normal cells, the tricarboxylic acid (TCA) cycle drives the generation of ATP in the presence of oxygen, a process known as oxidative phosphorylation. However, under conditions of limiting O_2 or when energy is needed rapidly, glycolysis becomes the preferred route of energy production [83]. The preference of cancer cells to employ glycolysis may be a sign of response to hypoxia, which occurs as the tumor outgrows the blood supply. Although p53 can be activated during many stressors, it is speculated that hypoxia is indeed one of them. In cellular responses to hypoxia that involve the transcription factor hypoxia inducible factor (HIF), it has been shown that induction of p53 under low oxygen concentration may trigger HIF-p53 interaction [84]. In addition, reduced nutrient or energy levels fail to stimulate both the AKT-mTOR pathway and AMP-activated protein kinase, which responds to an increased AMP/ATP ratio, resulting in p53 activation [85,86]. Furthermore, AKT activates MDM2 that regulates p53 stability; therefore, reduced AKT function will preclude MDM2 negative regulation of p53, leading to p53 activation under low-nutrient conditions. Malate dehydrogenase that converts malate to oxaloacetic acid in the TCA cycle has been shown to interact with p53 during deprivation of cellular glucose levels [87]. Okorokov and Milner noted that ADP promoted and ATP inhibited the ability of p53 to bind DNA [88]. Several studies have also documented that p53 has the capability of slowing down the glycolytic pathway to control the growth of cancerous cells by inhibiting the expression of glucose transporters [89]. Furthermore, although the underlying mechanism is still unclear, p53 can also inhibit NF-κB-mediated pro-survival pathways by limiting the activity of IκB kinase α and IκB kinase β functions [90]. Collectively, these findings indicate that a lack of nutrients and deregulation of nutrient-sensing pathways can each modulate a p53 response and that combinations of these abnormalities during tumor progression amplify the protective p53 response.

ACETYLATION AND REGULATION OF METABOLIC ENZYMES

Our analysis by functional clustering of the "lysine acetylome" revealed that at least 92 proteins were involved in metabolic events and energy production, including the TCA cycle, glycolysis, pyruvate metabolism, and fatty acid metabolism (Figure 5). Furthermore, as shown in Figure 6, a significant number of enzymes and the respective pathways in which they are implicated could be regulated by acetylation. For instance, 24 proteins in the TCA pathway can be acetylated. Recently, it was reported that the activities of key enzymes regulating the choice of glycolysis versus gluconeogenesis and the branching between the TCA cycle and glyoxylate bypass could possibly be regulated by acetylation [25]. In this context, acetyl coenzyme A (Ac-coA) is particularly important owing to its unique role as the acetate donor for all cellular acetylation reactions. In mammalian cells, Ac-coA is synthesized in two pathways. In the first pathway, Ac-coA synthetase condenses acetate and coenzyme A into Ac-coA. In the second pathway, energy from hydrolyzed ATP is utilized by ATP-citrate lyase (ACL) to convert citrate, a TCA cycle intermediate, and coenzyme A into Ac-coA and oxaloacetate. Cytoplasmic Ac-coA serves as a building block for lipids, whereas nuclear Ac-coA contributes to acetylation of chromatin and its associated proteins. This demarcation within Ac-coA metabolism is necessitated by cellular need of the coenzyme. Very recently, it was demonstrated that histone acetylation in several human cell lines relies primarily on ACL activity, thus linking the TCA cycle, glycolysis and intracellular energy status to gene activity [91]. Abrogation of ACL activity results in alteration of global histone acetylation and gene transcription. Using gene knockdown strategies, it was determined that ACL is the major source of Ac-coA for histone acetylation under normal growth conditions. However, acetate supplementation following cellular deprivation of ACL is able to rescue histone acetylation, suggesting that Ac-coA production by Ac-coA synthetase can compensate for the decrease in ACL activity, which is dependent upon the availability of acetate. This mechanism may allow the acetate that is generated during histone deacetylation to be recycled back to Ac-coA. Hence, because the pool of Ac-coA for epigenetic control arises from the TCA cycle and is influenced by energy flux, the metabolic status of cells is intertwined with gene transcription. Importantly, given the continuous requirement for the high-energy compounds Ac-coA, S-adenosyl methionine (SAM) and coenzyme NAD (or NAD+) for chromatin modifications, it is of high priority to investigate whether the abundance of the latter two compounds also contributes to transcriptional regulation via epigenetic mechanisms.

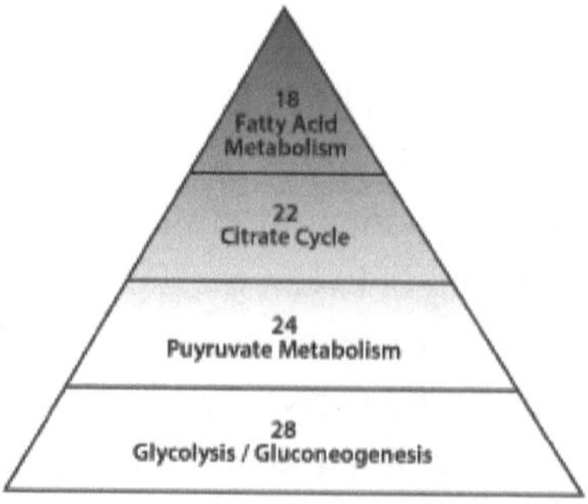

FIGURE 5. Role of lysine acetylation in enzymes involved in intermediary metabolism. The pyramid provides an overview of the involvement as well as overlapping roles of metabolic enzymes such as pyruvate dehydrogenase, acetyl coA synthetase and pyruvate kinase in four different metabolic pathways.

Similar to Ac-coA, NAD is a key compound that captures electrons in the form of hydride during glycolysis and the TCA cycle. In contrast to many reactions in which NAD is the essential coenzyme and only undergoes a change in redox state, in sirtuin-mediated deacetylation reactions NAD is hydrolyzed into nicotinamide and *O*-acetyl-ribose. The former compound is a potent inhibitor of sirtuin KDAC activity, whereas the latter is a signaling molecule [92]. Because of the obligatory need of sirtuins for NAD in catalysis and their susceptibility to nicotinamide inhibition, the activity of sirtuins is controlled by the intracellular ratio of NAD to NADH [93,94]. During each cycle of glycolysis and TCA, in which cells derive energy from glucose and pyruvate breakdown, NAD is reduced into NADH, thus decreasing the NAD:NADH ratio that inhibits sirtuin activity. Conceptually, this reduction of sirtuin function may then be compensated for by downregulation of Ac-coA synthetase, due to its inactivation by acetylation. Fluctuation of NAD abundance modulates the activity of sirtuins that act on acetylated chromatin and transcription factors.

Histone deacetylation not only represses transcription, but also inhibits recombination. One of the functions of yeast sirtuin Sir2p is to suppress the formation of rDNA extrachromosomal circles that have been postulated to be related to cellular senescence [95]. Thus, the metabolism and availability of NAD may impact both the genome and cellular physiology in a multitude of ways, including global and local changes in nucleosomal organization and the functions of transcription factors regulated by lysine acetylation.

POTENTIAL OF TARGETING ACETYLATION

The molecular events that follow acetylation could lead to recruitment of either bromodomain-containing proteins or KDACs [9,35]. Therefore, on the one hand, the enzymes that catalyze acetylation are obvious targets of intervention; on the other hand, proteins that interact with the acetylated-lysine moiety could also be potential targets [35,96]. With respect to KATs, two natural products, anacardic acid and garcinol (a polyprenylated benzophenone), are reported to inhibit both KAT3A/3B and KAT2B in the 5-10 micromolar range *in vitro* [97-99]. In contrast, curcumin displays selective activity against KAT3A/3B but not KAT2B [99]. Subsequent studies suggest that anacardic acid may be a broad-spectrum HAT inhibitor, as it also interferes with the KAT5 [100]. Isothiazolones were identified in a high-throughput screen as inhibitors of KAT2B and KAT3A/3B[101]. These compounds could be broadly useful as biological tools for evaluating the roles of HATs in transcriptional studies and may serve as lead agents for the development of novel anti-neoplastic therapeutics. Recent studies show that small molecules designed against the acetyl-lysine-binding hydrophobic pocket of conserved bromodomains affect transcriptional regulation and other cellular processes in cancer cells [102,103]. Furthermore, small-molecule modulators of KDACs have already emerged as promising therapeutic agents for cancer, cardiac illness, diabetes, and neurodegenerative disorders. Hence, studies focusing on lysine acetylation as well as molecular events that follow acetylation could identify non-histone

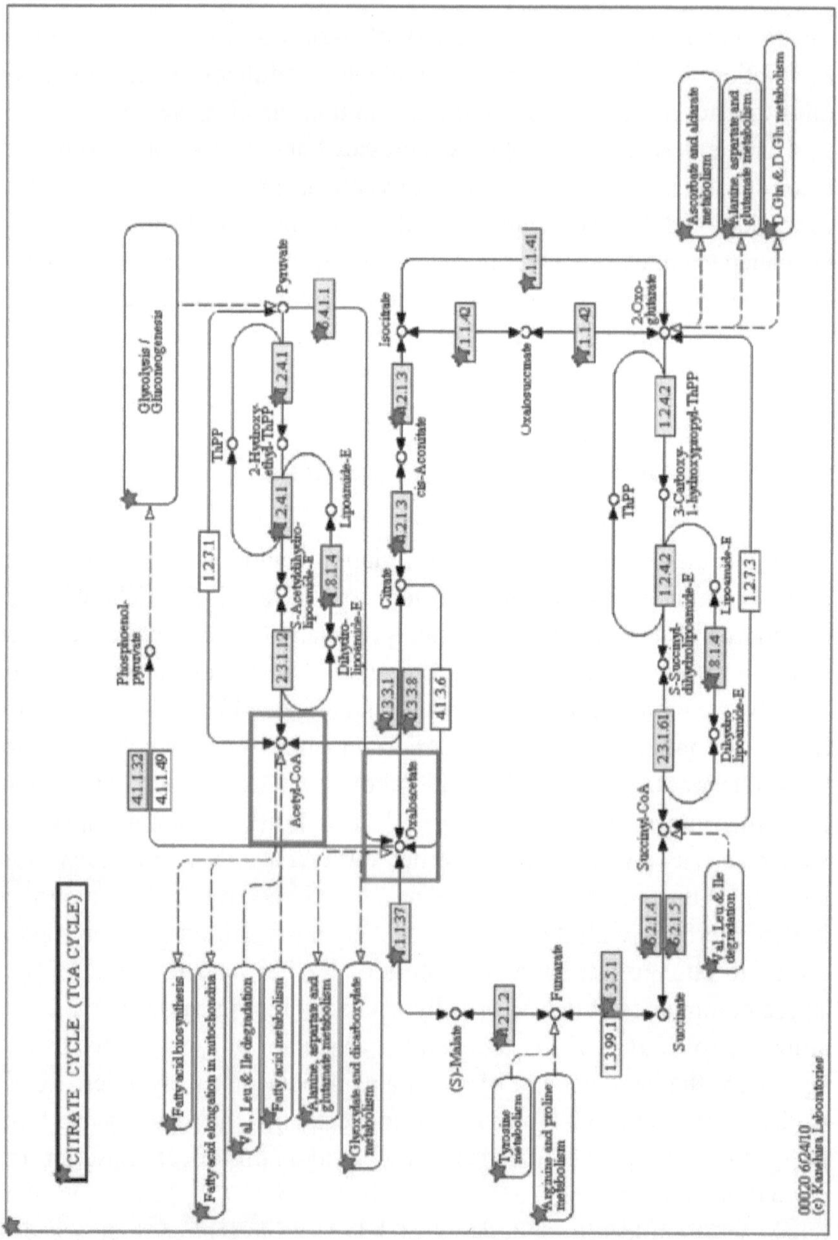

FIGURE 6. An integrated view of the metabolic processes (fatty acid biosynthesis, glycolysis, gluconeogenesis and amino acid metabolism) that converge on the citric acid cycle. Red stars represent lysine-acetylated metabolic enzymes and the red box highlights the critical metabolic products integrated into various metabolic pathways.

FIGURE 6 continued...
Highlighted enzymes with their respective EC numbers are: pyruvate dehydrogenase (EC 1. 2. 4. 1), dihydrolipoyl dehydrogenase (EC 1. 8. 1. 4), pyruvate carboxylase (EC 6. 4. 1. 1), malate dehydrogenase (EC 1. 1. 1. 37), fumarate hydratase (EC 4. 2. 1. 2), citrate (Si)-synthase (EC 2. 3. 3. 1), ATP citrate synthase (EC 2. 3. 3. 8), aconitate hydratase (EC 4. 2. 1. 3), isocitrate dehydrogenase (NADP+) (1. 1. 1. 42), isocitrate dehydrogenase (NAD+) (1. 1. 1. 41), succinate dehydrogenase (1. 3. 5. 1), succinate--CoA ligase (6. 2. 1. 4; GDP forming), and succinate--CoA ligase (6. 2. 1. 5; ADP forming).

targets for KAT- and KDAC-modulating compounds as well as illuminate molecular basis of signaling on chromatin and unravel new avenues to improve the efficacy of related therapeutic agents.

CONCLUSIONS AND FUTURE PERSPECTIVES

Evidently, many acetylated proteins are not only key components within nuclear processes, but also play crucial roles in signaling pathways, such as DNA damage response, immune network and inflammation. This has propelled the idea that instead of phosphorylation being the major contributor, signaling pathways are possibly controlled by the synchronized actions of phosphorylation, acetylation, and several other PTMs. Although acetylation regulates the activity of metabolic enzymes, the role of phosphorylation in conjunction with acetylation in metabolic pathways is not clear. However, what is clear is that lysine acetylation definitely expands the plasticity within the metabolic and cellular signaling networks. This notion is reinforced by the recent analyses of the "lysine acetylome" explicated above, which broadened the scope of acetylation-mediated regulation through an expansive clustering into diverse functional groups. This list offers new insights into the role of acetylation and possible routes to dissect its mechanism, especially in regulation of diseases like cancer and neurodegenerative disorders.

Meanwhile, proteomic surveys by MS will continue to identify new acetylated proteins, which along with efficient mapping of acetylation sites by MS should reveal additional sites [104]. For instance, accumulating studies on p53 acetylation indicate that, subsequent to *in vitro* biochemical characterization, cell lines and genetically altered mouse models will be especially effective for characterizing the biological functions associated with particular lysine PTMs [105]. Most importantly, such approaches will facilitate mapping within signaling pathways that are regulated by reversible acetylation. One pertinent question is how such modifications interact with other PTMs within the same or different protein(s) and form dynamic programs for regulating cellular functions under normal and pathological settings. Within the acetyl-proteome, the functional impact of lysine acetylation is context-dependent and varies from protein to protein. As in histones, the molecular interplay of lysine acetylation with other PTMs, either agonistically or antagonistically, generates codified molecular signaling programs that are crucial for governing the functions of various nuclear and cytoplasmic proteins [9,106-108].

REFERENCES

1. Vaissiere T, Sawan C, Herceg Z, Epigenetic interplay between histone modifications and DNA methylation in gene silencing. *Mutat Res* **2008**, 659, 40-48.
2. Shukla V, Vaissiere T, Herceg Z, Histone acetylation and chromatin signature in stem cell identity and cancer. *Mutat Res* **2008**, 637, 1-15.
3. Kondo Y, Epigenetic cross-talk between DNA methylation and histone modifications in human cancers. *Yonsei Med J* **2009**, 50, 455-463.
4. Campos EI, Reinberg D, Histones, annotating chromatin. *Annu Rev Genet* **2009**, 43, 559-599.
5. Kouzarides T, Chromatin modifications and their function. *Cell* **2007**, 128, 693-705.
6. Ruthenburg AJ, Li H, Patel DJ, Allis CD, Multivalent engagement of chromatin modifications by linked binding modules. *Nat Rev Mol Cell Biol* **2007**, 8, 983-994.
7. Sims RJ, Reinberg D, Is there a code embedded in proteins that is based on post-translational modifications? *Nat Rev Mol Cell Biol* **2008**, 9, 815-820.
8. Jenuwein T, Allis CD, Translating the histone code. *Science* **2001**, 293, 1074-1080._
9. Strahl BD, Allis CD, The language of covalent histone modifications. *Nature* **2000**, 403, 41-45._
10. Cheung P, Allis CD, Sassone-Corsi P, Signaling to chromatin through histone modifications. *Cell* **2000**, 103, 263-271.
11. Prives C, Manley JL, Why is p53 acetylated? *Cell* **2001**, 107, 815-818.
12. Mujtaba S, He Y, Zeng L, Farooq A, Carlson JE, *et al.*, Structural basis of lysine-acetylated HIV-1 Tat recognition by PCAF bromodomain. *Mol Cell* **2002**, 9, 575-586.
13. Mujtaba S, He Y, Zeng L, Yan S, Plotnikova O, *et al.*, Structural mechanism of the bromodomain of the coactivator CBP in p53 transcriptional activation. *Mol Cell* **2004**, 13, 251-263.
14. Barlev NA, Liu L, Chehab NH, Mansfield K, Harris KG, *et al.*, Acetylation of p53 activates transcription through recruitment of coactivators/histone acetyltransferases. *Mol Cell* **2001**, 8, 1243-1254.

15. Brownell JE, Zhou J, Ranalli T, Kobayashi R, Edmondson DG, *et al.*, Tetrahymena histone acetyltransferase A, a homolog to yeast Gcn5p linking histone acetylation to gene activation. *Cell* 1996, 84, 843-851.
16. Choudhary C, Kumar C, Gnad F, Nielsen ML, Rehman M, *et al.*, Lysine acetylation targets protein complexes and co-regulates major cellular functions. *Science* **2009**, 325, 834-840.
17. Zhao S, Xu W, Jiang W, Yu W, Lin Y, *et al.*, Regulation of cellular metabolism by protein lysine acetylation. *Science* **2010**, 327, 1000-1004.
18. Winston F, Allis CD, The bromodomain, a chromatin-targeting module? *Nat Struct Biol* 1999, 6, 601-604.
19. Glozak MA, Sengupta N, Zhang X, Seto E, Acetylation and deacetylation of non-histone proteins. *Gene* **2005**, 363, 15-23.
20. Marson CM, Histone deacetylase inhibitors, design, structure-activity relationships and therapeutic implications for cancer. *Anticancer Agents Med Chem* **2009**, 9, 661-692.
21. Peh KH, Wan BY, Assem ES, Middleton JB, Dines J, *et al.*, Mode of action of histone deacetylase inhibitors on mast cell histamine release and colon muscle contraction. *Inflamm Res* **2009**, 58(Suppl 1), 24-25.
22. Wang Z, Zang C, Cui K, Schones DE, Barski A, *et al.*, Genome-wide mapping of HATs and HDACs reveals distinct functions in active and inactive genes. *Cell* **2009**, 138, 1019-1031.
23. Kim SC, Sprung R, Chen Y, Xu Y, Ball H, *et al.*, Substrate and functional diversity of lysine acetylation revealed by a proteomics survey. *Mol Cell* **2006**, 23, 607-618.
24. Schwer B, Eckersdorff M, Li Y, Silva JC, Fermin D, *et al.*, Calorie restriction alters mitochondrial protein acetylation. *Aging Cell* **2009**, 8, 604-606.
25. Wang Q, Zhang Y, Yang C, Xiong H, Lin Y, *et al.*, Acetylation of metabolic enzymes coordinates carbon source utilization and metabolic flux. *Science* **2010**, 327, 1004-1007.
26. Yu W, Lin Y, Yao J, Huang W, Lei Q, *et al.*, Lysine 88 acetylation negatively regulates ornithine carbamoyltransferase activity in response to nutrient signals. *J Biol Chem* **2009**, 284, 13669-13675.
27. Gottlieb E, Vousden KH, p53 regulation of metabolic pathways. *Cold Spring Harb Perspect Biol* **2010**, 2, a001040.
28. Vousden KH, Ryan KM, p53 and metabolism. *Nat Rev Cancer* **2009**, 9, 691-700.
29. Lee KK, Workman JL, Histone acetyltransferase complexes, one size doesn't fit all. *Nat Rev Mol Cell Biol* **2007**, 8, 284-295.
30. Yang XJ, Seto E, The Rpd3/Hda1 family of lysine deacetylases, from bacteria and yeast to mice and men. *Nat Rev Mol Cell Biol* **2008**, 9, 206-218.
31. Haigis MC, Guarente LP, Mammalian sirtuins--emerging roles in physiology, aging, and calorie restriction. *Genes Dev* **2006**, 20, 2913-2921.
32. Huang da W, Sherman BT, Lempicki RA, Systematic and integrative analysis of large gene lists using DAVID bioinformatics resources. *Nat Protoc* **2009**, 4, 44-57.
33. Dennis G, Sherman BT, Hosack DA, Yang J, Gao W, *et al.*, DAVID, Database for Annotation, Visualization, and Integrated Discovery. *Genome Biol* **2003**, 4, P3.
34. 'Davie JR, Covalent modifications of histones, expression from chromatin templates. *Curr Opin Genet Dev* 1998, 8, 173-178.

35. Zeng L, Zhou MM, Bromodomain, an acetyl-lysine binding domain. *FEBS Lett* **2002**, 513, 124-128.
36. Clayton AL, Hazzalin CA, Mahadevan LC, Enhanced histone acetylation and transcription, a dynamic perspective. *Mol Cell* **2006**, 23, 289-296.
37. Shogren-Knaak M, Ishii H, Sun JM, Pazin MJ, Davie JR, *et al.*, Histone H4-K16 acetylation controls chromatin structure and protein interactions. *Science* **2006**, 311, 844-847.
38. Shogren-Knaak M, Peterson CL, Switching on chromatin, mechanistic role of histone H4-K16 acetylation. *Cell Cycle* **2006**, 5, 1361-1365.
39. Schubeler D, MacAlpine DM, Scalzo D, Wirbelauer C, Kooperberg C, *et al.*, The histone modification pattern of active genes revealed through genome-wide chromatin analysis of a higher eukaryote. *Genes Dev* **2004**, 18, 1263-1271.
40. Pokholok DK, Harbison CT, Levine S, Cole M, Hannett NM, *et al.*, Genome-wide map of nucleosome acetylation and methylation in yeast. *Cell* **2005**, 122, 517-527.
41. Milne TA, Briggs SD, Brock HW, Martin ME, Gibbs D, *et al.*, MLL targets SET domain methyltransferase activity to Hox gene promoters. *Mol Cell* **2002**, 10, 1107-1117.
42. Pray-Grant MG, Daniel JA, Schieltz D, Yates JR, Grant PA, Chd1 chromodomain links histone H3 methylation with SAGA- and SLIK-dependent acetylation. *Nature* **2005**, 433, 434-438.
43. Sims RJ, Chen CF, Santos-Rosa H, Kouzarides T, Patel SS, *et al.*, Human but not yeast CHD1 binds directly and selectively to histone H3 methylated at lysine 4 via its tandem chromodomains. *J Biol Chem* **2005**, 280, 41789-41792.
44. Schneider J, Wood A, Lee JS, Schuster R, Dueker J, *et al.*, Molecular regulation of histone H3 trimethylation by COMPASS and the regulation of gene expression. *Mol Cell* **2005**, 19, 849-856.
45. Xiao A, Li H, Shechter D, Ahn SH, Fabrizio LA, *et al.*, WSTF regulates the H2A.X DNA damage response via a novel tyrosine kinase activity. *Nature* **2009**, 457, 57-62.
46. Zippo A, De Robertis A, Serafini R, Oliviero S, PIM1-dependent phosphorylation of histone H3 at serine 10 is required for MYC-dependent transcriptional activation and oncogenic transformation. *Nat Cell Biol* **2007**, 9, 932-944.
47. Gupta A, Sharma GG, Young CS, Agarwal M, Smith ER, *et al.*, Involvement of human MOF in ATM function. *Mol Cell Biol* **2005**, 25, 5292-5305.
48. Karam CS, Kellner WA, Takenaka N, Clemmons AW, Corces VG, 14-3-3 mediates histone cross-talk during transcription elongation in Drosophila. *PLoS Genet* **2010**, 6, e1000975.
49. Zippo A, Serafini R, Rocchigiani M, Pennacchini S, Krepelova A, *et al.*, Histone crosstalk between H3S10ph and H4K16ac generates a histone code that mediates transcription elongation. *Cell* **2009**, 138, 1122-1136.
50. Mizzen CA, Allis CD, Transcription. New insights into an old modification. *Science* **2000**, 289, 2290-2291.
51. Rea S, Eisenhaber F, O'Carroll D, Strahl BD, Sun ZW, *et al.*, Regulation of chromatin structure by site-specific histone H3 methyltransferases. *Nature* **2000**, 406, 593-599.
52. Vogelstein B, Lane D, Levine AJ, Surfing the p53 network. *Nature* **2000**, 408, 307-310.
53. Prives C, Hall PA, The p53 pathway. *J Pathol* 1999, 187, 112-126.

54. Kruse JP, Gu W, SnapShot, p53 posttranslational modifications. *Cell* **2008**, 133, 930-930.
55. Farmer G, Colgan J, Nakatani Y, Manley JL, Prives C, Functional interaction between p53, the TATA-binding protein (TBP), andTBP-associated factors in vivo. *Mol Cell Biol* 1996, 16, 4295-4304.
56. Farmer G, Friedlander P, Colgan J, Manley JL, Prives C, Transcriptional repression by p53 involves molecular interactions distinct from those with the TATA box binding protein. *Nucleic Acids Res* 1996, 24, 4281-4288.
57. Thut CJ, Chen JL, Klemm R, Tjian R, p53 transcriptional activation mediated by coactivators TAFII40 and TAFII60. *Science* 1995, 267, 100-104.
58. Tang Y, Luo J, Zhang W, Gu W, Tip60-dependent acetylation of p53 modulates the decision between cell-cycle arrest and apoptosis. *Mol Cell* **2006**, 24, 827-839.
59. Gu W, Roeder RG, Activation of p53 sequence-specific DNA binding by acetylation of the p53 C-terminal domain. *Cell* 1997, 90, 595-606.
60. Gu W, Shi XL, Roeder RG, Synergistic activation of transcription by CBP and p53. *Nature* 1997, 387, 819-823.
61. Liu L, Scolnick DM, Trievel RC, Zhang HB, Marmorstein R, *et al.*, p53 sites acetylated in vitro by PCAF and p300 are acetylated in vivo in response to DNA damage. *Mol Cell Biol* 1999, 19, 1202-1209.
62. Sakaguchi K, Herrera JE, Saito S, Miki T, Bustin M, *et al.*, DNA damage activates p53 through a phosphorylation-acetylation cascade. *Genes Dev* 1998, 12, 2831-2841.
63. Pearson M, Carbone R, Sebastiani C, Cioce M, Fagioli M, *et al.*, PML regulates p53 acetylation and premature senescence induced by oncogenic Ras. *Nature* **2000**, 406, 207-210.
64. Ashcroft M, Kubbutat MH, Vousden KH, Regulation of p53 function and stability by phosphorylation. *Mol Cell Biol* 1999, 19, 1751-1758.
65. Haupt Y, Maya R, Kazaz A, Oren M, Mdm2 promotes the rapid degradation of p53. *Nature* 1997, 387, 296-299.
66. Sykes SM, Mellert HS, Holbert MA, Li K, Marmorstein R, *et al.*, Acetylation of the p53 DNA-binding domain regulates apoptosis induction. *Mol Cell* **2006**, 24, 841-851.
67. Huang J, Dorsey J, Chuikov S, Perez-Burgos L, Zhang X, *et al.*, G9a and Glp methylate lysine 373 in the tumor suppressor p53. *J Biol Chem* **2010**, 285, 9636-9641.
68. Krummel KA, Lee CJ, Toledo F, Wahl GM, The C-terminal lysines fine-tune P53 stress responses in a mouse model but are not required for stability control or transactivation. *Proc Natl Acad Sci USA* **2005**, 102, 10188-10193.
69. Feng L, Lin T, Uranishi H, Gu W, Xu Y, Functional analysis of the roles of posttranslational modifications at the p53 C terminus in regulating p53 stability and activity. *Mol Cell Biol* **2005**, 25, 5389-5395.
70. Langley E, Pearson M, Faretta M, Bauer UM, Frye RA, *et al.*, Human SIR2 deacetylates p53 and antagonizes PML/p53-induced cellular senescence. *EMBO J* **2002**, 21, 2383-2396.
71. Natoli G, Saccani S, Bosisio D, Marazzi I, Interactions of NF-kappaB with chromatin, the art of being at the right place at the right time. *Nat Immunol* **2005**, 6, 439-445.
72. Zhong H, May MJ, Jimi E, Ghosh S, The phosphorylation status of nuclear NF-kappa B determines its association with CBP/p300 or HDAC-1. *Mol Cell* **2002**, 9, 625-636.

73. Zhong H, Voll RE, Ghosh S, Phosphorylation of NF-kappa B p65 by PKA stimulates transcriptional activity by promoting a novel bivalent interaction with the coactivator CBP/p300. *Mol Cell* 1998, 1, 661-671.
74. Chen LF, Greene WC, Shaping the nuclear action of NF-kappaB. *Nat Rev Mol Cell Biol* **2004**, 5, 392-401.
75. Chen L, Fischle W, Verdin E, Greene WC, Duration of nuclear NF-kappaB action regulated by reversible acetylation. *Science* **2001**, 293, 1653-1657.
76. Buerki C, Rothgiesser KM, Valovka T, Owen HR, Rehrauer H, *et al*., Functional relevance of novel p300-mediated lysine 314 and 315 acetylation of RelA/p65. *Nucleic Acids Res* **2008**, 36, 1665-1680.
77. Baek SH, Ohgi KA, Rose DW, Koo EH, Glass CK, *et al*., Exchange of N-CoR corepressor and Tip60 coactivator complexes links gene expression by NF-kappaB and beta-amyloid precursor protein. *Cell* **2002**, 110, 55-67.
78. Yamamoto Y, Verma UN, Prajapati S, Kwak YT, Gaynor RB, Histone H3 phosphorylation by IKK-alpha is critical for cytokine-induced gene expression. *Nature* **2003**, 423, 655-659.
79. Ott M, Schnolzer M, Garnica J, Fischle W, Emiliani S, *et al*., Acetylation of the HIV-1 Tat protein by p300 is important for its transcriptional activity. *Curr Biol* 1999, 9, 1489-1492.
80. Mujtaba S, Zhou MM, Anti-viral opportunities during transcriptional activation of latent HIV in the host chromatin. *Methods* **2011**, 1(53), 97-101.
81. Hardie DG, Biochemistry. Balancing cellular energy. *Science* **2007**, 315, 1671-1672.
82. Tennant DA, Duran RV, Boulahbel H, Gottlieb E, Metabolic transformation in cancer. *Carcinogenesis* **2009**, 30, 1269-1280.
83. Pfeiffer T, Schuster S, Bonhoeffer S, Cooperation and competition in the evolution of ATP-producing pathways. *Science* **2001**, 292, 504-507.
84. An WG, Kanekal M, Simon MC, Maltepe E, Blagosklonny MV, *et al*., Stabilization of wild-type p53 by hypoxia-inducible factor 1alpha. *Nature* 1998, 392, 405-408.
85. Hardie DG, AMPK and SNF1, Snuffing Out Stress. *Cell Metab* **2007**, 6, 339-340.
86. Hardie DG, AMP-activated/SNF1 protein kinases, conserved guardians of cellular energy. *Nat Rev Mol Cell Biol* **2007**, 8, 774-785.
87. Lee SM, Kim JH, Cho EJ, Youn HD, A nucleocytoplasmic malate dehydrogenase regulates p53 transcriptional activity in response to metabolic stress. *Cell Death Differ* **2009**, 16, 738-748.
88. Okorokov AL, Milner J, An ATP/ADP-dependent molecular switch regulates the stability of p53-DNA complexes. *Mol Cell Biol* 1999, 19, 7501-7510.
89. Schwartzenberg-Bar-Yoseph F, Armoni M, Karnieli E, The tumor suppressor p53 down-regulates glucose transporters GLUT1 and GLUT4 gene expression. *Cancer Res* **2004**, 64, 2627-2633.
90. Kawauchi K, Araki K, Tobiume K, Tanaka N, p53 regulates glucose metabolism through an IKK-NF-kappaB pathway and inhibits cell transformation. *Nat Cell Biol* **2008**, 10, 611-618.
91. Wellen KE, Hatzivassiliou G, Sachdeva UM, Bui TV, Cross JR, *et al*., ATP-citrate lyase links cellular metabolism to histone acetylation. *Science* **2009**, 324, 1076-1080.
92. Sauve AA, Wolberger C, Schramm VL, Boeke JD, The biochemistry of sirtuins. *Annu Rev Biochem* **2006**, 75, 435-465.

93. Fulco M, Schiltz RL, Iezzi S, King MT, Zhao P, *et al.*, Sir2 regulates skeletal muscle differentiation as a potential sensor of the redox state. *Mol Cell* **2003**, 12, 51-62.
94. Imai S, Armstrong CM, Kaeberlein M, Guarente L, Transcriptional silencing and longevity protein Sir2 is an NAD-dependent histone deacetylase. *Nature* **2000**, 403, 795-800.
95. Imai S, Johnson FB, Marciniak RA, McVey M, Park PU, *et al.*, Sir2, an NAD-dependent histone deacetylase that connects chromatin silencing, metabolism, and aging. *Cold Spring Harb Symp Quant Biol* **2000**, 65, 297-302.
96. Dhalluin C, Carlson JE, Zeng L, He C, Aggarwal AK, *et al.*, Structure and ligand of a histone acetyltransferase bromodomain. *Nature* 1999, 399, 491-496.
97. Balasubramanyam K, Altaf M, Varier RA, Swaminathan V, Ravindran A, *et al.*, Polyisoprenylated benzophenone, garcinol, a natural histone acetyltransferase inhibitor, represses chromatin transcription and alters global gene expression. *J Biol Chem* **2004**, 279, 33716-33726.
98. Balasubramanyam K, Swaminathan V, Ranganathan A, Kundu TK, Small molecule modulators of histone acetyltransferase p300. *J Biol Chem* **2003**, 278, 19134-19140.
99. Balasubramanyam K, Varier RA, Altaf M, Swaminathan V, Siddappa NB, *et al.*, Curcumin, a novel p300/CREB-binding protein-specific inhibitor of acetyltransferase, represses the acetylation of histone/nonhistone proteins and histone acetyltransferase-dependent chromatin transcription. *J Biol Chem* **2004**, 279, 51163-51171.
100. Sun Y, Jiang X, Chen S, Price BD, Inhibition of histone acetyltransferase activity by anacardic acid sensitizes tumor cells to ionizing radiation. *FEBS Lett* **2006**, 580, 4353-4356.
101. Stimson L, Rowlands MG, Newbatt YM, Smith NF, Raynaud FI, *et al.*, Isothiazolones as inhibitors of PCAF and p300 histone acetyltransferase activity. *Mol Cancer Ther* **2005**, 4, 1521-1532.
102. Filippakopoulos P, Qi J, Picaud S, Shen Y, Smith WB, *et al.*, Selective inhibition of BET bromodomains. *Nature* **2010**, 468, 1067-1073.
103. Mujtaba S, Zeng L, Zhou MM, Structure and acetyl-lysine recognition of the bromodomain. *Oncogene* **2007**, 26, 5521-5527.
104. Huq MD, Wei LN, Post-translational modification of nuclear co-repressor receptor-interacting protein 140 by acetylation. *Mol Cell Proteomics* **2005**, 4, 975-983.
105. Toledo F, Wahl GM, Regulating the p53 pathway, in vitro hypotheses, in vivo veritas. *Nat Rev Cancer* **2006**, 6, 909-923.
106. Margueron R, Trojer P, Reinberg D, The key to development, interpreting the histone code? *Curr Opin Genet Dev* **2005**, 15, 163-176.
107. Berger SL, The complex language of chromatin regulation during transcription. *Nature* **2007**, 447, 407-412.
108. Latham JA, Dent SY, Cross-regulation of histone modifications. *Nat Struct Mol Biol* **2007**, 14, 1017-1024.

CHAPTER 6

ROLES OF HISTONE DEACETYLASES IN EPIGENETIC REGULATION: EMERGING PARADIGMS FROM STUDIES WITH INHIBITORS

GENEVIÈVE P. DELCUVE, DILSHAD H. KHAN, and JAMES R. DAVIE

INTRODUCTION

Acetylation of the lysine ε-amino group, first discovered on histones, is a dynamic posttranslational modification (PTM) regulated by the opposing activities of lysine acetyltransferases (KATs) and histone deacetylases (HDACs). Histone acetylation is a modulator of chromatin structure involved in DNA replication, DNA repair, heterochromatin silencing and gene transcription [1,2]. Hyperacetylation usually marks transcriptionally active genes, as it contributes to the decondensed chromatin state and maintains the unfolded structure of the transcribed nucleosome [2-6]. Moreover, specific acetylated sites on core histones are read by bromodomain modules found in proteins, and sometimes in KATs, which are components of chromatin-remodeling complexes involved in transcriptional activation [7]. Conversely, HDACs are found in corepressor complexes and, by removing acetyl groups from histones, induce the formation of a compacted, transcriptionally repressed chromatin structure. As discussed below, however, this model reflects quite an oversimplification of the role of HDACs in transcription regulation.

Many nonhistone proteins (transcription factors, regulators of DNA repair, recombination and replication, chaperones, viral proteins and others) are also subject to acetylation [8-10]. Investigators in a recent study used high-resolution mass spectrometry to identify 3,600 acetylation sites in 1,750 human proteins and showed that lysine acetylation is implicated in the regulation of nearly all nuclear functions and many cytoplasmic processes [11]. Furthermore, acetylation is regulated by and/or regulates other PTMs. Through either recruitment or occlusion of binding proteins, PTMs may lead to or prevent a secondary PTM on histones and nonhistone proteins [12,13]. In particular, histone H3 phosphorylation on serine 10 or 28, rapid and transient PTMs in response to the stimulation of signaling pathways such as the mitogen-activated protein kinase (MAPK) pathways, are associated with histone acetylation and transcriptional activation of specific genes [14]. A crosstalk also exists between histone acetylation and H3 methylation. Although acetylation is generally linked to transcription activation, the effect of methylation depends on which amino acid residue is modified and the degree to which this residue is methylated (mono-, di- or trimethylation of lysine). Methylation of H3 lysine 4 or 36 is associated with transcription activation, but methylation of lysine 9 or 27 is linked to transcription repression [15,16].

To date, 18 different mammalian HDACs have been identified and divided into four classes based on their sequence similarity to yeast counterparts [17,18]. HDACs from the classical family are dependent on Zn^{2+} for deacetylase activity and constitute classes I, II and IV. Class I HDACs, closely related to yeast RPD3, comprise HDAC1, HDAC2, HDAC3 and HDAC8. Class II HDACs, related to yeast HDA1, are divided into subclass IIa (HDAC4, HDAC5, HDAC7 and HDAC9) and subclass IIb (HDAC6 and HDAC10). Class IV contains only HDAC11. Class III HDACs consist of seven sirtuins, which require the NAD^+ cofactor for activity.

Inhibitors of Zn^{2+}-dependent HDACs were originally discovered as inducers of transformed cell growth arrest and cell death and only later were identified as inhibitors of HDAC activity [19]. It was recognized that HDACs are upregulated in many cancers or aberrantly recruited to DNA following chromosomal translocations, particularly in hematologic malignancies [20,21]. The specificity of HDAC inhibitors toward tumor cells, although poorly understood, has led to their development as anticancer drugs. More recently, clinical studies using HDAC inhibitors have been extended to a range of nononcologic diseases, such as sickle cell anemia, HIV infection, cystic fibrosis, muscular dystrophy and neurodegenerative and inflammatory disorders [21-23]. The use of HDAC inhibitors also constitutes a chemical approach to the study of HDAC cellular functions. In addition, crucial developmental and physiological roles of HDACs have been elucidated by knockout studies [21,24].

CLASS I HDAC COMPLEXES

Class I HDACs are ubiquitously expressed nuclear enzymes, although HDAC8 is generally poorly expressed [17]. Except for HDAC8, class I HDACs are components of multiprotein complexes. Knockout studies have shown that class I HDACs are involved in cell proliferation and survival [21,24]. As products of an evolutionary recent gene duplication, HDAC1 and HDAC2 exhibit a high degree of homology (85% identity for human proteins) [18,25] and have undergone little functional divergence, although specific and distinct roles have also been identified for each of them [26,27]. For example, targeted disruption of the *Hdac1* alleles in mouse results in cell proliferation defects and embryonic lethality by embryonic day E9. 5 [28], whereas mice lacking HDAC2 survive at least until the perinatal period [29-31]. On the basis of their differential distributions in the brain at distinct stages of neuroglial development, HDAC1 and HDAC2 appear to have different functions during the development of the central nervous system [32]. Moreover, HDAC2, but not HDAC1, negatively regulates memory formation and synaptic plasticity [33]. Surprisingly, researchers in a recent study suggested that HDAC1 has a protective role against the formation of immature teratomas with high malignant potential in both mouse studies and human patients [34].

HDAC1 and HDAC2 form homo- and heterodimers between each other [26,35,36], which presumably allows them to act together or separately from each other. The dimer is a requirement for HDAC activity [36]. Dissociation of the dimer with a HDAC1 N-terminal peptide will inhibit HDAC activity [36]. Viruses have capitalized on this mechanism to inhibit HDAC activity. The adenoviral protein GAM1 inhibits HDAC1 activity by binding to the N-terminal region of HDAC1, which likely dissociates the dimer [37].

HDAC1 and HDAC2 heterodimer levels seem to depend on the cell type, because it was shown that 80% to 90% of HDAC1 and HDAC2 proteins were associated with each other in the nucleus of human breast cancer MCF-7 cells [38], whereas 40% to 60% of HDAC1 and HDAC2 proteins were found to be free from each other in mouse embryonic fibroblasts [39]. Moreover, a genomewide mapping study in primary human $CD4^+$ T cells revealed a differential distribution of HDAC1 and

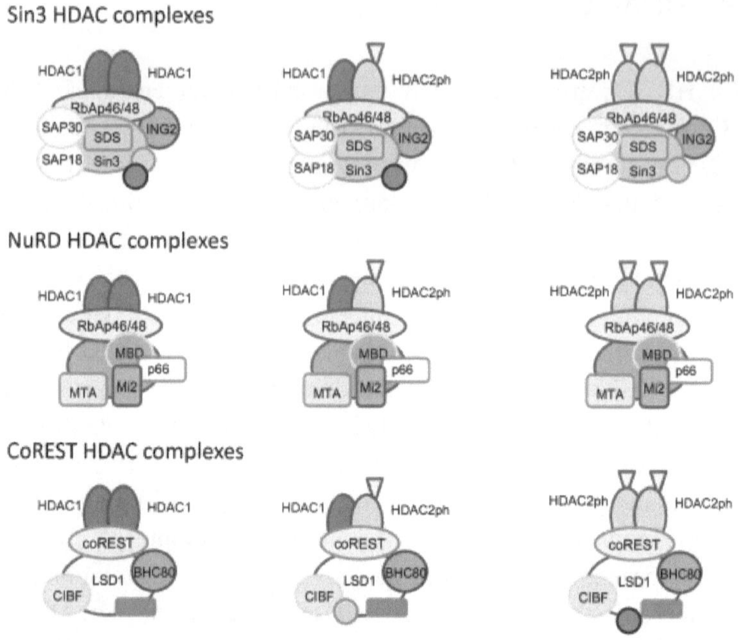

FIGURE 1. Class I HDAC1-HDAC2 multiprotein complexes. Multiprotein complexes containing HDAC1-HDAC2 homo- or heterodimers are shown. HDAC2 is shown as phosphorylated, which is a requirement for multiprotein complex formation. Phosphorylation is indicated by a the triangles.

HDAC2 along regulatory and coding regions [40]. Conversely, HDAC1 and HDAC2 were both associated with regulatory and coding regions in MCF-7 cells [38,41]. HDAC1 and HDAC2 relative expression levels also vary with cell types. For example, T-lymphocyte Jurkat cells express negligible levels of HDAC2 compared to HDAC1 levels [42], and throughout the adult brain HDAC2 is preferentially expressed in neurons, whereas HDAC1 is more abundant in glial cells [32,33]. It is likely that, at least in cells expressing markedly different relative levels of HDAC1 and HDAC2, homodimer formation would prevail over heterodimer formation.

HDAC1 and HDAC2 are both found in multiprotein corepressor complexes Sin3, nucleosome-remodeling HDAC (NuRD) and CoREST, which are recruited to chromatin regulatory regions by transcription factors (for example, Sp1, Sp3, p53, NF-κB and YY1) and have very diverse, often cell-specific, roles (Figure 1) [17,43]. Although it is generally assumed that both HDACs can be paired within the same complex, to the best of our knowledge, it has been demonstrated only in studies using exogenously expressed, tagged HDAC1 and not in studies characterizing endogenous HDAC corepressor complexes [36]. The Sin3 core complex contains Sin3A or Sin3B, HDAC1 and/or HDAC2, SAP18, SAP30 and retinoblastoma-associated proteins (RbAps) RbAp46 and RbAp48 and serves as a platform for the addition of other modules with enzymatic functions such as nucleosome remodeling, DNA methylation, histone methylation and N-acetylglucosamine transferase activity [44,45].

The NuRD complex has a variable composition that is dependent on the cell type and external stimuli. It is the only complex holding both HDAC- and ATP-dependent chromatin-remodeling activities, which are carried out by HDAC1 and/or HDAC2 and Mi-2α and/or Mi-2β, respectively. The other known components of NuRD are structural and/or regulatory proteins RbAp46/RbAp48 and, in some instances, also p66α or p66β, the methyl-CpG-binding domain-containing proteins (MBD2 or MBD3), with only MBD2 being able to recognize methylated DNA and the three members of the metastasis-associated protein family (MTA1, MTA2 or MTA3), with different MTA proteins allowing distinct downstream responses to the activation of different signaling pathways [45,46]. Lysine-specific demethylase 1 (KDM1/LSD1) has also been identified as a component of NuRD [47].

HDAC1 and HDAC2 are also components of the Nanog- and Oct4-associated deacetylase (NODE) complex, a NuRD-related repression complex, also comprising MTA1 or MTA2, p66α or p66β, but not the histone-binding proteins RbAp46/RbAp48 and the helicase-like ATPase Mi-2. NODE is involved in the control of embryonic stem cell fate by repressing Nanog and Oct4 target genes [48].

Also including HDAC1 and HDAC2, but composed of proteins distinct from those of Sin3 and NuRD, the CoREST complex is recruited by the RE1 silencing transcription (REST) factor, also known as the "neuronal restricted silencing factor" (NRSF), to the RE1 DNA motif associated with many genes encoding fundamental neuronal traits. As a component of the CoREST complex and as a consequence of histone H3 deacetylation, KDM1/LSD1 promotes demethylation of H3 dimethylated on lysine 4 (H3K4me2), an event that facilitates the formation of a repressive chromatin structure [49,50]. Although CoREST acts as a corepressor in terminally differentiating nonneuronal cells by recruiting KDM1/LSD1 to demethylate H3K4me2 and the methyltransferase G9a to methylate H3K9 at the RE1 sites of target genes, it acts as a coactivator of transcription in embryonic stem cells and neural stem cells by recruiting an H3K4 methyltransferase to the RE1 sites of target genes [51]. CoREST can also form larger complexes by association with ZNF217, a Krüppel-like zinc finger protein and strong candidate oncogene product found in breast cancer, or with other complexes, such as the chromatin-remodeling complex SWI/SNF or the C-terminal binding protein (CtBP) complex [45,52]. Interestingly, CoREST appears to be involved in the negative regulation of synaptic plasticity and memory formation by HDAC2 [33]. Chromatin immunoprecipitation (ChIP) and immunocoprecipitation experiments performed on mouse forebrains showed that, though both HDAC1 and HDAC2 were incorporated in Sin3 and NuRD complexes and were found to be enriched at the promoters of cell-cycle genes, HDAC2 was preferentially associated with the CoREST complex to repress neuronal gene expression [33].

A novel HDAC complex, MiDAC, is specific to mitotic cells and includes HDAC1, HDAC2, either one of the related ELM-SANT proteins MIDEAS or TRERF1, and DNTTIP1 (terminal deoxynucleotidyl transferase (TdT)-interacting protein), although the authors who published these findings suggested that the MiDAC complex has a TdT-independent function in cell division. Whether the putative histone acetylase CDYL is also a MiDAC component is presently unclear [53].

The discussion above illustrates that the HDAC1 and HDAC2 homo- or heterodimer can exist with different proteins. The combination of these proteins likely determines the overall activity, substrate specificity and genomic location of the HDAC1 and/or HDAC2 containing complex.

The two highly related complexes nuclear receptor corepressor (NCoR or NCOR1) and silencing mediator of retinoic acid and thyroid hormone receptor (SMRT or NCOR2) consist of HDAC3, transducin β-like 1 (TBL1), TBL-related 1 (TBLR1) and G protein pathway suppressor 2 (GPS2)[45,54]. NCoR and SMRT also interact with class IIa HDACs, which exhibit no deacetylase activity of their own but are believed to recruit NCoR/SMRT HDAC3 activity to distinct promoters through their associated factors, such as myocyte enhancer factor 2 (MEF2) [55]. NCoR, but not SMRT, interacts with zinc finger and BTB domain-containing 33 (ZBTB33 or Kaiso), which is a protein that binds to methylated DNA. NCoR and SMRT are regulated by different kinase pathways and play different roles in development. Although NCoR binds preferentially to the thyroid hormone receptor, SMRT prefers the retinoic acid receptor [45,54]. It is noteworthy that repression by NCoR/SMRT is an integral phase of the cyclical process that is the transcriptional activation of genes controlled by liganded receptors. NCoR/SMRT repression is necessary to prime chromatin for subsequent transcription initiation [56]. TBL1 and TBLR1 are involved in the active dismissal of corepressor complexes [54]. Besides its role in transcriptional control, the HDAC3-NCoR-SMRT axis is critical to the maintenance of heterochromatin content and genomic stability [57].

Histone deacetylation also acts in concert with the Polycomb repressive complexes (PRC1/PRC2) or the G9a complex, which catalyze the trimethylation of H3K27 or H3K9, respectively. H3K27me3 is a repressive mark that can easily be reverted, thus conferring plasticity to the chromatin structure. The DNA of Polycomb target genes is generally unmethylated, but some genes can undergo *de novo* DNA methylation under certain circumstances, notably in cancer cells. H3K9me3, on the other hand, is associated with DNA methylation and is a stable mark, denoting permanent silencing [58,59]. As mentioned above, the cooperation of DNA methylation and histone deacetylation in gene silencing is also established by the recruitment of complexes such as Sin3 or NuRD via proteins that bind methylated DNA such as MeCP2 or MBD2.

CLASS II HISTONE DEACETYLASES

Class II HDACs shuttle between nucleus and cytoplasm and have tissue-specific expression and functions [21,24,43,60]. Class IIa HDACs (HDAC4, HDAC5, HDAC7 and HDAC9) are signal transducers characterized by the presence in their regulatory N-terminal domains of two or three conserved serine residues subject to reversible phosphorylation.

Phosphorylation leads to the binding of the 14-3-3 proteins, the nuclear export of HDACs and the derepression of their target genes. A range of kinases and phosphatases acting downstream of diverse biological pathways have been shown to regulate the nucleocytoplasmic trafficking of class IIa HDACs [61]. Because of a substitution of Tyr with His in their catalytic site, class IIa HDACs have negligible intrinsic deacetylase activity but are able to bind acetylated lysine. It has been suggested that, under some circumstances, class IIa HDACs may act as bromodomains, recognizing acetylated lysine in a sequence-dependent context and recruiting chromatin-modifying enzymes to regulate transcription [62]. As mentioned above, class IIa HDAC association with MEF2 provides additional targeting for the SMRT-NCoR complex [55]. Class IIa HDACs also interact with numerous other transcription factors. However, the biological relevance of these associations has been established only for the MEF2-regulated processes [63,64]. Class IIa HDACs are not affected by most HDAC inhibitors at pharmacologically relevant concentrations [62].

Class IIb HDACs (HDAC6 and HDAC10) have duplicated catalytic domains, albeit the duplication is partial in the case of HDAC10. HDAC6 and HDAC10 shuttle between nucleus and cytoplasm, but their location is primarily cytoplasmic. Little is known of the role of HDAC10. HDAC6 is an α-tubulin deacetylase as well as a cortactin deacetylase and thus is involved in the control of microtubule- and actin-dependent cell motility. Chaperone protein HSP90 is another substrate of HDAC6. Moreover, HDAC6 plays a critical role in the cellular clearance of misfolded proteins via formation of aggresomes or autophagy [43]. Thus HDAC6 is a potential therapeutic target for the treatment of an array of diseases, including neurodegenerative diseases and cancer [65-67].

CLASS IV HISTONE DEACETYLASE 11

HDAC11 has sequence similarity to classes I and II HDACs. Aside from its evolutionary conservation's implying a vital role across species [43] and a study suggesting a role in the decision between immune activation and immune tolerance [68], little is known of HDAC11 functions.

SELECTIVITY OF HISTONE DEACETYLASE INHIBITORS

The active site of Zn^{2+}-dependent HDACs consists of a tubular pocket with two adjacent histidine residues, two aspartic acid residues, one tyrosine residue (substituted with histidine in class IIa HDACs, as mentioned above) and a Zn^{2+} ion at the bottom of the pocket, all forming a charge-relay system [69]. The HDAC inhibitors currently used in clinical trials or approved by the US Food and Drug Administration fit into this active site pocket, owing to a pharmacophore featuring a Zn^{2+}-chelating group and a linker spanning the length of the tubular pocket and connected to a cap that blocks the active site by interacting with the external surface of HDACs. Depending on their chemical Zn^{2+}-binding group, HDAC inhibitors belong to different classes, including hydroxamic acids, carboxylic acids, benzamides and cyclic tetrapeptides [19]. A central theme in the literature on HDAC inhibitors is their isoform selectivity or, rather, their perceived lack of isoform selectivity. HDAC inhibitors have generally been considered pan-inhibitors, inhibiting all HDACs from the classical family or class I-specific inhibitors. This view has recently been dispelled, however, by a study revealing no targeting of class IIa HDACs by most HDAC inhibitors [62]. Although it is not known which HDAC isoform's inhibition is responsible for the therapeutic or toxic effects observed in clinical trials, it has generally been assumed that the development of isoform-selective inhibitors would result in preferable clinical outcomes. This theory is unproven to date [20,23]. However, researchers who have performed conventional assays have analyzed the affinities of HDAC inhibitors for different HDACs by using purified HDACs, whereas HDAC activity is mostly associated with multiprotein complexes, the role and composition of which are often cell type-specific. This fact was taken into consideration in a pioneering study in which the investigators carried out the chemoproteomic profiling of 16 HDAC inhibitors with different chemical structures across six human cell lines and six mouse tissues [53]. In that study, a nonselective HDAC inhibitor bound to sepharose beads was added to cell lysates under conditions that preserved the integrity of protein complexes. In a competition assay, the mixture was spiked with a range of concentrations of a free inhibitor interfering with the capture of

HDAC complexes by the immobilized inhibitor. Captured proteins were analyzed by quantitative mass spectrometry, and target complexes were reconstituted by matching half-maximal inhibitory concentration values. This initial complex identification was further confirmed by quantitative immunoprecipitation experiments. Although the results collected in this study confirmed that class IIa HDACs were not targeted by any of the studied inhibitors, they mostly conflicted with the isoform selectivity data previously obtained in assays using purified HDACs [62,70]. This is not surprising, in view of a previous kinetic study suggesting that the *in vitro* mode of action of the HDAC inhibitor trichostatin A (TSA) depended on whether the assay conditions preserved HDAC complexes or resulted in their dissociation [71]. Incidentally, it was also shown that TSA did not disrupt HDAC1 and HDAC2 interaction with Sin3A [71]. However, it was shown that TSA and suberoylanilide hydroxamic acid (SAHA, also known as vorinostat), but not less bulky inhibitors such as sodium butyrate or valproic acid, dissociated inhibitor of growth 2 (ING2) from the Sin3 complex, thus disrupting the ING2-mediated recruitment of Sin3 to chromatin [72]. Bantscheff *et al.* [53] found that some inhibitors had different affinities for different complexes. In particular, inhibitors from the benzamide class displayed a higher affinity for the HDAC3-NCoR complex than for NuRD and CoREST complexes, whereas they did not target the Sin3 complex. The affinity of valproic acid, an inhibitor from the carboxylic acid class with moderate potency for class I HDACs, was highest for the CoREST complex, decreased gradually for the NuRD and NCoR complexes and was lowest for the Sin3 complex [53]. The different affinities of HDAC inhibitors detected for different complexes are in agreement with the previous observation that proteins in close proximity to the HDAC active site could interact with the cap of HDAC inhibitors, leading to the suggestion that HDAC-associated proteins could specify inhibitor selectivity [73]. The class IIb enzymes HDAC6 and HDAC10 were inhibited only by hydroxamate compounds. The anti-inflammatory drug bufexamac was identified as a class IIb HDAC inhibitor (HDAC6 and HDAC10), however, and accordingly promoted tubulin hyperacetylation at pharmacologically relevant concentrations. It was shown that hyperacetylation of different HDAC substrates corresponded to the HDAC inhibitor selectivity. For example, class I-selective HDAC inhibitors induced histone but not tubulin hyperacetylation, whereas nonselective HDAC inhibitors stimulated both histone and tubulin hyperacetylation. The authors also identified a novel mitotic HDAC complex, which they called MiDAC [53]. This study provides a new and promising path toward understanding the mode of action of class I-specific HDAC inhibitors and to develop isoform-selective compounds. For example, the combination of this methodology with genomic profiling of the different complexes by ChIP assay (specifically large-scale variants ChIP-on-chip and ChIP-seq) would be highly informative.

TRANSCRIPTIONAL REPROGRAMMING BY HISTONE DEACETYLASE INHIBITORS

Inhibition of HDAC activity results in transcriptional reprogramming, which is believed to contribute largely to the therapeutic benefits of HDAC inhibitors on cancers, cardiovascular diseases, neurodegenerative disorders and pulmonary diseases [24]. Inhibition of HDAC enzymatic activity affects the expression of only 5% to 20% of genes, however, with equal numbers of genes being upregulated and downregulated [19]. Only a fraction of these changes are direct effects of HDAC inhibitors, and others are downstream effects, necessitating new protein synthesis. Only some of the direct effects can be inferred as direct consequences of inhibition of histone deacetylation, and others are the results of other mechanisms, such as the inhibition of transcription factor deacetylation, resulting in an altered affinity for DNA binding sites on target gene regulatory regions, an altered interaction with other factors or an altered half-life [74]. Gene expression changes and biological functions targeted by HDAC inhibitors have been thoroughly addressed in recent comprehensive reviews [21,75]. Thus only a few examples of HDAC inhibitor-induced transcriptional changes involving novel or unexpected mechanisms are presented below.

p21 (*CDKN1A*), which encodes the cyclin-dependent kinase inhibitor p21, mediating cell cycle arrest, differentiation or apoptosis, is a model gene. Its transcription is directly upregulated in different cell types by different HDAC inhibitors, thus contributing to the antitumor effect of HDAC inhibitors. In parallel with transcriptional activation, a reorganization of chromatin, including histone hyperacetylation, takes place in both the proximal and distal promoter regions [76]. *p21* is regulated by a variety of factors, including p53. HDAC inhibitor-mediated transcriptional activation is independent of p53, however, and consequently can occur in tumor cells lacking a functional p53. Researchers in a recent study demonstrated that the nucleosomal response to the stimulation of the MAPK signaling pathway was required for *p21* induction by the HDAC inhibitor TSA. As part of the nucleosomal response, histone H3 in the *p21* proximal promoter region was phosphorylated on serine 10 by the mitogen- and stress-activated protein kinase 1 (MSK1). It was shown that this phosphorylation event was crucial to the acetylation of neighboring lysine 14. The phosphoacetylation mark was recognized by the 14-3-3ζ protein, reader of phosphoserine marks, and was thus protected from removal by PP2A phosphatase [77]. Presumably, 14-3-3 also acts as a scaffold for the recruitment of chromatin remodeler, leading to initiation of transcription [78]. Additionally, treatment with the HDAC inhibitor depsipeptide, also known as romidepsin, can induce*p21* expression by causing acetylation of p53, protecting it from ubiquitination-induced degradation and allowing the recruitment of the p300 KAT to the p53-responsive *p21* promoter [79]. The *p21*gene may generate several alternate variants [80,81]. The impact of HDAC inhibitors on the genesis of these variants remains to be determined. Some HDAC inhibitors alter pre-mRNA splicing by changing the expression of splicing factors, which are components of the spliceosome. As an example, butyrate, but not TSA, increases the expression of SFRS2 [82]. SFRS2 is required for the expression of *p21* [82].

The induction of the *Fos* and *Jun* immediate-early genes following the activation of the MAPK pathway is also dependent on MSK-mediated phosphorylation of histone H3 in the promoter region. However, the outcome of HDAC inhibition by TSA on these genes was opposite to the one on *p21*and opposite to the common belief that histone hyperacetylation is linked to transcription activation. Treatment with TSA resulted in rapid enhancement of H3 acetylation at the promoter region of these genes, but transcription was inhibited [83]. Furthermore, it was shown that continuous dynamic turnover of acetylation was characteristic of genes carrying the active methylation mark on H3K4, but not of genes carrying the repressive methylation mark on H3K9. The authors concluded that acetylation turnover rather than stably enhanced acetylation was crucial to the induction of the *Fos* and *Jun* genes [83]. A similar cyclical process that entails alternating activating and repressive epigenetic events during the hormone-dependent activation of genes has been described [56]. Nonetheless, other scenarios are possible; for example, the transcription activation of *Fos* and *Jun* could require the deacetylation of a nonhistone protein associated with their regulatory region. Investigators in several studies have suggested a role for deacetylation of transcription factors or other proteins in gene induction [84,85]. A proposal for the role of HDACs in the basal transcription from the mouse mammary tumor virus (MMTV) promoter and some other TATA/Inr-containing core promoters is that deacetylation of protein components of the reinitiation complex would allow the recruitment of RNA polymerase II [86].

In another example contrasting with the predominant view of HDACs' being transcriptional repressors, it was shown that HDAC1 served as a coactivator for the glucocorticoid receptor (GR) and that this function was dynamically modulated by acetylation of HDAC1 itself [87]. Researchers in a more recent study reported that HDAC2 was also required for GR-mediated transcriptional activation, and mechanistic insight was presented regarding collaborative regulation by HDAC1 and HDAC2 [36]. HDAC1 and HDAC2 act synergistically in the GR-mediated transcriptional activation at the MMTV promoter. So far, neither their acetylated target nor the nature of the coactivator complex with which they associate has been identified. It was shown that the coactivation function of HDAC1 and HDAC2 is dynamically regulated by acetylation of the HDAC1 C-terminal tail, with K432 being an important site among the six lysine residues that may be acetylated. It is of interest to note that the C-terminal domain of HDAC1 and HDAC2 is a region carrying modifications (phosphorylation and acetylation) that regulate HDAC activity. The acetylation event represses the deacetylase activity of both the HDAC1 homodimer and the HDAC1/HDAC2 heterodimer, even though only HDAC1 is acetylated [36,87]. It was suggested that HDAC1 and HDAC2 form homo- and heterodimers with the catalytic domains facing each other and that this arrangement was required for catalytic activity [36]. A modification of one of the HDACs is all that is required to inhibit the activity of the HDAC dimer.

EPIGENETIC MECHANISMS DIRECTED BY NONCODING RNAS

A widespread effect of HDAC inhibitors on microRNAs (miRNAs) levels was first reported in the breast cancer cell line SKBr3. Following a short exposure of these cells to a proapoptotic dose of the HDAC inhibitor LAQ824 (also known as dacinostat), significant changes in the levels of 40% of the miRNA population were detected. The majority of miRNAs were downregulated, but some were upregulated [88]. miRNAs are short, noncoding RNAs (ncRNAs) of about 23 nucleotides that regulate gene expression at the posttranscriptional level by binding to the 3'-UTRs of target mRNAs, leading to their degradation or translation repression. Although the biogenesis of miRNAs is well understood, little is known of the regulation of miRNA expression, but there is increasing evidence that miRNA expression is widely misregulated in tumors, with tumor suppressor miRNAs targeting growth-inducing genes being downregulated and oncogenic miRNAs targeting growth-inhibiting genes being upregulated [89,90].

Similarly, misregulation of miRNA expression is characteristic of metastasis [91]. miRNAs are initially transcribed by RNA polymerase II into long-coding or noncoding intragenic or intergenic RNAs. They can overlap RNA transcripts in the same, opposite or both directions. Noncoding intragenic RNAs transcribed in the same direction as the coding RNA can be transcribed from the same promoter as the host RNA or can have their own promoter embedded in an intron. Promoters used by other ncRNAs are mostly unknown [90]. Some promoters are associated with CpG islands, which can become hypo- or hypermethylated during tumorigenesis, resulting in transcriptional activation or silencing, respectively. Researchers in a number of studies have demonstrated the reversal of specific miRNA transcriptional silencing following treatment with DNA methyltransferase (DNMT) and/or HDAC inhibitors, indicating that these anticancer agents can indirectly induce posttranscriptional repression of target genes [89,90,92-94]. Investigators in one study showed that transcriptional silencing of miR-22 in acute lymphoblastic leukemia cells was independent of DNA methylation of the CpG island within the promoter region, but entailed K27 trimethylation of associated H3 histone. miR-22 transcriptional silencing could be reversed by TSA treatment [95]. In acute myeloid leukemia (AML), miR-223 was shown to be a direct transcriptional target of the AML1/ETO fusion oncoprotein resulting from the t(8;21) translocation. By recruiting HDAC1, DNMT and MeCP2 activities, AML1/ETO induces heterochromatic silencing of miR-223 [96].

Alternatively, it has been shown that miRNAs can regulate genes encoding epigenetic regulators such as DNMT3a and DNMT3b, Polycomb-associated K-methyltransferase EZH2 (also known as KMT6), HDAC1 and HDAC4 [89,90]. It was determined that the *HDAC1* gene, a direct target of miR-449a, was upregulated in prostate cancer cells and tissues as a consequence of miR-449a downregulation [97]. Similarly, *HDAC4* targeted by miR-1-1 was upregulated in human hepatocellular carcinoma cells and primary hepatocellular carcinoma [98]. During mouse development, HDAC4 plays a crucial role in the regulation of skeletogenesis and myogenesis, and its expression has been shown to be controlled at the posttranscriptional level by miR-140 and miR-1, respectively [99,100].

More recently, the existence of another class of miRNAs mediating transcriptional gene silencing (TGS) was demonstrated [101,102]. Contrary to miRNA-mediated posttranscriptional silencing, which is transient and dependent on the sustained presence of the effector miRNA, TGS targets promoter regions and triggers heterochromatin formation by inducing DNA and histone methylation (H3K27me3 and H3K9me2), which leads to long-term silencing. This process can be inhibited by TSA, indicating a role for HDACs. The mechanism implicated in the chromatin reorganization in the promoter region of the target gene is not fully understood, but it is known to involve RNA-RNA pairing between the small RNA and the nascent RNA transcript and, among other factors, EZH2, DNMT3A and HDAC1 [103,104]. Moreover, in miRNA-induced TGS of HIV-1, heterochromatin formation was initiated at the nucleation center, as directed by the miRNA, and was further extended both upstream and downstream to include adjacent genes. It was directly shown that HDAC1 was involved in this process [105]. As a side note, miRNAs are also able to activate transcription, notably of the gene *p21*, for example, perhaps by targeting and repressing a repressor miRNA [103,104].

Long, noncoding RNAs (lncRNAs) have been known to recruit Polycomb proteins and associated HDACs to initiate and maintain heterochromatin formation in the silencing of developmentally important genes, such as in X-chromosome inactivation and parental imprinting [104,106]. It has been proposed that long, single-stranded ncRNAs are integral components of chromatin, which may stabilize binding of nonhistone proteins to chromatin, such as the heterochromatin protein 1 (HP1), and/or may play a role similar to that of H1, a linker histone contributing to the formation of higher-order chromatin structures. It is possible that RNA could facilitate the folding of the chromatin fiber and the formation of loops involving long-distance interactions [107].

The role of ncRNAs in the regulation of gene expression is beyond the scope of this review, but it has become evident that the majority of protein-coding genes are regulated by antisense RNAs and, moreover, that a large part of the genomes of humans and other complex organisms consist of

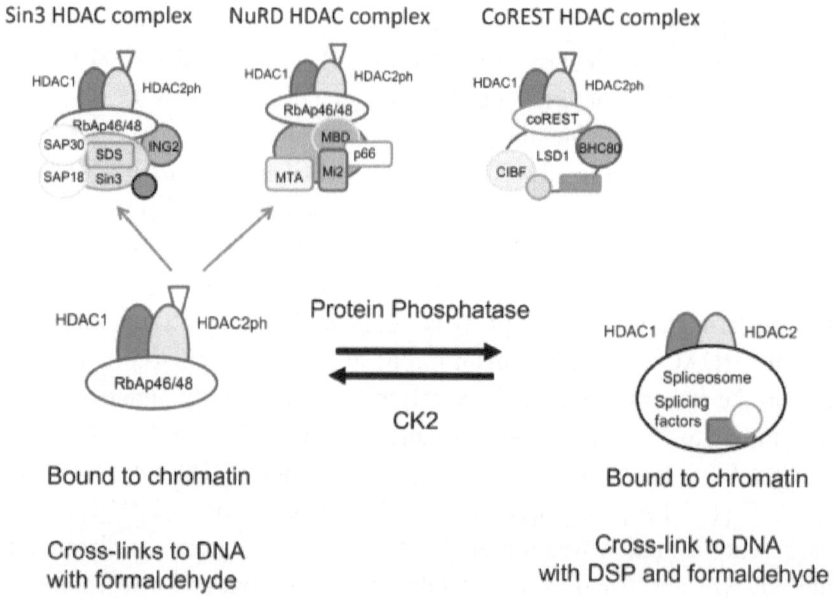

FIGURE 2. Model for the regulation of HDAC1-HDAC2 complex formation by phosphorylation. When phosphorylated by CK2, HDAC2 binds to the core components of Sin3, NuRD and CoREST complexes as homodimer or heterodimer with HDAC1. In the low or unphosphorylated states, HDAC1 and HDAC2 bind to proteins such as the serine/arginine (SR)-rich proteins and RNA-binding protein Hu antigen R (HuR/ELAVL1) which interact with the spliceosome.

non-protein-coding DNA sequences, which are transcribed. It has been suggested that these noncoding regions, previously termed "junk DNA," supply a vast array of ncRNAs which control the different epigenomes supporting development and generated throughout life in response to diet and environment. There is increasing evidence that misregulation by ncRNAs is also responsible for cancers and other diseases. Furthermore, it has been proposed that ncRNAs "provide the regulatory power and plasticity required to program our ontogeny and cognition" [106, p1610].

ROLE OF PHOSPHORYLATION IN TARGETING OF CLASS I HISTONE DEACETYLASES

Phosphorylation of HDAC1 (S393, S421, S423), HDAC2 (S394, S422, S424) and HDAC3 (S424) stimulates enzyme activity [108,109]. *In vitro* HDAC2 is phosphorylated by casein kinase (CK2), whereas HDAC1 can be phosphorylated by CK2, cAMP-dependent protein kinase and protein kinase G [25]. This difference constitutes more evidence that, although they share a high degree of homology, HDAC1 and HDAC2 have distinct and separately regulated functions. The regulation of HDAC1 and HDAC2 activities by PTMs has recently been reviewed [109]. Phosphorylation of HDAC1 and HDAC2 is also required for their incorporation into the Sin3, NuRD and CoREST corepressor complexes [25,41,110]. Recruitment of HDAC2 to regulatory regions of target genes by transcription factors also depends on its highly phosphorylated state [41]. On the other hand, the nonphosphorylated or monophosphorylated HDAC2 is associated with coding regions of transcribed genes [41] (Figure 2). Although unmodified and monophosphorylated HDAC2 are more abundant than highly phosphorylated HDAC2, it is the highly phosphorylated form that is preferentially cross-linked to chromatin with formaldehyde or cisplatin [111]. Thus, under conditions typically used in ChIP, highly phosphorylated HDAC2, but not unmodified or monophosphorylated HDAC2, is preferentially cross-linked to nuclear DNA *in situ* with formaldehyde. Through the use of a dual cross-linking ChIP assay, however, all isoforms of HDAC1 and HDAC2 could be mapped along regulatory and coding regions of transcribed genes, with the monophosphorylated or unmodified HDAC2 being associated with the coding region [41].

It should be noted that HDAC phosphorylation is dynamic and dependent on the balance of opposing activities of involved kinases and phosphatases. Treatment of cultured cells with the protein phosphatase inhibitor okadaic acid resulted in HDAC1 and HDAC2 hyperphosphorylation concomitant with the dissociation of HDAC1 and HDAC2, as well as the dissociation of HDAC1 from mSin3A or YY1. On the other hand, the HDAC1 and HDAC2 interactions with RbAp46 or RbAp48 were not disrupted [112]. In view of the above-described results [25,41,110], however, it appears that the observed dissociation of the HDAC-corepressor complexes subsequent to okadaic acid treatment are due not to the hyperphosphorylation of HDAC1 and HDAC2, but rather to the hyperphosphorylation of other unidentified factors.

A role of dynamic phosphorylation was also demonstrated for HDAC3. HDAC3 activity not only depends on its binding to the NCoR/SMRT complex but also is increased by CK2-mediated phosphorylation on S424. Conversely, HDAC3 interaction with protein phosphatase 4 results in decreased activity [113]. Moreover, it was shown that the phosphorylation state of HDAC3 had no effect on its association with NCoR or its subcellular localization [113]. In a recent study in which human HCT116 colon cancer cells were treated with sulforaphane (SFN), a cancer chemoprotective agent abundant in broccoli, a link between CK2-mediated phosphorylation of HDAC3 and SMRT and the disruption of the nuclear HDAC3-SMRT corepressor complex, followed by the export of HDAC3 into the cytoplasm, was suggested [114]. The proposed model is that 14-3-3 and Pin1 compete with each other for binding of cytoplasmic phosphorylated HDAC3, with binding to 14-3-3 eventually resulting in the reentry of HDAC3 into the nucleus and binding to Pin1 directing HDAC3 degradation. Thus extended exposure to SFN would lead to HDAC3 degradation [114]. Again, to reconcile these results with previous ones [113], it must be assumed that HDAC3 dissociation from SMRT and export to the cytoplasm are due to events other than CK2-mediated HDAC3 phosphorylation.

In the specific context of oxidative and/or nitrative stress, such as cigarette smoke, in chronic obstructive pulmonary disease (COPD), HDAC2 phosphorylation or nitration was linked to its ubiquitination and degradation by proteasome [115-117]. Corticosteroids are used to reduce inflammation in the airways of patients with COPD. The mode of action of corticosteroids includes the recruitment of HDAC2 and silencing of proinflammatory genes. However, resistance to the anti-inflammatory actions of corticosteroids occurs under oxidative and/or nitrative stress. Among a number of mechanisms contributing to this corticosteroid insensitivity, phosphoinositide 3-kinase-δ- and CK2-mediated HDAC2 phosphorylation were implicated [115,116,118]. Of note, administration of HDAC inhibitors affecting class I HDACs should not be given to COPD patients in the treatment of other diseases.

ROLE OF CLASS I HISTONE DEACETYLASES ASSOCIATED WITH CODING REGIONS

Several studies have addressed the role and mode of recruitment of class I HDACs to the coding region of active genes. In *Saccharomyces cerevisiae*, the Rpd3S HDAC complex is recruited by the chromodomain of its EAF3 subunit and the plant homeobox domain (PHD) of its RCO1 subunit to the H3K36me mark to deacetylate nucleosomes behind the elongating RNA polymerase II, thus preventing cryptic initiation of transcription within the coding region [119-121]. In mammals, a novel complex composed of SIN3B, HDAC1, the EAF3 ortholog MRG15 and the PHD finger-containing PF1 (a homolog of yeast RCO1) was recently identified. This complex is associated with discrete loci of constitutively active genes and is believed to regulate RNA polymerase II progression within transcribed regions [122].

A role of HDAC1 in alternative pre-mRNA splicing has recently been reported in HeLa cells [123]. Alternative splicing of pre-mRNA gives rise to mature mRNA isoforms coding for functionally different proteins. This alternative splicing plays essential roles in differentiation and development as well as in diseases. There has been increasing evidence that transcription and splicing are coupled, and recent studies have revealed the contribution of chromatin structure and histone modifications to the selection of splice sites [124,125]. One of these studies addressed the role of HDACs in the regulation of alternative splicing [123]. It was shown that following treatment of HeLa cells with the HDAC inhibitor sodium butyrate, about 4% of the genes exhibited altered splicing. Further characterization of one of these genes, fibronectin (*FN1*), indicated that HDAC inhibition resulted in alternative exon skipping and was associated with increased histone H4 acetylation, increased RNA polymerase II processivity and reduced cotranscriptional association of the splicing regulator SRp40 at the target exon. Moreover, knockdown studies demonstrated that HDAC1, but not HDAC2, activity was required for the alternative splicing [123]. Although that study provided mechanistic insight into the role of HDAC1 in alternative splicing, the question of how HDAC1 is targeted to a particular intron-exon junction remains. It is possible that a newly identified class of small ncRNAs, the splice site RNAs, whose 3' ends map precisely to the splice donor site of internal exons in animals, serve as markers [126]. It is noteworthy that the genes whose splicing was affected by HDAC inhibition were all genes involved in cell fate and differentiation [123]. One of these genes encodes the tau protein, which is highly expressed in the central nervous system. It turned out that the expression of a splice variant of the tau protein upregulated in some neurodegenerative diseases was reduced following treatment with sodium butyrate. This indicates that some of the therapeutic benefits of HDAC inhibitors may be due to the modulation of alternative splicing. Furthermore, that study [123] represents another example of the different functions of HDAC1 and HDAC2. A recent study demonstrated that Hu proteins (for example, HuR/ELAVL1), which are splicing regulators, bind to HDAC2 and inhibit its enzyme activity. The Hu proteins are recruited to transcribed genes by an interaction with RNA polymerase II and transferred to pre-mRNA at Hu target sites. Inhibition of HDAC2 activity by the Hu proteins results in localized increase in histone acetylation at specific exons [127], increasing the transcriptional elongation rate.

Questions remain regarding the role of HDAC2 or the presence of HDAC3 on coding regions. A genomewide mapping study in primary human $CD4^+$ T cells showed that HDAC1, HDAC2, HDAC3 and HDAC6 were enriched in active genes. Moreover, HDAC1 and HDAC3 were present mostly in promoter regions, whereas HDAC2 and HDAC6 were localized to both promoter and coding regions of active genes [40]. However, these results conflict with those of studies in which other cell types were studied. In MCF-7 cells, HDAC1 and HDAC2 were both associated with regulatory and coding regions [38,41]. Moreover, bufexamac, a class IIb-specific HDAC inhibitor, did not affect the acetylation levels of histones in HeLa cells, suggesting that histones are not substrates of HDAC6 [53]. Nonetheless, the study showed that dynamic acetylation was associated with active chromatin characterized by the H3K4 methylation mark [40]. This observation was in agreement with results obtained in mouse fibroblast cells [83,128]. A recent study in *S. cerevisiae* reported that dynamic acetylation was required for the recruitment of splicing factors during co-transcriptional spliceosome assembly [129].

CONCLUSION

One major issue with HDAC inhibitors often referred to in the literature is their lack of specificity, in particular their lack of isoform selectivity [130]. Another issue is the lack of known targets. There is still much to learn about the many ways that HDAC inhibitors affect gene expression. Indeed, we have just started to comprehend the breadth of their actions in changing gene expression in normal and abnormal states. New tools and methods are continually being developed, however, leading to discoveries that challenge our paradigms. For example, the specificity of HDAC inhibitors might be directed not toward HDAC isoforms but toward HDAC complexes. This finding underlines the importance of resolving HDAC interactions with other proteins and identifying genomic targets of HDAC complexes in relevant tissues in normal and disease states.

Classes I and II HDACs often exist as dimers. For example, class I HDACs form heterodimers with class II HDACs (for example, HDAC3-HDAC4). Furthermore, class I HDAC1 and HDAC2 form either homodimers or heterodimers. Multiprotein complexes with either an HDAC1-HDAC2 heterodimer versus an HDAC1 or HDAC2 homodimer may have different properties and substrate preferences. Discovery of the mechanism regulating HDAC1-HDAC 2 homo- versus heterodimer formation in cells will be important to understanding the biology of these enzymes.

There is still much to be learned about the mechanistic linkages between class I HDACs, transcription and RNA splicing. Whether other splicing regulators, such as Hu proteins, regulate HDAC activity will be an important question to address. The profound impact of HDAC inhibitors on the alternate splicing of RNAs and miRNAs requires further investigation to understand the full impact of the inhibitors on the cellular spectrum of RNAs and proteins. Recently developed mass spectrometry approaches are beginning to sort out the effects of miRNAs on protein profiles. Such approaches are required to understand the effect of HDAC inhibitor-altered miRNA profiles on the proteome [131].

REFERENCES

1. Groth A, Rocha W, Verreault A, Almouzni G: Chromatin challenges during DNA replication and repair. *Cell* **2007**, 128:721-733.
2. Shahbazian MD, Grunstein M: Functions of site-specific histone acetylation and deacetylation. *Annu Rev Biochem* **2007**, 76:75-100.
3. Tse C, Sera T, Wolffe AP, Hansen JC: Disruption of higher order folding by core histone acetylation dramatically enhances transcription of nucleosomal arrays by RNA polymerase III. *Mol Cell Biol* 1998, 18:4629-4638.
4. Wang X, He C, Moore SC, Ausio J: Effects of histone acetylation on the solubility and folding of the chromatin fiber. *J Biol Chem* **2001**, 276:12764-12768.
5. Shogren-Knaak M, Ishii H, Sun JM, Pazin MJ, Davie JR, Peterson CL: Histone H4-K16 acetylation controls chromatin structure and protein interactions. *Science* **2006**, 311:844-847.
6. Davie JR, He S, Li L, Sekhavat A, Espino P, Drobic B, Dunn KL, Sun JM, Chen HY, Yu J, Pritchard S, Wang X: Nuclear organization and chromatin dynamics: Sp1, Sp3 and histone deacetylases. *Adv Enzyme Regul* **2008**, 48:189-208.
7. Lee KK, Workman JL: Histone acetyltransferase complexes: one size doesn't fit all. Nat Rev *Mol Cell Biol* **2007**, 8:284-295.
8. Yang XJ, Seto E: HATs and HDACs: from structure, function and regulation to novel strategies for therapy and prevention. *Oncogene* **2007**, 26:5310-5318.
9. Wang Q, Zhang Y, Yang C, Xiong H, Lin Y, Yao J, Li H, Xie L, Zhao W, Yao Y, Ning ZB, Zeng R, Xiong Y, Guan KL, Zhao S, Zhao GP: Acetylation of metabolic enzymes coordinates carbon source utilization and metabolic flux. *Science* **2010**, 327:1004-1007.
10. Patel J, Pathak RR, Mujtaba S: The biology of lysine acetylation integrates transcriptional programming and metabolism. *Nutr Metab* (Lond) **2011**, 8:12.
11. Choudhary C, Kumar C, Gnad F, Nielsen ML, Rehman M, Walther TC, Olsen JV, Mann M:Lysine acetylation targets protein complexes and co-regulates major cellular functions. *Science* **2009**, 325:834-840.

12. Latham JA, Dent SY: Cross-regulation of histone modifications. *Nat Struct Mol Biol* **2007**, 14:1017-1024.
13. Yang XJ, Seto E: Lysine acetylation: codified crosstalk with other posttranslational modifications. *Mol Cell* **2008**, 31:449-461.
14. Hazzalin CA, Mahadevan LC: MAPK-regulated transcription: a continuously variable gene switch? *Nat Rev Mol Cell Biol* **2002**, 3:30-40.
15. Kouzarides T: Chromatin modifications and their function. *Cell* **2007**, 128:693-705.
16. Allis CD, Berger SL, Cote J, Dent S, Jenuwien T, Kouzarides T, Pillus L, Reinberg D, Shi Y, Shiekhattar R, Shilatifard A, Workman J, Zhang Y: New nomenclature for chromatin-modifying enzymes. *Cell* **2007**, 131:633-636.
17. de Ruijter AJ, van Gennip AH, Caron HN, Kemp S, van Kuilenburg AB: Histone deacetylases (HDACs): characterization of the classical HDAC family. *Biochem J* **2003**, 370:737-749.
18. Gregoretti IV, Lee YM, Goodson HV: Molecular evolution of the histone deacetylase family: functional implications of phylogenetic analysis. *J Mol Biol* **2004**, 338:17-31.
19. Smith KT, Workman JL: Histone deacetylase inhibitors: anticancer compounds. *Int J Biochem Cell* Biol **2009**, 41:21-25.
20. Witt O, Deubzer HE, Milde T, Oehme I: HDAC family: what are the cancer relevant targets? *Cancer Lett* **2009**, 277:8-21.
21. Marks PA: Histone deacetylase inhibitors: a chemical genetics approach to understanding cellular functions. *Biochim Biophys Acta* **2010**, 1799:717-725.
22. Wiech NL, Fisher JF, Helquist P, Wiest O: Inhibition of histone deacetylases: a pharmacological approach to the treatment of non-cancer disorders. *Curr Top Med Chem* **2009**, 9:257-271.
23. Wagner JM, Hackanson B, Lübbert M, Jung M: Histone deacetylase (HDAC) inhibitors in recent clinical trials for cancer therapy. *Clin Epigenetics* **2010**, 1:117-136.
24. Haberland M, Montgomery RL, Olson EN: The many roles of histone deacetylases in development and physiology: implications for disease and therapy. *Nat Rev Genet* **2009**, 10:32-42.
25. Tsai SC, Seto E: Regulation of histone deacetylase 2 by protein kinase CK2. *J Biol Chem* **2002**, 277:31826-31833.
26. Brunmeir R, Lagger S, Seiser C: Histone deacetylase HDAC1/HDAC2-controlled embryonic development and cell differentiation. *Int J Dev Biol* **2009**, 53:275-289.
27. Jurkin J, Zupkovitz G, Lagger S, Grausenburger R, Hagelkruys A, Kenner L, Seiser C:Distinct and redundant functions of histone deacetylases HDAC1 and HDAC2 in proliferation and tumorigenesis. *Cell Cycle* **2011**, 10:406-412.
28. Lagger G, O'Carroll D, Rembold M, Khier H, Tischler J, Weitzer G, Schuettengruber B, Hauser C, Brunmeir R, Jenuwein T, Seiser C: Essential function of histone deacetylase 1 in proliferation control and CDK inhibitor repression. *EMBO J* **2002**, 21:2672-2681.
29. Montgomery RL, Davis CA, Potthoff MJ, Haberland M, Fielitz J, Qi X, Hill JA, Richardson JA, Olson EN: Histone deacetylases 1 and 2 redundantly regulate cardiac morphogenesis, growth, and contractility. *Genes Dev* **2007**, 21:1790-1802.
30. Trivedi CM, Luo Y, Yin Z, Zhang M, Zhu W, Wang T, Floss T, Goettlicher M, Noppinger PR, Wurst W, Ferrari VA, Abrams CS, Gruber PJ, Epstein JA: Hdac2

regulates the cardiac hypertrophic response by modulating Gsk3β activity. *Nat Med* **2007**, 13:324-331.
31. Zimmermann S, Kiefer F, Prudenziati M, Spiller C, Hansen J, Floss T, Wurst W, Minucci S, Göttlicher M: Reduced body size and decreased intestinal tumor rates in HDAC2-mutant mice. *Cancer Res* **2007**, 67:9047-9054.
32. MacDonald JL, Roskams AJ: Histone deacetylases 1 and 2 are expressed at distinct stages of neuro-glial development. *Dev Dyn* **2008**, 237:2256-2267.
33. Guan JS, Haggarty SJ, Giacometti E, Dannenberg JH, Joseph N, Gao J, Nieland TJ, Zhou Y, Wang X, Mazitschek R, Bradner JE, DePinho RA, Jaenisch R, Tsai LH: HDAC2 negatively regulates memory formation and synaptic plasticity. *Nature* **2009**, 459:55-60.
34. Lagger S, Meunier D, Mikula M, Brunmeir R, Schlederer M, Artaker M, Pusch O, Egger G, Hagelkruys A, Mikulits W, Weitzer G, Muellner EW, Susani M, Kenner L, Seiser C: Crucial function of histone deacetylase 1 for differentiation of teratomas in mice and humans. *EMBO J* **2010**, 29:3992-4007.
35. Taplick J, Kurtev V, Kroboth K, Posch M, Lechner T, Seiser C: Homo-oligomerisation and nuclear localisation of mouse histone deacetylase 1. *J Mol Biol* **2001**, 308:27-38.
36. Luo Y, Jian W, Stavreva D, Fu X, Hager G, Bungert J, Huang S, Qiu Y: Trans-regulation of histone deacetylase activities through acetylation. *J Biol Chem* **2009**, 284:34901-34910.
37. Chiocca S, Kurtev V, Colombo R, Boggio R, Sciurpi MT, Brosch G, Seiser C, Draetta GF, Cotten M: Histone deacetylase 1 inactivation by an adenovirus early gene product. *Curr Biol* **2002**, 12:594-598.
38. He S, Sun JM, Li L, Davie JR: Differential intranuclear organization of transcription factors Sp1 and Sp3. *Mol Biol Cell* **2005**, 16:4073-4083.
39. Yamaguchi T, Cubizolles F, Zhang Y, Reichert N, Kohler H, Seiser C, Matthias P: Histone deacetylases 1 and 2 act in concert to promote the G1-to-S progression. Genes Dev **2010**, 24:455-469.
40. Wang Z, Zang C, Cui K, Schones DE, Barski A, Peng W, Zhao K: Genome-wide mapping of HATs and HDACs reveals distinct functions in active and inactive genes. *Cell* **2009**, 138:1019-1031.
41. Sun JM, Chen HY, Davie JR: Differential distribution of unmodified and phosphorylated histone deacetylase 2 in chromatin. *J Biol Chem* **2007**, 282:33227-33236.
42. Hassig CA, Tong JK, Fleischer TC, Owa T, Grable PG, Ayer DE, Schreiber SL: A role for histone deacetylase activity in HDAC1-mediated transcriptional repression. *Proc Natl Acad Sci USA* **1998**, 95:3519-3524.
43. Yang XJ, Seto E: The Rpd3/Hda1 family of lysine deacetylases: from bacteria and yeast to mice and men. Nat Rev *Mol Cell Biol* **2008**, 9:206-218.
44. Silverstein RA, Ekwall K: Sin3: a flexible regulator of global gene expression and genome stability. *Curr Genet* **2005**, 47:1-17.
45. Hayakawa T, Nakayama J: Physiological roles of class I HDAC complex and histone demethylase. *J Biomed Biotechnol* **2011**, **2011**:129383.
46. Denslow SA, Wade PA: The human Mi-2/NuRD complex and gene regulation. *Oncogene* **2007**, 26:5433-5438.
47. Wang Y, Zhang H, Chen YP, Sun YM, Yang F, Yu WH, Liang J, Sun LY, Yang XH, Shi L, Li RF, Li YY, Zhang Y, Li Q, Yi X, Shang YF: LSD1 is a subunit of the NuRD

complex and targets the metastasis programs in breast cancer. *Cell* **2009**, 138:660-672.

48. Liang J, Wan M, Zhang Y, Gu P, Xin H, Jung SY, Qin J, Wong J, Cooney AJ, Liu D, Songyang Z: Nanog and Oct4 associate with unique transcriptional repression complexes in embryonic stem cells. *Nat Cell Biol* **2008**, 10:731-739.
49. Lee MG, Wynder C, Cooch N, Shiekhattar R: An essential role for CoREST in nucleosomal histone 3 lysine 4 demethylation. *Nature* **2005**, 437:432-435.
50. Shi YJ, Matson C, Lan F, Iwase S, Baba T, Shi Y: Regulation of LSD1 histone demethylase activity by its associated factors. *Mol Cell* **2005**, 19:857-864.
51. Cunliffe VT: Eloquent silence: developmental functions of class I histone deacetylases. *Curr Opin Genet Dev* **2008**, 18:404-410.
52. Battaglia S, Maguire O, Campbell MJ: Transcription factor co-repressors in cancer biology: roles and targeting. *Int J Cancer* **2010**, 126:2511-2519.
53. Bantscheff M, Hopf C, Savitski MM, Dittmann A, Grandi P, Michon AM, Schlegl J, Abraham Y, Becher I, Bergamini G, Boesche M, Delling M, Dümpelfeld B, Eberhard D, Huthmacher C, Mathieson T, Poeckel D, Reader V, Strunk K, Sweetman G, Kruse U, Neubauer G, Ramsden NG, Drewes G: Chemoproteomics profiling of HDAC inhibitors reveals selective targeting of HDAC complexes. *Nat Biotechnol* **2011**, 29:255-265.
54. Perissi V, Jepsen K, Glass CK, Rosenfeld MG: Deconstructing repression: evolving models of co-repressor action. *Nat Rev Genet* **2010**, 11:109-123.
55. Fischle W, Dequiedt F, Hendzel MJ, Guenther MG, Lazar MA, Voelter W, Verdin E: Enzymatic activity associated with class II HDACs is dependent on a multiprotein complex containing HDAC3 and SMRT/N-CoR. *Mol Cell* **2002**, 9:45-57.
56. Métivier R, Penot G, Hübner MR, Reid G, Brand H, Kos M, Gannon F: Estrogen receptor-alpha directs ordered, cyclical, and combinatorial recruitment of cofactors on a natural target promoter. *Cell* **2003**, 115:751-763.
57. Bhaskara S, Knutson SK, Jiang G, Chandrasekharan MB, Wilson AJ, Zheng S, Yenamandra A, Locke K, Yuan JL, Bonine-Summers AR, Wells CE, Kaiser JF, Washington MK, Zhao Z, Wagner FF, Sun ZW, Xia F, Holson EB, Khabele D, Hiebert SW: Hdac3 is essential for the maintenance of chromatin structure and genome stability. *Cancer Cell* **2010**, 18:436-447.
58. Cedar H, Bergman Y: Linking DNA methylation and histone modification: patterns and paradigms. *Nat Rev Genet* **2009**, 10:295-304.
59. Sharma S, Kelly TK, Jones PA: Epigenetics in cancer. *Carcinogenesis* **2010**, 31:27-36.
60. Verdin E, Dequiedt F, Kasler HG: Class II histone deacetylases: versatile regulators. *Trends Genet* **2003**, 19:286-293.
61. Parra M, Verdin E: Regulatory signal transduction pathways for class IIa histone deacetylases. *Curr Opin Pharmacol* **2010**, 10:454-460.
62. Bradner JE, West N, Grachan ML, Greenberg EF, Haggarty SJ, Warnow T, Mazitschek R: Chemical phylogenetics of histone deacetylases. *Nat Chem Biol* **2010**, 6:238-243.
63. Martin M, Kettmann R, Dequiedt F: Class IIa histone deacetylases: regulating the regulators. *Oncogene* **2007**, 26:5450-5467.
64. Martin M, Kettmann R, Dequiedt F: Class IIa histone deacetylases: conducting development and differentiation. *Int J Dev Biol* **2009**, 53:291-301.

65. Valenzuela-Fernández A, Cabrero JR, Serrador JM, Sánchez-Madrid F: HDAC6: a key regulator of cytoskeleton, cell migration and cell-cell interactions. Trends *Cell* Biol **2008**, 18:291-297.
66. Li G, Jiang H, Chang M, Xie H, Hu L: HDAC6 α-tubulin deacetylase: a potential therapeutic target in neurodegenerative diseases. J Neurol Sci **2011**, 304:1-8.
67. Aldana-Masangkay GI, Sakamoto KM: The role of HDAC6 in cancer. *J Biomed Biotechnol* **2011**, **2011**:875824.
68. Villagra A, Sotomayor EM, Seto E: Histone deacetylases and the immunological network: implications in cancer and inflammation. *Oncogene* **2010**, 29:157-173.
69. Finnin MS, Donigian JR, Cohen A, Richon VM, Rifkind RA, Marks PA, Breslow R, Pavletich NP:Structures of a histone deacetylase homologue bound to the TSA and SAHA inhibitors. *Nature* 1999, 401:188-193.
70. Bertrand P: Inside HDAC with HDAC inhibitors. Eur J Med Chem **2010**, 45:2095-2116.
71. Sekhavat A, Sun JM, Davie JR: Competitive inhibition of histone deacetylase activity by trichostatin A and butyrate. *Biochem Cell Biol* **2007**, 85:751-758.
72. Smith KT, Martin-Brown SA, Florens L, Washburn MP, Workman JL: Deacetylase inhibitors dissociate the histone-targeting ING2 subunit from the Sin3 complex. *Chem Biol* **2010**, 17:65-74.
73. Salisbury CM, Cravatt BF: Activity-based probes for proteomic profiling of histone deacetylase complexes. *Proc Natl Acad Sci USA* **2007**, 104:1171-1176.
74. Glozak MA, Sengupta N, Zhang X, Seto E: Acetylation and deacetylation of nonhistone proteins. *Gene* **2005**, 363:15-23.
75. Wanczyk M, Roszczenko K, Marcinkiewicz K, Bojarczuk K, Kowara M, Winiarska M: HDACi: going through the mechanisms. *Front Biosci* **2011**, 16:340-359.
76. Gui CY, Ngo L, Xu WS, Richon VM, Marks PA: Histone deacetylase (HDAC) inhibitor activation of $p21^{WAF1}$ involves changes in promoter-associated proteins, including HDAC1. *Proc Natl Acad Sci USA* **2004**, 101:1241-1246.
77. Simboeck E, Sawicka A, Zupkovitz G, Senese S, Winter S, Dequiedt F, Ogris E, Di Croce L, Chiocca S, Seiser C: A phosphorylation switch regulates the transcriptional activation of cell cycle regulator p21 by histone deacetylase inhibitors. *J Biol Chem* **2010**, 285:41062-41073.
78. Drobic B, Pérez-Cadahía B, Yu J, Kung SK, Davie JR: Promoter chromatin remodeling of immediate-early genes is mediated through H3 phosphorylation at either serine 28 or 10 by the MSK1 multi-protein complex. *Nucleic Acids Res* **2010**, 38:3196-3208.
79. Zhao Y, Lu S, Wu L, Chai G, Wang H, Chen Y, Sun J, Yu Y, Zhou W, Zheng Q, Wu M, Otterson GA, Zhu WG: Acetylation of p53 at lysine 373/382 by the histone deacetylase inhibitor depsipeptide induces expression of $p21^{Waf1/Cip1}$. *Mol Cell Biol* **2006**, 26:2782-2790.
80. Radhakrishnan SK, Gierut J, Gartel AL: Multiple alternate p21 transcripts are regulated by p53 in human cells. *Oncogene* **2006**, 25:1812-1815.
81. Chen Y, Zhang L, Jones KA: SKIP counteracts p53-mediated apoptosis via selective regulation of $p21^{Cip1}$ mRNA splicing. *Genes Dev* **2011**, 25:701-716.
82. Edmond V, Brambilla C, Brambilla E, Gazzeri S, Eymin B: SRSF2 is required for sodium butyrate-mediated $p21^{WAF1}$ induction and premature senescence in human lung carcinoma cell lines. *Cell Cycle* **2011**, 10:1968-1977.

83. Hazzalin CA, Mahadevan LC: Dynamic acetylation of all lysine 4-methylated histone H3 in the mouse nucleus: analysis at c-fos and c-jun. *PLoS Biol* **2005**, 3:e393.
84. Zupkovitz G, Tischler J, Posch M, Sadzak I, Ramsauer K, Egger G, Grausenburger R, Schweifer N, Chiocca S, Decker T, Seiser C: Negative and positive regulation of gene expression by mouse histone deacetylase 1. *Mol Cell Biol* **2006**, 26:7913-7928.
85. Smith CL: A shifting paradigm: histone deacetylases and transcriptional activation. *Bioessays* **2008**, 30:15-24.
86. Lee SC, Magklara A, Smith CL: HDAC activity is required for efficient core promoter function at the mouse mammary tumor virus promoter. *J Biomed Biotechnol* **2011**, 2011:416905.
87. Qiu Y, Zhao Y, Becker M, John S, Parekh BS, Huang S, Hendarwanto A, Martinez ED, Chen Y, Lu H, Adkins NL, Stavreva DA, Wiench M, Georgel PT, Schiltz RL, Hager GL: HDAC1 acetylation is linked to progressive modulation of steroid receptor-induced gene transcription. *Mol Cell* **2006**, 22:669-679.
88. Scott GK, Mattie MD, Berger CE, Benz SC, Benz CC: Rapid alteration of microRNA levels by histone deacetylase inhibition. *Cancer Res* **2006**, 66:1277-1281.
89. Iorio MV, Piovan C, Croce CM: Interplay between microRNAs and the epigenetic machinery: an intricate network. *Biochim Biophys Acta* **2010**, 1799:694-701.
90. Sato F, Tsuchiya S, Meltzer SJ, Shimizu K: MicroRNAs and epigenetics. *FEBS J* **2011**, 278:1598-1609.
91. Lujambio A, Esteller M: How epigenetics can explain human metastasis: a new role for microRNAs. *Cell Cycle* **2009**, 8:377-382.
92. Bandres E, Agirre X, Bitarte N, Ramirez N, Zarate R, Roman-Gomez J, Prosper F, Garcia-Foncillas J: Epigenetic regulation of microRNA expression in colorectal cancer. *Int J Cancer* **2009**, 125:2737-2743.
93. Lee EM, Shin S, Cha HJ, Yoon Y, Bae S, Jung JH, Lee SM, Lee SJ, Park IC, Jin YW, An S:Suberoylanilide hydroxamic acid (SAHA) changes microRNA expression profiles in A549 human non-small cell lung cancer cells. *Int J Mol Med* **2009**, 24:45-50.
94. Zhang S, Cai X, Huang F, Zhong W, Yu Z: Effect of trichostatin A on viability and microRNA expression in human pancreatic cancer cell line BxPC-3. *Exp Oncol* **2008**, 30:265-268.
95. Li X, Liu J, Zhou R, Huang S, Huang S, Chen XM: Gene silencing of MIR22 in acute lymphoblastic leukaemia involves histone modifications independent of promoter DNA methylation. *Br J Haematol* **2010**, 148:69-79.
96. Fazi F, Racanicchi S, Zardo G, Starnes LM, Mancini M, Travaglini L, Diverio D, Ammatuna E, Cimino G, Lo-Coco F, Grignani F, Nervi C: Epigenetic silencing of the myelopoiesis regulator microRNA-223 by the AML1/ETO oncoprotein. *Cancer Cell* **2007**, 12:457-466.
97. Noonan EJ, Place RF, Pookot D, Basak S, Whitson JM, Hirata H, Giardina C, Dahiya R: miR-449a targets HDAC-1 and induces growth arrest in prostate cancer. *Oncogene* **2009**, 28:1714-1724.
98. Datta J, Kutay H, Nasser MW, Nuovo GJ, Wang B, Majumder S, Liu CG, Volinia S, Croce CM, Schmittgen TD, Ghoshal K, Jacob ST: Methylation mediated silencing of microRNA-1 gene and its role in hepatocellular carcinogenesis. *Cancer Res* **2008**, 68:5049-5058.

99. Tuddenham L, Wheeler G, Ntounia-Fousara S, Waters J, Hajihosseini MK, Clark I, Dalmay T:The cartilage specific microRNA-140 targets histone deacetylase 4 in mouse cells. *FEBS Lett* **2006**, 580:4214-4217.
100. Chen JF, Mandel EM, Thomson JM, Wu Q, Callis TE, Hammond SM, Conlon FL, Wang DZ: The role of microRNA-1 and microRNA-133 in skeletal muscle proliferation and differentiation. *Nat Genet* **2006**, 38:228-233.
101. Gonzalez S, Pisano DG, Serrano M: Mechanistic principles of chromatin remodeling guided by siRNAs and miRNAs. *Cell Cycle* **2008**, 7:2601-2608.
102. Kim DH, Saetrom P, Snøve O Jr, Rossi JJ: MicroRNA-directed transcriptional gene silencing in mammalian cells. *Proc Natl Acad Sci USA* **2008**, 105:16230-16235.
103. 103. Suzuki K, Kelleher AD: Transcriptional regulation by promoter targeted RNAs. *Curr Top Med Chem* **2009**, 9:1079-1087.
104. Turner AM, Morris KV: Controlling transcription with noncoding RNAs in mammalian cells. *Biotechniques* **2010**, 48:ix-xvi.
105. Suzuki K, Juelich T, Lim H, Ishida T, Watanebe T, Cooper DA, Rao S, Kelleher AD: Closed chromatin architecture is induced by an RNA duplex targeting the HIV-1 promoter region. *J Biol Chem* **2008**, 283:23353-23363.
106. Mattick JS: The central role of RNA in human development and cognition. *FEBS Lett* **2011**, 585:1600-1616.
107. Rodríguez-Campos A, Azorín F: RNA is an integral component of chromatin that contributes to its structural organization. *PLoS One* **2007**, 2:e1182.
108. Brandl A, Heinzel T, Krämer OH: Histone deacetylases: salesmen and customers in the post-translational modification market. *Biol Cell* **2009**, 101:193-205.
109. Segré CV, Chiocca S: Regulating the regulators: the post-translational code of class I HDAC1 and HDAC2. *J Biomed Biotechnol* **2011**, 2011:690848.
110. Pflum MK, Tong JK, Lane WS, Schreiber SL: Histone deacetylase 1 phosphorylation promotes enzymatic activity and complex formation. *J Biol Chem* **2001**, 276:47733-47741.
111. Sun JM, Chen HY, Moniwa M, Litchfield DW, Seto E, Davie JR: The transcriptional repressor Sp3 is associated with CK2 phosphorylated histone deacetylase 2. *J Biol Chem* **2002**, 277:35783-35786.
112. Galasinski SC, Resing KA, Goodrich JA, Ahn NG: Phosphatase inhibition leads to histone deacetylases 1 and 2 phosphorylation and disruption of corepressor interactions. *J Biol Chem* **2002**, 277:19618-19626.
113. Zhang X, Ozawa Y, Lee H, Wen YD, Tan TH, Wadzinski BE, Seto E: Histone deacetylase 3 (HDAC3) activity is regulated by interaction with protein serine/threonine phosphatase 4. *Genes Dev* **2005**, 19:827-839.
114. Rajendran P, Delage B, Dashwood WM, Yu TW, Wuth B, Williams DE, Ho E, Dashwood RH:Histone deacetylase turnover and recovery in sulforaphane-treated colon cancer cells: competing actions of 14-3-3 and Pin1 in HDAC3/SMRT corepressor complex dissociation/reassembly. *Mol Cancer* **2011**, 10:68.
115. Barnes PJ: Role of HDAC2 in the pathophysiology of COPD. *Annu Rev Physiol* **2009**, 71:451-464.
116. Adenuga D, Yao H, March TH, Seagrave J, Rahman I: Histone deacetylase 2 is phosphorylated, ubiquitinated, and degraded by cigarette smoke. *Am J Respir Cell Mol Biol* **2009**, 40:464-473.

117. Roche N, Marthan R, Berger P, Chambellan A, Chanez P, Aguilaniu B, Brillet PY, Burgel PR, Chaouat A, Devillier P, Escamilla R, Louis R, Mal H, Muir JF, Pérez T, Similowski T, Wallaert B, Aubier M: Beyond corticosteroids: future prospects in the management of inflammation in COPD. *Eur Respir Rev* **2011**, 20:175-182.
118. Chung KF, Marwick JA: Molecular mechanisms of oxidative stress in airways and lungs with reference to asthma and chronic obstructive pulmonary disease. *Ann N Y Acad Sci* **2010**, 1203:85-91.
119. Carrozza MJ, Li B, Florens L, Suganuma T, Swanson SK, Lee KK, Shia WJ, Anderson S, Yates J, Washburn MP, Workman JL: Histone H3 methylation by Set2 directs deacetylation of coding regions by Rpd3S to suppress spurious intragenic transcription. *Cell* **2005**, 123:581-592.
120. Keogh MC, Kurdistani SK, Morris SA, Ahn SH, Podolny V, Collins SR, Schuldiner M, Chin K, Punna T, Thompson NJ, Boone C, Emili A, Weissman JS, Hughes TR, Strahl BD, Grunstein M, Greenblatt JF, Buratowski S, Krogan NJ: Cotranscriptional set2 methylation of histone H3 lysine 36 recruits a repressive Rpd3 complex. *Cell* **2005**, 123:593-605.
121. Li B, Gogol M, Carey M, Lee D, Seidel C, Workman JL: Combined action of PHD and chromo domains directs the Rpd3S HDAC to transcribed chromatin. *Science* **2007**, 316:1050-1054.
122. Jelinic P, Pellegrino J, David G: A novel mammalian complex containing Sin3B mitigates histone acetylation and RNA polymerase II progression within transcribed loci. *Mol Cell Biol* **2011**, 31:54-62.
123. Hnilicová J, Hozeifi S, Dušková E, Icha J, Tománková T, Staněk D: Histone deacetylase activity modulates alternative splicing. *PLoS One* **2011**, 6:e16727.
124. Fox-Walsh K, Fu XD: Chromatin: the final frontier in splicing regulation? *Dev Cell* **2010**, 18:336-338.
125. Luco RF, Allo M, Schor IE, Kornblihtt AR, Misteli T: Epigenetics in alternative pre-mRNA splicing. *Cell* **2011**, 144:16-26.
126. Taft RJ, Simons C, Nahkuri S, Oey H, Korbie DJ, Mercer TR, Holst J, Ritchie W, Wong JJ, Rasko JE, Rokhsar DS, Degnan BM, Mattick JS: Nuclear-localized tiny RNAs are associated with transcription initiation and splice sites in metazoans. *Nat Struct Mol Biol* **2010**, 17:1030-1034.
127. Zhou HL, Hinman MN, Barron VA, Geng C, Zhou G, Luo G, Siegel RE, Lou H: Hu proteins regulate alternative splicing by inducing localized histone hyperacetylation in an RNA-dependent manner. *Proc Natl Acad Sci USA* **2011**, 108:E627-E635.
128. Clayton AL, Hazzalin CA, Mahadevan LC: Enhanced histone acetylation and transcription: a dynamic perspective. *Mol Cell* **2006**, 23:289-296.
129. Gunderson FQ, Merkhofer EC, Johnson TL: Dynamic histone acetylation is critical for cotranscriptional spliceosome assembly and spliceosomal rearrangements. *Proc Natl Acad Sci USA* **2011**, 108:2004-2009.
130. Chang J, Varghese DS, Gillam MC, Peyton M, Modi B, Schiltz RL, Girard L, Martinez ED:Differential response of cancer cells to HDAC inhibitors trichostatin A and depsipeptide. *Br J Cancer* **2012**, 106:116-125.
131. Lössner C, Meier J, Warnken U, Rogers MA, Lichter P, Pscherer A, Schnölzer M:Quantitative proteomics identify novel miR-155 target proteins. *PLoS One* **2011**, 6:e22146.

CHAPTER 7

ABERRANT EPIGENETIC SILENCING IS TRIGGERED BY A TRANSIENT REDUCTION IN GENE EXPRESSION

JON A. OYER, ADRIAN CHU, SUKHMANI BRAR, and MITCHELL S. TURKER

INTRODUCTION

Aberrant epigenetic silencing is a common and significant mechanism in cancer development and progression [1]. Like mutational events, aberrant silencing frequently inactivates tumor suppressor genes in both sporadic tumors and human cancer cell lines [2]. Unlike mutations, however, silencing is a stepwise process [3,4] with potential for reversal [5]. These observations have led to research to identify the molecular changes that accompany silencing. Such changes include promoter region DNA methylation, histone deacetylation, histone methylation at specific residues (e. g. H3K9, H3K27), and densely packed nucleosomes that create a closed chromatin structure [6]. However, a caveat is that these changes are most often documented at stably silenced alleles that were under continuous selective pressure within the tumor microenvironment for maintenance of the silenced state. Therefore, reported epigenetic modifications represent an ultimate endpoint and do not reveal how silencing initiates, nor do they reveal the order of epigenetic modifications that occur during the transition from active expression to stable silencing. Such information is required to create strategies to prevent the initiation or progression of aberrant epigenetic silencing.

Many models designed to examine initiation of silencing track normal epigenetic changes during development at imprinted genes [7] or during X chromosome inactivation [8], but developmentally programmed silencing may progress differently than aberrant silencing occurring in cancer.

Promoter DNA methylation is the most common modification associated with epigenetic silencing, and has previously been thought to play a causal role [9], but evidence is accumulating to suggest DNA methylation as a late step in the silencing process. For example, DNA methylation occurs after histone modifications for silenced, stably integrated transgenes [10]. A similar progression of epigenetic modifications occurs for silencing of the endogenous tumor suppressor gene RASSF1A [11]. Previous studies in our laboratory showed that silencing of an integrated Aprt transgene allows the spread of DNA methylation into a promoter region, which stabilizes the silenced transcriptional state [4]. Although DNA methylation has been the most common modification associated with cancer-related silencing, examples of epigenetic silencing occurring independent of DNA methylation show it is not an absolute requirement [12–14]. Collectively these data suggest that DNA methylation primarily functions to maintain and stabilize the silenced state and that other epigenetic processes are required to initiate silencing.

If DNA methylation is neither a required nor an initiating step for aberrant silencing, how is this process triggered? Recent studies suggest reduced expression as one possibility. For example, in ovarian cancer, loss of the GATA6 transcription factor results in reduced expression and subsequent epigenetic silencing of the downstream target Disabled-2 [15]. Also, inhibition of ERα (estrogen receptor α) signaling in breast cancer cell lines reduces expression and induces silencing of the downstream target gene, PR(progesterone receptor) [16]. These are two instances that involve loss of transcriptional activators, but evidence also exists that reducing expression by inappropriate recruitment of transcriptional repressors can lead to silencing. An inherited mutation in the DAPK1 promoter apparently causes B-cell chronic lymphocytic leukemia by increased localization of a transcriptional repressor that reduces expression and correlates with silencing [17]. In addition to altered signaling pathways, some environmental changes accompanying tumor progression also reduce gene expression, which could initiate silencing. For example, hypoxia, a common feature of tumor microenvironments, represses expression of tumor suppressor genes (e.g., E-CAD, BRCA1, MLH1, and RUNX) [18–21] frequently silenced in cancer [2,22,23].

RESULTS

A SYSTEM TO STUDY TRANSIENT REDUCTIONS IN GENE EXPRESSION

To test the hypothesis that a transient reduction in gene expression can initiate epigenetic silencing, we used the tet-off system [24] to control transcription levels of human HPRT cDNA in the mouse Dif-6 cell line, which lacks expression of endogenous Hprt [25]. In this system the tet-Transcriptional Activator (tTA) localizes to the tet-responsive promoter (PTRE) and promotes HPRT expression (Fig. 1A). Adding the tetracycline analog doxycycline (Dox) to the growth medium reduces HPRT expression by directly binding tTA and inhibiting its localization to the promoter (Fig. 1B). Three stable transfectants, HP11, HP13, and HP14, expressing high levels of HPRT were established. After 48 hours growth in Dox medium, HPRT expression was reduced by more than 90% relative to untreated controls, with the HP11 cell line exhibiting the strongest Dox response (Fig. 1C). Although HPRT expression is significantly reduced, cell cultures growing in Dox media remain sensitive to selection against HPRT (Fig. S1) and can grow under conditions that require HPRT expression (data not shown). A concentration of 1 μg/ml Dox showed a maximum effect on expression without causing toxicity (data not shown) and was used in all subsequent Dox treatments.

A TRANSIENT REDUCTION IN GENE EXPRESSION INDUCES PHENOTYPIC GENE INACTIVATION

Following removal of Dox from the culture medium, the tTA protein can again bind to PTRE and restoreHPRT expression. However, our hypothesis predicts that during the period of reduced expression a fraction of alleles will become epigenetically silenced and thus will be unable to restore HPRT expression upon removal of Dox. To test this hypothesis, cells were grown in Dox media for a week to reduce expression and allow adequate time for HPRT protein turnover before removing Dox and selecting for

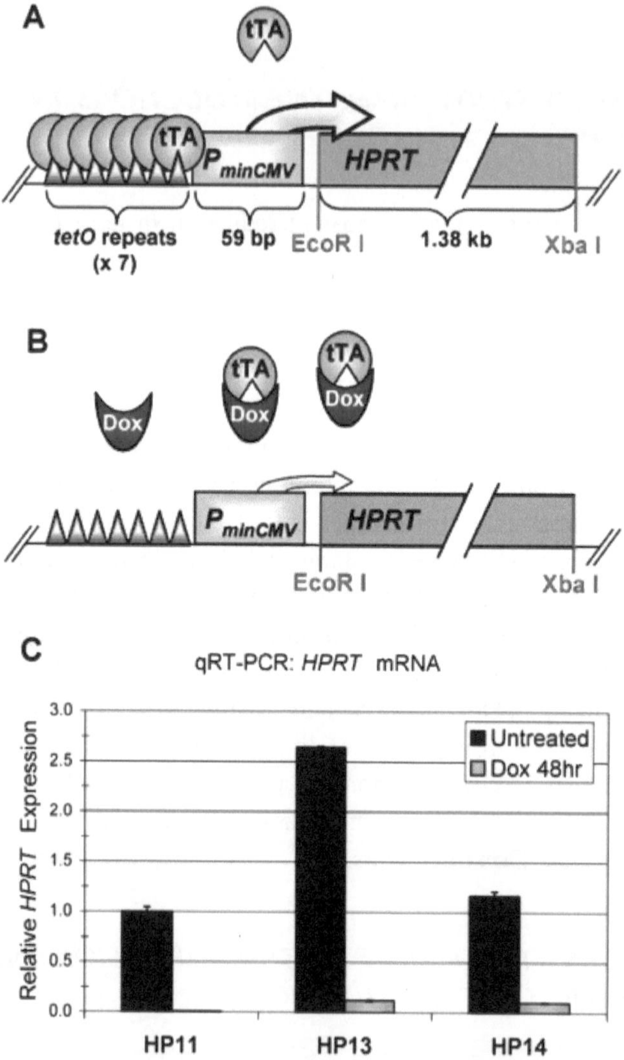

FIGURE 1. Treatment with doxycycline (Dox) reduces PTRE-HPRTexpression. (A) The tet-off system was used to express *HPRT*. The 1.38 kb *HPRT* cDNA was cloned into the 5′ EcoR I and 3′ Xba I restriction sites downstream of the minimal CMV promoter (PminCMV). In the absence of Dox, tTA binds to the seven 19 bp tet operator (tetO) sequence repeats in the promoter and activates high expression of *HPRT*. (B) Adding Dox reduces *HPRT* expression by direct binding of tTA, though minimal expression levels remain. (C) *HPRT* mRNA levels are significantly reduced within 48 hours of growth in media containing 1 μg/ml Dox. *HPRT* expression was measured by quantitative RT-PCR (qRT-PCR) and normalized to *Gapdh* expression levels. Each bar represents the average of triplicate reactions with error bars indicating minimum and maximum fold change.

HPRT deficient cells with the purine analog 6-thioguanine (TG). The fraction of surviving TG-resistant cells reflects the gene inactivation frequency for PTRE-HPRT. Dox exposure was found to induce TG-resistant cells for all three cell lines, at frequencies ranging from 1. 4×10−3 to 9. 4×10−2, which were several orders of magnitude higher than untreated cultures (Fig. 2A). Moreover, PTRE-HPRTinactivation frequency increased with longer durations of initial Dox exposure (Fig. 2B). These results demonstrate that transient reductions in gene expression correlate with greatly increased frequencies ofPTRE-HPRT inactivation.

EPIGENETIC MODIFICATIONS CONSISTENT WITH SILENCING CHARACTERIZE INACTIVATED PTRE-HPRT ALLELES

Individual TG-resistant clones were characterized to identify molecular changes correlating with PTRE-HPRT inactivation. TG-resistant clones were isolated from the HP13 and HP14 parental lines following one week of Dox treatment. HPRT mRNA levels in all TG-resistant clones were substantially lower than those observed in the Dox treated parental cells (Fig. 3). DNA methylation at the PTRE-HPRT promoters of HP14-derived TG-resistant cells was measured via bisulfite sequencing (Fig. 4). As expected, all CpG sites within the PCMVmin, tetO repeats, and nearby regions were unmethylated in actively expressing HP14 cells. Moreover, these sites remained unmethylated in the same cells grown in the presence of Dox for one week. In contrast, all TG-resistant clones analyzed exhibited DNA methylation in the promoter region, though the density of DNA methylation varied (Fig. 4). TG1 and TG2 both exhibited low levels of DNA methylation and contained some alleles without any methylated CpG sites in the minimal CMV promoter (PminCMV). In contrast, TG5 and TG6 exhibited substantially more DNA methylation within the promoter, including the core PminCMV region. The other two TG-resistant lines, TG3 and TG4, contained intermediate to high levels of DNA methylation within the promoter relative to the other cell lines.

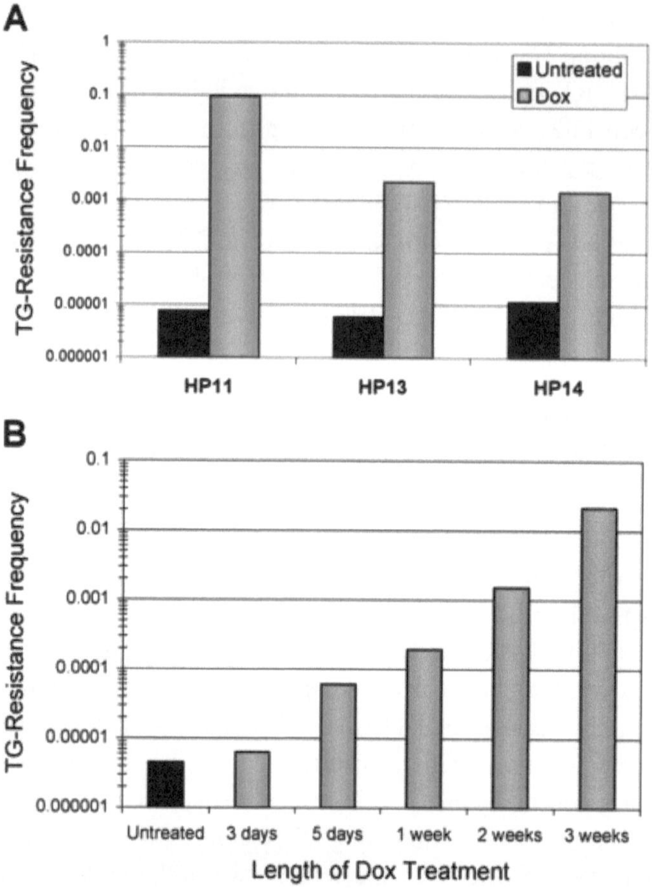

FIGURE 2. Dox exposure induces PTRE-HPRT inactivation. (A) Reducing expression of *PTRE-HPRT* by growing cells for 1 week in medium containing 1 µg/ml Dox increased the frequency of gene inactivation, as measured by TG-resistance. TG-resistance was measured by washing out Dox, selecting cells with 5 µg/ml TG, and counting surviving colonies after 2 weeks of continuous selection. During Dox treatment or the equivalent period without treatment, cells were maintained in medium containing puromycin and G418 to maintain the *PTRE-HPRT* and tTA constructs respectively, but without azaserine/hypoxanthine (AzHx) selection. Only cells that express *HPRT* can grow in AzHx selection. (B) TG-resistance frequencies increased as a function of time HP13 cells were exposed to Dox before starting TG selection. HP13 cells were continuously cultured in 1 µg/ml Dox for 3 weeks, and TG-resistance was measured at different points during the Dox treatment (3, 5, 7, 14, and 21 days). A parallel control culture was maintained in medium containing puromycin and G418 without Dox, and TG-resistance was measured after 21 days (Untreated).

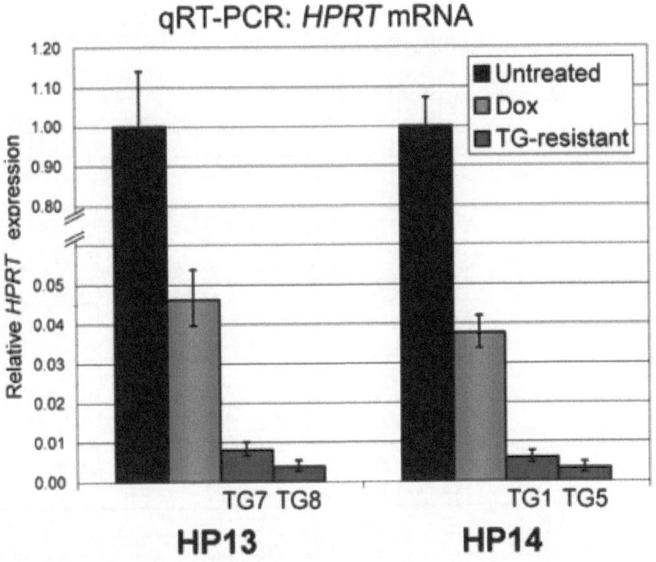

FIGURE 3. PTRE-HPRT inactivation correlates with reduced mRNA levels. Dox-induced TG-resistant clones were isolated and expanded. *HPRT* mRNA levels in TG-resistant clones from the HP13 (TG7 and TG8) and HP14 (TG1 and TG5) parental cells are lower than both active expression levels (Untreated) and reduced expression levels after exposure to Dox for one week (Dox). *HPRT* mRNA was measured by qRT-PCR and normalized to *Gapdh* expression levels. Each bar represents the average of triplicate reactions with error bars indicating minimum and maximum fold change.

The absence of dense promoter DNA methylation in some clones suggested additional mechanisms were contributing to PTRE-HPRT inactivation. Chromatin immunoprecipitation (ChIP) analysis was used to measure specific histone modifications associated with either active transcription (methyl-K4, acetyl-K9, and acetyl-K14 of histone H3) or silenced transcription (dimethyl-K9 of histone H3) at the PTRE-HPRTpromoter (Fig. 5). As expected, cells expressing high levels of HPRT have a histone modification pattern at the promoter consistent with active transcription. Specifically, the actively transcribed PTRE–HPRTpromoter was associated with high levels of H3 acetylation (Fig. 5A) and methylation at lysine 4 (methyl-K4 H3) (Fig. 5B) relative to modification levels measured at the active Gapdh promoter (P-Gapdh). The repressive modification dimethyl-K9 H3 was low in the HPRT expressing cells (Fig. 5C), measured relative to the silenced Mage-a locus (P-Mage) [26,27]. Reducing HPRT

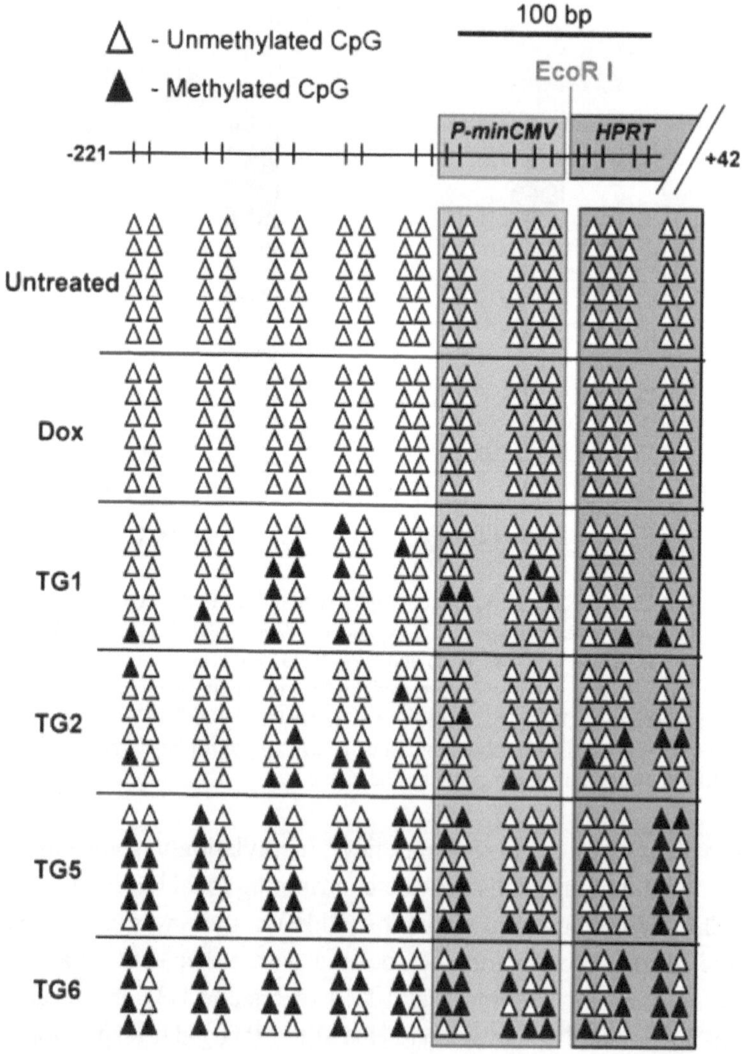

FIGURE 4. PTRE-HPRT inactivation correlates with increased promoter DNA methylation. Allelic methylation patterns for parental HP14 cells expressing high levels of *HPRT* (Untreated), reduced levels of *HPRT* after an 1 week Dox treatment (Dox), and HP14-derived TG-resistant clones (TG1, TG2, TG5, and TG6). Bisulfite sequencing identified methylated (closed triangles) and unmethylated (open triangles) CpG sites within individual alleles. Schematic of the promoter shows approximate positioning of CpG sites (vertical bars) within the minimal CMV promoter and the 5′ region (~42 bp) of the *HPRT* cDNA sequence. The start of the *HPRT* cDNA sequence, EcoR I restriction site, has been designated base position +1.

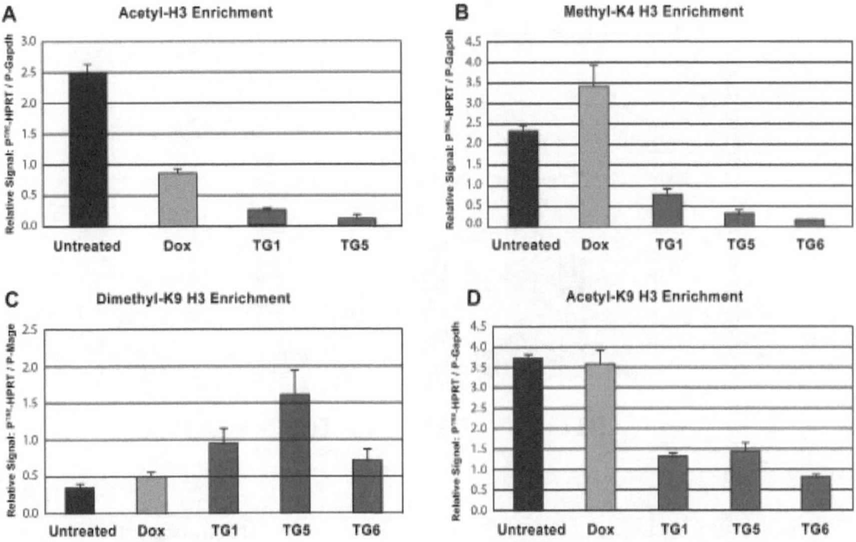

FIGURE 5. PTRE-HPRT inactivation correlates with repressive histone modifications. ChIP analysis measuring histone H3 modifications at the *PTRE-HPRT* promoter in HP14 cells expressing high levels of *HPRT* (Untreated), reduced levels of *HPRT* after 1 week Dox treatment (Dox), and HP14-derived TG-resistant cell lines (TG1, TG5, and TG6). (A) ChIP analysis measuring acetylated histone H3 using a polyclonal antibody raised against a peptide corresponding to acetyl-K9 and acetyl-K14. (B) ChIP analysis measuring methylation at lysine 4 of histone H3 (methyl-K4 H3). The antibody used for immunoprecipitation recognizes all three forms of methylation at K4, mono-, di-, and tri-methyl. (C) ChIP analysis measuring the repressive modification of dimethylation at lysine 9 of histone H3 (di-methyl-K9 H3). (D) ChIP analysis measuring acetylation at lysine 9 of histone H3 (acetyl-K9 H3). Immunoprecipitated DNA levels were quantified by qRT-PCR. Specific signal was calculated by measuring fold change between pull down and input at the *Hprt* promoter (*PTRE-HPRT*), *Gapdh* promoter (*P-Gapdh*), and *Mage-a* promoter (*P-Mage-a*). For activating modifications, levels at *PTRE-HPRT* are displayed relative to the *Gapdh* promoter; for the repressive modification, dimethyl-K9 H3, results are displayed relative to the *Mage* promoter. Error bars indicate the SD from triplicate reactions.

expression by treating cells with Dox did not reduce levels of methyl-K4 H3 (Fig. 5B) or significantly change the levels of dimethyl-K9 H3 (Fig. 5C) at the PTRE-HPRT promoter. However, H3 acetylation decreased significantly after reducing HPRT expression by Dox treatment (Fig. 5A). The antibody used for the acetyl-H3 ChIP recognizes both acetyl-K9 H3 and acetyl-K14 H3 [28]. To probe this decrease further, an additional ChIP

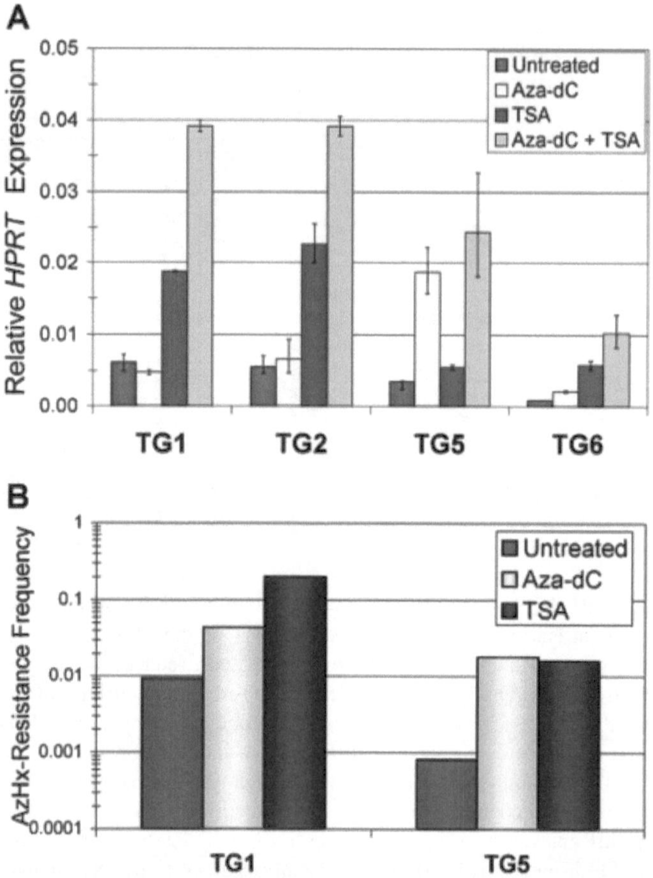

FIGURE 6. Silenced PTRE-HPRT alleles are reactivated by inhibiting histone deacetylation or DNA methylation. (A) Inhibition of DNA methylation and histone deacetylation increased *HPRT* mRNA levels. HP14-derived TG-resistant cell lines (TG1, TG2, TG5, and TG6) were treated with 300 nM 5-aza-dC (Aza-dC), inhibiting histone deacetylation with 100 nM trichostatin A (TSA), or a combination of the 300 nM 5-aza-dC and 100 nM TSA treatments (Aza-dC+TSA). Cells were treated with inhibitors overnight (~16 hours), and RNA was harvested 24 hours later. The units shown along the Y-axis are relative to those measured in the untreated parental HP14 cells (see Figure 3). *HPRT* expression was measured by qRT-PCR and normalized to *Gapdh* expression levels. Each bar represents the average of duplicate reactions with error bars indicating minimum and maximum fold change. (B) TG-resistant cell lines were capable of reactivating *PTRE-HPRT* expression. Cells were plated with azaserine/hypoxanthine (AzHx) selection, which requires HPRT enzyme activity for cell survival, to isolate and measure the number of cells that reactivated *HPRT* expression. Before plating and selection, cells were treated overnight with 300 nM 5-aza-dC (Aza-dC), 100 nM TSA (TSA), or vehicle control (untreated) and allowed to recover for 24 hours. Frequencies represent the fraction of cell colonies surviving after two weeks of continuous AzHx selection.

was conducted with antibody directed specifically against acetyl-K9 H3. In this case, no decrease was observed after the one-week exposure to Dox (Fig. 5D). Therefore, reducing expression by Dox treatment caused loss of acetyl-K14 H3 without decreasing other modifications associated with transcriptional activity, i.e., methyl-K4 H3 or acetyl-K9 H3. Dox treatment had no effect on histone modifications measured at the control promoters (P-Gapdh and P-Mage) used for normalization.

In contrast to the observations for histone modifications in the presence of Dox, ChIP analysis for the Dox-independent, TG resistant clones revealed markedly reduced levels of methyl-K4 H3 and acetyl-K9 H3. Increased dimethyl-K9 H3 was also observed at the PTRE-HPRT promoter in the TG-resistant cells, ranging from a 2-fold increase in TG6 to a nearly 5-fold increase in TG5 (Fig. 5C). These results demonstrate that the transition from reduced expression to gene inactivation is associated with a shift from activating to repressive histone modifications consistent with epigenetic silencing.

SILENCED PTRE-HPRT ALLELES ARE REACTIVATED BY INHIBITING HISTONE DEACETYLASES OR DNA METHYLATION

One of the hallmarks of epigenetic silencing is reversibility. To confirm definitively that the inducedPTRE–HPRT inactivation was due to silencing, we measured the effects of inhibiting histone deacetylation and/or DNA methylation on gene reactivation in the TG-resistant cells. First, changes inHPRT mRNA levels were measured for the TG-resistant cells after inhibiting histone deacetylation with trichostatin A (TSA) treatment or inhibiting DNA methylation with 5-aza-2′-deoxycytidine (5-aza-dC) (Fig. 6A). The different TG-resistant clones had varied responses to histone deacetylase (HDAC) inhibition ranging from an approximately 3-fold increase in HPRT mRNA (TG1 and TG2) to no response (TG5). Inhibiting DNA methylation gave a nearly reciprocal result with the TG5 cell line showing the largest 5-aza-dC induction of HPRT mRNA, an approximately 5-fold increase, and little response in the TG1 and TG2 clones, which exhibited the strongest response after HDAC inhibition. Combining inhibition of histone deacetylases and DNA methylation by treating the cells with 5-aza-dC and TSA simultaneously resulted in synergistic induction of HPRT expression for every TG-resistant cell lines except for TG5, which exhibited at best an additive effect (Fig. 6A).

Next we determined if the silenced alleles could phenotypically reactivate by selecting for revertant cells in media requiring HPRT expression for survival (azaserine / hypoxanthine or AzHx). Two TG-resistant cell lines, TG1and TG5, spontaneously gave rise to AzHx-resistant colonies at frequencies of 9. 3×10−3and 8. 2×10−4, respectively (Fig. 6B). TSA and 5-aza-dC treatments were used to determine if inhibiting histone deacetylases or DNA methylation, respectively, would induce phenotypic reactivants similar to their effects on induced reactivation at the RNA level (Fig. 6B). Phenotypic reactivants were induced, though the results did not mimic precisely those obtained by measuring HPRT mRNA levels. For example, TSA treatment increased the frequency of phenotypic reactivation of the TG5 cell line despite the apparent lack of induction when measuring HPRT mRNA one day after TSA treatment. While these discrepancies reveal differences between the two assays, the combined results clearly demonstrate that TG-resistance was due to epigenetic mechanisms.

REACTIVATED ALLELES EXHIBIT MEMORY OF TRANSCRIPTIONAL SILENCING

Several laboratories have reported that 5-aza-dC reactivated promoters exhibit rapid re-silencing [29,30]; however these experiments could not use selection to maintain expression. Our system allowed continuous selection to ensure maintenance of the reactivated promoter state by growing the cells in AzHx medium, which requires HPRT enzyme activity for cell survival. We therefore asked whether promoter reactivation stabilized under selective conditions or alternatively whether the reactivated promoters retained a memory of silencing, as defined by high frequency re-silencing. Although selection ensures HPRT expression, the absolute expression levels were variable ranging from 14% to 90% of HPRT expression in the parental cells. Two clones, reactivants 1 and 2, were isolated from TG-resistant HP13 cells that had spontaneously reactivated HPRT expression

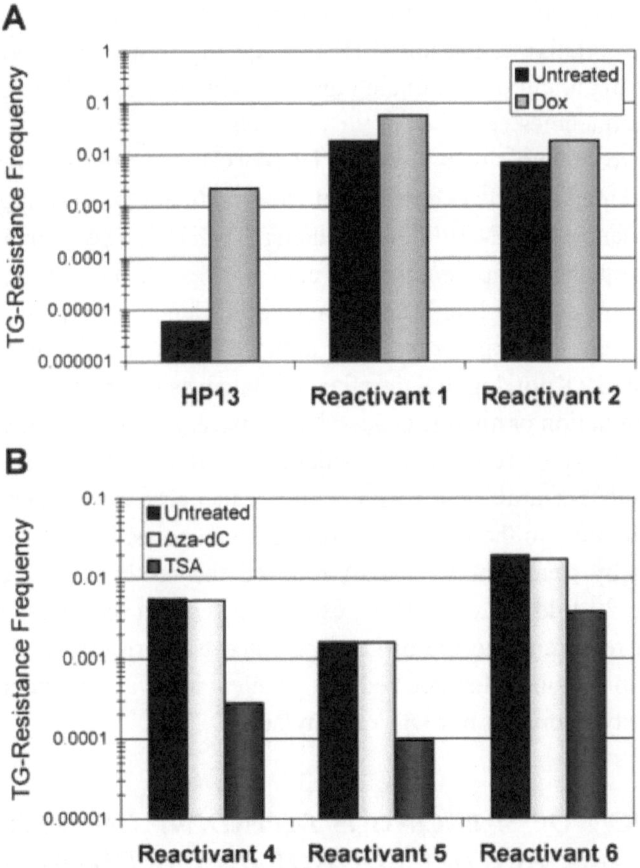

FIGURE 7. Reactivation of silenced PTRE-HPRT alleles is unstable. (A) *PTRE-HPRT* inactivation frequencies for HP13-derived clones with epigenetically silenced and then reactivated *HPRT* expression (Reactivant 1 and 2) were measured after one week without Dox exposure (Untreated) or after one week Dox treatment (Dox). The silencing frequency for the parental HP13 cell line (HP13) is shown for comparison. (B) *PTRE-HPRT* inactivation frequencies for HP14-derived clones with epigenetically silenced and then reactivated *HPRT* expression (Reactivants 4–6) were measured after overnight treatment with 300 nM 5-aza-dC (Aza-dC), 100 nM TSA (TSA), or vehicle control (Untreated) before selecting against *HPRT* activity with TG. Frequencies represent the fraction of cell colonies surviving after two weeks of TG selection.

and grew well in AzHx medium. Spontaneous and Dox-induced silencing frequencies were determined for both clones (Fig. 7A). These reactivant cell lines spontaneously re-silenced at high frequencies, 1. 8×10−2 and 6. 8×10−3, relative to the initial HP13 silencing frequency of 4. 5×10−6, with Dox treatment only inducing an approximately 3-fold increase in silencing frequencies.

Three reactivant clones from the HP14 TG1 cell line (reactivants 4–6) were examined and also showed that Dox treatment was not required for high frequency PTRE-HPRT re-silencing (Fig. 7B). The spontaneous silencing frequency for parental HP14 cells was less than 6×10−6 (Fig. 2A), but all reactivant cell lines had spontaneous silencing frequencies (10−3 to 10−2) equal to or higher than that induced by the week-long Dox treatment (~10−3). Knowing that the silenced state was reversible by inhibiting DNA methylation or histone deacetylation, we examined whether either of these events were required for re-silencing. After inhibiting DNA methylation with 5-aza-dC, the re-silencing frequencies were essentially unchanged relative to the high spontaneous frequencies. In contrast, HDAC inhibition by treatment with TSA reduced the re-silencing frequencies from 5- to 20-fold. In total, these results showed that the reactivated cells no longer required a period of Dox-mediated transcriptional reduction to silence expression and suggested that re-silencing was dependent on histone deacetylation, but not DNA methylation.

INITIATION OF SILENCING IS DEPENDENT ON HISTONE DEACETYLASE ACTIVITY BUT NOT DNA METHYLATION

After demonstrating that HDAC inhibition reduced re-silencing of reactivated alleles, we tested whether inhibiting HDACs or DNA methylation would affect initial silencing induced by Dox treatment. Induced silencing frequencies were measured again for the HP13 and HP14 parental HPRT expressing cell lines, with the modification of adding TSA or 5-aza-dC for the last 16 hours the cells were in Dox media. Inhibiting DNA methylation did not significantly affect the Dox-induced silencing frequency, but HDAC inhibition drastically reduced the silencing frequency (Fig. 8). These results show that HDAC activity is an early requirement for

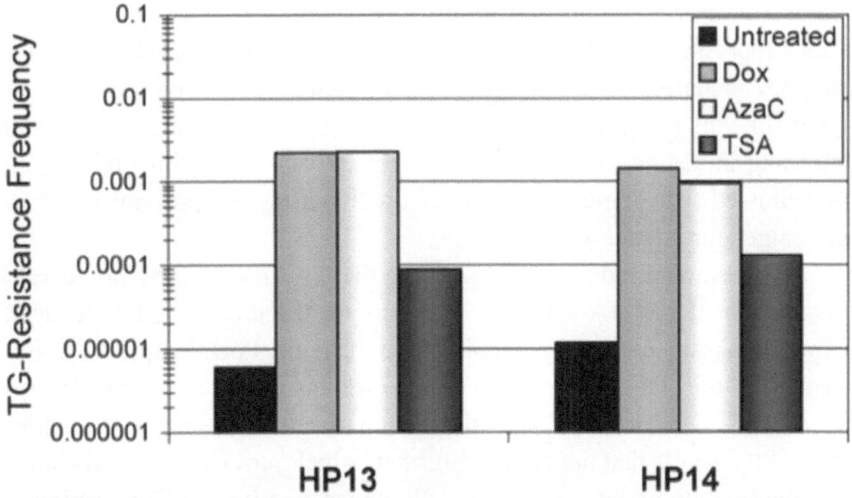

FIGURE 8. Histone deacetylase inhibition prevents Dox-inducedPTRE-HPRT silencing. *PTRE-HPRT* inactivation frequencies for HP13 and HP14 cells were measured after treatment exposure to 1 μg/ml Dox for one week (Dox), exposure to Dox for one week plus 300 nM 5-aza-dC for the last 16 hours (Aza-dC), exposure to Dox for one week plus 100 nM TSA for the last 16 hours (TSA), or no treatments (Untreated). Frequencies represent the fraction of cell colonies surviving after two weeks of continuous TG selection.

silencing induced by decreased transcription in our model, but DNA methylation is not required.

DISCUSSION

Aberrant epigenetic silencing is a significant mechanism of tumor suppressor gene inactivation, but how this process initiates in mammalian cells is poorly understood. We used the tet-off system to test the hypothesis that a transient and reversible reduction in gene expression could sensitize a promoter to undergo silencing. This hypothesis was based on observations showing reduced gene expression correlates with subsequent tumor suppressor gene silencing (see Introduction) and results from our laboratory showing transcriptional silencing allowed DNA methylation to spread into a promoter region [4]. Moreover, some tumor suppressor genes that are frequently silenced in cancer are also repressed by specific environmental exposures. For example, the tumor microenvironment causes hypoxia, which represses the E-Cadherin [18], BRCA1 [19], and MLH1 [20] tumor suppressor genes. All are epigenetically silenced in one or more cancer types [2], which suggests a relation between transcriptional repression and silencing in cancer. Here we report results from an experimental system that allowed us to demonstrate that a reduction in gene expression can trigger epigenetic silencing.

Dox treatment reduces expression in the tet-off system by preventing association of the tTA activator protein with the promoter, but reduced expression is not equivalent to epigenetic silencing. For example, the Dox-treated cultures remained sensitive to TG while silenced clones are TG-resistant (Fig. S1). However, a small fraction of cells exposed to Dox exhibited HPRT levels that are reduced further, which provided TG-resistance, and the fraction of cells increases with longer durations of Dox exposure. The induced TG-resistance was also relatively stable because it did not require continued exposure to Dox. These observations demonstrate that the reduced expression in the presence of Dox sensitized some alleles to undergo epigenetic silencing, but was insufficient to confer TG-resistance by itself. All evidence obtained in these experiments supported the conclusion that Dox-induced TG-resistance was due to epigenetic silencing as opposed to mutational events. The best evidence was the ability of TG-resistant cells to reactivate expression and restore functional HPRT activity, which was evident by growth of the cells in AzHx media. Besides the nine different TG-resistant clones described in this paper, we examined an additional fifteen TG-resistant clones induced by Dox treatment. At least one characteristic of epigenetic silencing (i.e., TSA or 5-aza-dC induction of HPRT mRNA or reactivant cell clones) was measured in each of these TG-resistant clones. In total, all twenty-four of the examined Dox-induced clones were shown to have silenced PTRE-HPRT alleles. Additionally, the silencing frequencies induced by Dox treatment were orders of magnitude higher than that expected for HPRT inactivating mutations (<10−6), and a previous study that characterized base-pair substitutions in the Dif-6 cells showed they do not have a mutator phenotype [31]. Dox treatment has also been used extensively in cell culture without having displayed mutagenic properties.

High-level promoter expression in the tet-off system occurs via localization of the tTA protein and activity of its VP16 activation domain; this domain promotes expression through recruitment of TBP, TFIIB, and the SAGA complex [32]. Then reduced expression during Dox treatment likely results from losing recruitment of these factors. The resultant disruption in recruitment of the SAGA complex and its associated histone acetyltransferases may therefore cause the concurrent decrease in acetyl-K14 H3. In contrast, acetyl-K9 H3 did not decrease when gene expression was reduced by Dox treatment, which demonstrates that the acetylation state of K9 and K14 of H3 may be regulated independently. Previous studies have also observed acetylation at K9 H3 can remain high despite decreased gene expression levels [33]. Levels of the repressive histone modification dimethyl-K9 H3 remained relatively low during reduced expression in the presence of Dox, which is not surprising considering the continuing presence of acetyl-K9 H3 should prevent the addition of methyl groups at K9 H3. Hypoxic conditions have been reported to increase dimethyl-K9 H3 upon repression of the mouse Mlh1 [34] and human RUNX3promoters [21]. Increased dimethyl-K9 H3 has also been reported to result from nickel exposure [35], which can induce silencing of a gpt transgene in hamster cells [36]. While reduced expression alone did not induce methylation of K9 H3 in our system, increased levels of dimethyl-K9 H3 were observed after alleles transitioned to the silenced state identified by TG-resistance.

Results provided by our experiments help establish specific distinctions between the states of transcriptional repression and epigenetic silencing. In our system epigenetic silencing was defined asHPRT expression that was reduced to levels that allowed growth in TG selection. Therefore, the most evident difference was that clones with silenced alleles were TG-resistant while cells growing in Dox remained sensitive to TG selection. The molecular basis of this phenotypic difference was demonstrated by showing TG-resistant cells had lower levels of HPRT mRNA than cells treated with Dox (Fig. 3) and molecular changes associated with epigenetic silencing (Figs. 4, 5). While the reduced expression after Dox treatment correlated with a loss of acetyl-K14 H3 at the PTRE-HPRTpromoter, TG-resistance and epigenetic silencing correlated with additional molecular changes including DNA methylation, reduced methyl-K4 H3, loss of acetyl-K9 H3, and increased dimethyl-K9 H3 at the PTRE-HPRT promoter. Although increased DNA methylation was one of the molecular changes observed at silenced promoters in our system, DNA methylation was not required for the initiation of silencing because 5-aza-dC treatment had no effect on the frequency of silenced clones induced by Dox treatment. Evidence that the 5-aza-dC treatment used here was sufficient to inhibit DNA methylation was provided with experiments showing 5-aza-dC treatment induced reactivation of silenced PTRE-HPRT promoters that were hypermethylated (Fig. 6). Additionally, bisulfite sequencing analysis showed HPRT silencing in the TG1 and TG2 cell lines did not require high levels of DNA methylation (Fig. 4). In contrast to inhibition of DNA methylation, inhibiting HDAC activity prevented most, but not all, of the Dox-dependent increase in HPRT silencing. This observation suggests the presence of two populations of silenced alleles at the end of the Dox treatment. One population would be silenced alleles that are readily reactivated by TSA treatment, and the second population would be alleles that are more stably silenced and fail to restore HPRT expression after TSA treatment. Presumably, the second population would have acquired additional repressive epigenetic modifications that cooperate with histone deacetylation to stabilize the silenced state.

A speculative model (Fig. 9) to explain the results obtained herein is that promoters with high transcriptional activity are resistant to silencing and are characterized by epigenetic modifications commonly associated with active expression (Fig. 9A). After transcriptional activity decreases at the promoter, acetyl-K14 H3 levels are reduced, and the promoter is more susceptible to epigenetic silencing (Fig. 9B). Although decreased acetyl-K14 H3 alone is not sufficient to induce epigenetic silencing, loss of this modification could decrease protection of the promoter from epigenetic silencing. Similarly, histone H3 acetylation has been shown to establish a protective boundary against spreading of DNA methylation [10]. The transition from reduced expression to epigenetic silencing initiates with histone deacetylation based on the observations that acetyl-K9 H3 levels were low at silenced PTRE-HPRT promoters and inhibiting class I and II HDACs reduced the frequency of epigenetic silencing (Fig. 9C). Initially the silenced alleles are unstable and can be reactivated by TSA treatment, but as additional epigenetic modifications occur the silenced state stabilizes and is resistant to TSA treatment alone

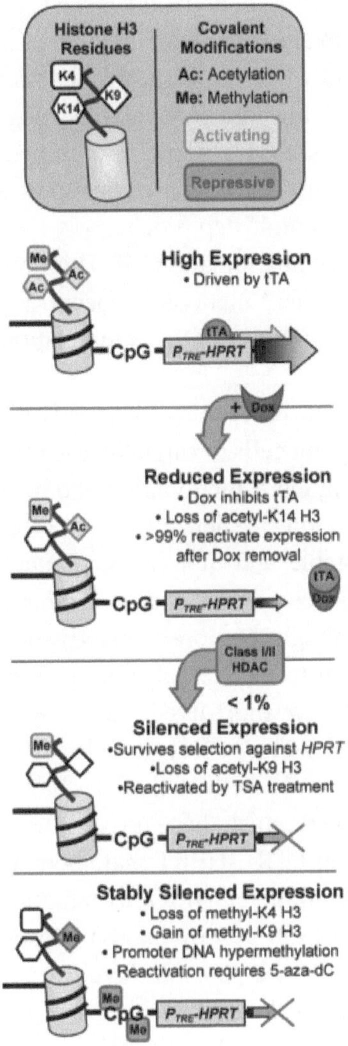

FIGURE 9. A model for induced silencing via reduced gene expression. (A) The VP16 activation domain promotes high levels of expression. DNA in the promoter region DNA is unmethylated and histone H3 is enriched for activating modifications (methyl-K4, acetyl-K9, and acetyl-K14). (B) Adding Dox reduces expression levels and acetylation at K14 of histone H3. (C) Reduced expression sensitizes alleles to undergo silencing; silenced alleles become unable to restore expression after Dox removal. The transition to silencing correlates with a further reduction in detectable mRNA and hypoacetylation at K9 H3 and is inhibited by TSA treatment (Class I/II HDAC inhibitor). (D) Additional epigenetic changes (loss of methylation at K4-H3, methylation at K9-H3, and DNA methylation) occur with continued TG selection against *HPRT* expression, as the silenced state stabilizes.

(Fig. 9D). We propose DNA methylation as a late step in epigenetic silencing because 5-aza-dC treatment did not affect the initiation of silencing. Although loss of methyl-K4 H3 is also shown as a secondary step, our results are not inconsistent with this loss being an early step in epigenetic silencing similar to loss of acetyl-K9 H3. While future experiments are required to test this model directly, aspects of it are consistent with prior observations. One is that silencing is a multistep process in which DNA methylation occurs downstream of silencing initiation [4]. This conclusion is supported by multiple observations of DNA methylation occurring after histone modification [4,10,11,37] and examples of DNA methylation-independent silencing [12–14].

A current focus in cancer treatment is reactivating silenced tumor suppressor genes in malignant cells through the use of pharmacological agents [5]. Although inhibiting DNA methylation and histone deacetylation usually reactivates expression of silenced alleles [38,39], such renewed expression is often unstable and quickly re-silences at a high frequency, possibly as a consequence of retention of some repressive histone modifications [29,30]. Although temporary reactivation of tumor suppressor genes may be sufficient to induce anti-tumor effects, re-silencing would ensure that these effects are short-lived. Thus, it would be helpful to know if high frequency re-silencing reflects a lack of prolonged expression, or alternatively if silenced and reactivated alleles have a persistent memory of the silenced state. To distinguish these possibilities, we isolated subclones from cells with silenced PTRE-HPRT that spontaneously reactivated expression and used selection for HPRT to maintain the reactivated state for at least one month (~50 cell divisions). Despite the prolonged time of reactivated expression, the absolute level of expression is not always restored to the original level, and the reactivatedPTRE-HPRT alleles still re-silence at a high frequency. Additionally, re-silencing did not require the Dox-mediated reduction in expression that was required for the initial silencing event. Thus, the memory of silencing was clearly persistent and likely reflects retention of epigenetic modifications. The inhibition of re-silencing with TSA suggests similarities with the initiation of silencing, which was also inhibited with TSA treatment.

We propose that the PTRE-HPRT system presented in this study represents a valid model for initiation and progression of aberrant silencing in cancer because silenced PTRE-HPRT alleles display the hallmarks of tumor suppressor gene silencing (promoter region DNA methylation, histone hypoacetylation, loss of methyl-K4 H3, and gain of methyl-K9 H3). In other words, we believe that the principle of reduced expression as a trigger for silencing will apply to bona fide mammalian promoters. Although our system utilized a non-mammalian promoter, endogenous levels of enzymes that control histone modifications and DNA methylation were responsible for the transition from repression to silencing. This is a unique and significant difference between our experimental system and previous systems that induced silencing by direct recruitment of repressive protein domains [40] or direct establishment of DNA methylation [41,42] at promoters. Hence, our system has the potential to detect multiple independent pathways of epigenetic silencing, which could be cell type specific. For example, histone modifications and DNA methylation are both observed at silenced promoters in colon cancer cells, whereas some of the same promoters only exhibit histone modifications when silenced in prostate cancer cells [14].

In summary, we used the tet-off system to provide a clear demonstration that reduced transcriptional potential can sensitize a promoter to undergo epigenetic silencing. Consistent with prior work, the results demonstrate that silencing is a multistep process in which promoter region DNA methylation is secondary to altered histone modification. We propose that these results are applicable to tumor suppressor promoters that are repressible by internal or external environmental exposures and that the model we created will be useful for identifying molecular determinants of aberrant silencing in mammalian cells.

MATERIALS AND METHODS

TET-OFF CONSTRUCTS

The Tet-Off system has been described previously [24]. The pTet-Off plasmid (Clontech) expresses the neomycin (Neo) resistance gene (Neor) and tTA, a fusion protein composed of the amino-terminus of the tetracycline repressor and the activation domain of the VP16 protein. The 1. 38 kb HPRT full-length cDNA sequence (Accesion # NM_000194. 1) was isolated by EcoRI and XbaI digestion of the TrueCloneHPRT cDNA expression vector (Origene, catalog #TC120047). pTRE-tight-HPRT was created by directionally cloning the HPRT fragment into the EcoRI and XbaI restriction sites within the pTRE-Tight (Clontech) multiple cloning site.

CELL CULTURE

Dif-6 cells were cultured in Dulbecco's modified Eagle's medium (Hyclone) supplemented with 5% fetal bovine serum (Hyclone) and 5% Serum Plus (SABC Biosciences). 5×106 Dif-6 cells were transfected by electroporation [25] with 4 μg of pTet-Off plasmid (Clontech) that expressed the tTA activating protein and selected for linked Neor with 500 μg/ml of G418. A transfectant showing high tTA expression was selected for a second transfection with 10 μg of PTRE-HPRT and 2 μg of a separate plasmid containing a bacterial puromycin (pur) resistance gene. Stable transfectants expressing functional HPRT were selected with media containing 10 μg/ml azaserine (Sigma) and 10 μg/ml hypoxanthine (Sigma) (AzHx medium). Selection for the pur gene was with 1. 5 μg/ml puromycin (Invitrogen). Individual clones were expanded and screened for physical linkage between pur and PTRE-HPRT by identifying clones with low frequency PTRE-HPRT inactivation (via TG selection) while retaining resistance to puromycin. Comparing the PTRE promoter signal to the Gapdh promoter signal from genomic DNA samples by quantitative-PCR measured PTRE-HPRT copy number in the cell lines. HP11 and HP14 contained single copies and HP13 contained two copies of PTRE-HPRT. These parental cells were routinely cultured in AzHx, G418, and puromycin to retain expression all constructs and expression of PTRE-HPRT. Doxycycline hyclate (Dox) (Sigma) was added to DMEM at a concentration of 1 μg/ml for silencing experiments. Dox medium also contained G418 and puromycin to retain the tTA and PTRE-HPRTconstructs, respectively. These drugs were also used during TG selection to retain both constructs in clones with silenced alleles. Cell exposed to Dox were not exposed to AzHx, unless indicated.

RNA PREPARATION AND ANALYSIS

Total RNA was isolated from cell cultures with the RNeasy Mini Kit (Qiagen) according to manufacturer's instructions. Total RNA samples were converted to cDNA using Quantitect Reverse Transcription Kit (Qiagen) with removal of genomic DNA contamination. 100 ng cDNA was used as input in subsequent quantitative-PCR analysis for either HPRT (TaqMan assay Hs99999909_m1, Applied Biosystems) or Gapdh (Mouse TaqMan Endogenous Control, Applied Biosystems) with iQ Supermix (Bio-Rad) and a Bio-Rad iCycler. HPRT Results were normalized in relation to Gapdh mRNA levels and displayed relative to an arbitrary value.

SILENCING AND REACTIVATION CELL CLONING ASSAYS

To measure PTRE-HPRT inactivation or reactivation, cells were plated into 100 mm plates at densities ranging from 1×104 to 1×105 cells per plate. The next day the medium was removed, cells were rinsed with DMEM, and TG or AzHx selective medium was to select against or for HPRT expression, respectively. Cells were cultured for approximately two weeks in the appropriate selective media before staining live colonies with crystal violet solution. To estimate cloning efficiencies, additional cells were plated under identical conditions as selective plates but at lower densities, 250 to 1000 cells per plate and without selection for or against HPRT expression. Silencing or reactivation frequencies were calculated by dividing the number of clones growing under selection by the effective number of cells plated (as determined with the cloning efficiency plates).

DRUG TREATMENTS

Cells were treated with media containing 100 nM TSA (Wako) overnight (~16 hours) to inhibit histone deacetylation. Cells were treated with media containing 300 nM 5-aza-dC (Sigma) overnight (~16 hours) to inhibit DNA methylation. For HPRT mRNA analysis, cells were allow to recover

24 hours in DMEM after drug treatment (TSA or 5-aza-dC) before harvesting for RNA purification.

DNA METHYLATION BISULFITE SEQUENCING ASSAY

Genomic DNA was isolated from cell cultures using DNAzol (Molecular Research Center) according to the manufacturer's instructions. For each treatment, 4 μg of genomic DNA was digested with Bsr I, and modified in a solution of 6. 24 M urea, 4 M sodium bisulfite, and 10 mM hydroquinone as described previously [4]. PCR amplification of modified DNA, cloning of PCR products, and sequence analysis were also described elsewhere [42], with the following exceptions. The primers used in the initial PCR reaction were the TRE-NaBis-S sense primer 5′-GTA TTT ATT AGG GTT ATT GTT TTA TGA G-3′ and the HPRT NaBis-A antisense primer 5′-CAA AAT AAA TCA AAA TCA TAA CCT AAT TC-3′. 1 μl of the PCR product was used as input in the subsequent semi-nested PCR reaction using the TRE-NaBis-NS primer 5′-GTA TTT AGA AAA ATA AAT AAA TAG GGG TTT-3′ and HPRT-NaBis-A for amplification. PCR products were cloned using Strataclone PCR cloning kit (Stratagene). Sequencing analysis showed all cytosine bases not present in the CpG dinucleotide context were converted to thymine indicating complete bisulfite modification of the genomic template occurred.

CHROMATIN IMMUNOPRECIPITATION

ChIP assays were carried out using EZ ChIP chromatin immunoprecipitation kit (Millipore) with the following specific details or modifications. Proteins were cross-linked to DNA in 5×107 cells by adding formaldehyde to a final concentration of 1% and incubating for 10 minutes at room temperature. The cross-linking reaction was stopped by addition of glycine to a final concentration of 125 mM and incubating for 5 minutes at room temperature. Cells were rinsed with cold PBS containing complete protease inhibitor cocktail (Roche) and resuspended in SDS lysis buffer. Lysates were sonicated using a Branson 450 microtip sonicator to shear

DNA into 100–1000 bp fragments. Protein-DNA complexes were immunoprecipitated using antibodies to acetyl-K9/K14 H3 (06-599, Millipore), acetyl-K9 H3 (07-352, Millipore), mono/di/trimethyl-K4 H3 (05-791, Millipore), and dimethyl-K9 H3 (ab1220, Abcam). 5 μl of each specific antibody was added to lysates from ~1×106 cells and incubated overnight at 4°C. Immunocomplexes were isolated by incubating for 3 hours at 4°C with a 3 1 mixture of Protein A and Protein G conjugated magnetic Dynabeads (Invitrogen) that had been blocked with salmon sperm DNA and BSA. Beads were washed once with each of the following: low salt buffer, high salt buffer, LiCl buffer, and 1× TE. Immunocomplexes were eluted by incubating beads at 65°C for 15 minutes in 200 μl elution buffer (50 mM Tris-HCl, 10 mM EDTA, 1% SDS), and the cross-links were reversed by incubating at 65°C overnight. After incubation with 0. 2 μg/ml RNase A at 37°C for 2 hours and 0. 2 μg/ml Proteinase K at 55°C for 2 hours, DNA was purified using QiaQuick PCR purification kit (Qiagen). Quantitative PCR using an Icycler and iQ SYBR Green Supermix (Bio-Rad) was used to analyze the immunoprecipitated DNA. The PTRE-HPRT promoter was amplified using the 5′-AAC GTA TGT CGA GGT AGG CGT GTA-3′ sense promoter and the 5′-ATC TCC TTC ATC ACA TCT CGA G-3′ antisense promoter. The active Gapdh promoter was amplified using the 5′-TTG AGC TAG GAC TGG ATA AGC AGG-3′ sense promoter and the 5′-AAG AAG ATG CGG CCG TCT CTG GAA-3′ antisense promoter. The silenced Mage-a promoter was amplified using the 5′-GTT CTA GTG TCC ATA TTG GTG-3′ sense promoter and the 5′-AAC TGG CAC AGC ATG GAG AC-3′ antisense promoter. The specific signal from each immunoprecipitation relative to signal from input was calculated for the three promoters, PTRE-HPRT, Gapdh, and Mage. For activating modifications, levels at PTRE-HPRT are displayed relative to the Gapdh promoter; for the repressive modification, dimethyl-K9 H3, results are displayed relative to the Mage promoter.

REFERENCES

1. Sawan C, Vaissière T, Murr R, Herceg Z. Epigenetic drivers and genetic passengers on the road to cancer. *Mutat Res* **2008**; 642:1–13.
2. Esteller M. Epigenetics in cancer. *N Engl J Med* **2008**; 358:1148–59.

3. Turker MS. Gene silencing in mammalian cells and the spread of DNA methylation. *Oncogene* **2002**; 21:5388–93.
4. Yates PA, Burman R, Simpson J, Ponomoreva ON, Thayer MJ, et al. Silencing of mouse Aprt is a gradual process in differentiated cells. *Mol Cell Biol* **2003**; 23:4461–70.
5. Mund C, Brueckner B, Lyko F. Reactivation of epigenetically silenced genes by DNA methyltransferase inhibitors: basic concepts and clinical applications. *Epigenetics : official journal of the DNA Methylation Society* **2006**; 1:7–13.
6. Gronbaek K, Hother C, Jones PA. Epigenetic changes in cancer. *APMIS* **2007**; 115:1039–59.
7. Reik W. Stability and flexibility of epigenetic gene regulation in mammalian development. *Nature* **2007**; 447:425–32..
8. Erwin JA, Lee JT. New twists in X-chromosome inactivation. *Curr Opin Cell Biol* **2008**; 20:349–55. doi:10.
9. Siegfried Z, Eden S, Mendelsohn M, Feng X, Tsuberi BZ, et al. DNA methylation represses transcription in vivo. *Nat Genet* **1999**; 22:203–6.
10. Mutskov V, Felsenfeld G. Silencing of transgene transcription precedes methylation of promoter DNA and histone H3 lysine 9. *EMBO J* **2004**; 23:138–49.
11. Strunnikova M, Schagdarsurengin U, Kehlen A, Garbe JC, Stampfer MR, et al. Chromatin inactivation precedes de novo DNA methylation during the progressive epigenetic silencing of the RASSF1A promoter. *Mol Cell Biol* **2005**; 25:3923–33.
12. Banelli B, Casciano I, Romani M. Methylation-independent silencing of the p73 gene in neuroblastoma. *Oncogene* **2000**; 19:4553–6. doi:10.
13. Markus J, Garin MT, Bies J, Galili N, Raza A, et al. Methylation-independent silencing of the tumor suppressor INK4b (p15) by CBFbeta-SMMHC in acute myelogenous leukemia with inv(16). *Cancer Res* **2007**; 67:992–1000.
14. Kondo Y, Shen L, Cheng AS, Ahmed S, Boumber Y, et al. Gene silencing in cancer by histone H3 lysine 27 trimethylation independent of promoter DNA methylation. *Nat Genet* **2008**; 40:741–50.
15. Caslini C, Capo-chichi CD, Roland IH, Nicolas E, Yeung AT, et al. Histone modifications silence the GATA transcription factor genes in ovarian cancer. *Oncogene* **2006**; 25:5446–61.
16. Leu Y, Yan PS, Fan M, Jin VX, Liu JC, et al. Loss of estrogen receptor signaling triggers epigenetic silencing of downstream targets in breast cancer. *Cancer Res* **2004**; 64:8184–92.
17. Raval A, Tanner SM, Byrd JC, Angerman EB, Perko JD, et al. Downregulation of death-associated protein kinase 1 (DAPK1) in chronic lymphocytic leukemia. *Cell* **2007**; 129:879–90.
18. Krishnamachary B, Zagzag D, Nagasawa H, Rainey K, Okuyama H, et al. Hypoxia-inducible factor-1-dependent repression of E-cadherin in von Hippel-Lindau tumor suppressor-null renal cell carcinoma mediated by TCF3, ZFHX1A, and ZFHX1B. *Cancer Res* **2006**; 66:2725–31.
19. Bindra RS, Gibson SL, Meng A, Westermark U, Jasin M, et al. Hypoxia-induced down-regulation of BRCA1 expression by E2Fs. *Cancer Res* **2005**; 65:11597–604.
20. Bindra RS, Crosby ME, Glazer PM. Regulation of DNA repair in hypoxic cancer cells. Cancer Metastasis Rev. **2007**; 26:249–60.

21. Lee S, Kim J, Kim W, Lee Y. Hypoxic silencing of tumor suppressor RUNX3 by histone modification in gastric cancer cells. *Oncogene* **2008.** Available: http://www.nature. com/onc/journal/vaop/ncurrent/abs/onc**2008**377a. html.
22. Li QL, Ito K, Sakakura C, Fukamachi H, Inoue KI, et al. Causal relationship between the loss of RUNX3 expression and gastric cancer. *Cell* **2002**; 109:113–24.
23. Li Q, Kim H, Kim W, Choi J, Lee YH, et al. Transcriptional silencing of the RUNX3 gene by CpG hypermethylation is associated with lung cancer. *Biochem Biophys Res Commun* **2004**; 314:223–8.
24. Gossen M, Bujard H. Tight control of gene expression in mammalian cells by tetracycline-responsive promoters. *Proc Natl Acad Sci USA* **1992**; 89:5547–51.
25. Turker MS, Mummaneni P, Bishop PL. Region- and cell type-specific de novo DNA methylation in cultured mammalian cells. *Somat Cell Mol Genet* **1991**; 17:151–7.
26. Dodge JE, Kang Y, Beppu H, Lei H, Li E. Histone H3-K9 methyltransferase ESET is essential for early development. *Mol Cell Biol* **2004**; 24:2478–86.
27. Tachibana M, Ueda J, Fukuda M, Takeda N, Ohta T, et al. Histone methyltransferases G9a and GLP form heteromeric complexes and are both crucial for methylation of euchromatin at H3-K9. *Genes Dev* **2005**; 19:815–26.
28. Yan C, Boyd DD. Histone H3 acetylation and H3 K4 methylation define distinct chromatin regions permissive for transgene expression. *Mol Cell Biol* **2006**; 26:6357–71.
29. McGarvey KM, Fahrner JA, Greene E, Martens J, Jenuwein T, et al. Silenced tumor suppressor genes reactivated by DNA demethylation do not return to a fully euchromatic chromatin state. *Cancer Res* **2006**; 66:3541–9.
30. Egger G, Aparicio A, Escobar S, Jones P. Inhibition of Histone Deacetylation Does Not Block Resilencing of p16 after 5-Aza-2′-Deoxycytidine \ldots. *Cancer Res* **2007.** Available: http://cancerres.aacrjournals.org/cgi/content/abstract/67/1/346.
31. Shin-Darlak CY, Skinner AM, Turker MS. A role for Pms2 in the prevention of tandem CC→TT substitutions induced by ultraviolet radiation and oxidative stress. DNA Repair (Amst) **2005**; 4:51–7. Available: http://www.ncbi.nlm.nih.gov/pubmed/15533837. Accessed 11 Feb 2009.
32. Hall DB, Struhl K. The VP16 activation domain interacts with multiple transcriptional components as determined by protein-protein cross-linking in vivo. *J Biol Chem* **2002**; 277:46043–50.
33. Stirzaker C, Song JZ, Davidson B, Clark SJ. Transcriptional gene silencing promotes DNA hypermethylation through a sequential change in chromatin modifications in cancer cells. *Cancer Res* **2004**; 64:3871–7.
34. Chen H, Yan Y, Davidson TL, Shinkai Y, Costa M. Hypoxic stress induces dimethylated histone H3 lysine 9 through histone methyltransferase G9a in mammalian cells. *Cancer Res* **2006**; 66:9009–16.
35. Chen H, Ke Q, Kluz T, Yan Y, Costa M. Nickel ions increase histone H3 lysine 9 dimethylation and induce transgene silencing. *Mol Cell Biol* **2006**; 26:3728–37.
36. Lee YW, Klein CB, Kargacin B, Salnikow K, Kitahara J, et al. Carcinogenic nickel silences gene expression by chromatin condensation and DNA methylation: a new model for epigenetic carcinogens. *Mol Cell Biol* 1995; 15:2547–57.
37. Bachman KE, Park BH, Rhee I, Rajagopalan H, Herman JG, et al. Histone modifications and silencing prior to DNA methylation of a tumor suppressor gene. *Cancer Cell* **2003**; 3:89–95.

38. Fahrner JA, Eguchi S, Herman JG, Baylin SB. Dependence of histone modifications and gene expression on DNA hypermethylation in cancer. *Cancer Res* **2002**; 62:7213–8.
39. Cameron EE, Bachman KE, Myöhänen S, Herman JG, Baylin SB. Synergy of demethylation and histone deacetylase inhibition in the re-expression of genes silenced in cancer. *Nat Genet* **1999**; 21:103–7.
40. Li F, Papworth M, Minczuk M, Rohde C, Zhang Y, et al. Chimeric DNA methyltransferases target DNA methylation to specific DNA sequences and repress expression of target genes. *Nucleic Acids Res* **2007**; 35:100–12.
41. Schübeler D, Lorincz MC, Cimbora DM, Telling A, Feng YQ, et al. Genomic targeting of methylated DNA: influence of methylation on transcription, replication, chromatin structure, and histone acetylation. *Mol Cell Biol* **2000**; 20:9103–12.
42. Hoffman AR, Hu JF. Directing DNA methylation to inhibit gene expression. *Cell Mol Neurobiol* **2006**; 26:425–38.
43. Yates PA, Burman RW, Mummaneni P, Krussel S, Turker MS. Tandem B1 Elements Located in a Mouse Methylation Center Provide a Target for de Novo DNA Methylation. *J Biol Chem* 1999; 274:36357–36361.

CHAPTER 8

THE SOUND OF SILENCE: RNAi IN POLY (ADP-RIBOSE) RESEARCH

CHRISTIAN BLENN, PHILIPPE WYRSCH, and FELIX R. ALTHAUS

THE LIFE CYCLE OF POLY (ADP-RIBOSE)

Poly(ADP-ribosyl)-ation belongs to the nonprotein posttranslational modifications and is a metabolite of the enzymatic cofactor nicotinamide adenine dinucleotide (NAD^+). Poly (ADP-ribose) polymerases (PARPs) cleave the nicotinic moiety from NAD^+ and convert the ADP-ribose (ADPR) units into long APD-ribose polymers (PAR). At the protein level, the E-D-loop-E-D NAD^+ fold is the most conserved region in PARPs and is therefore termed the PARP signature motif [1–3]. To date, 17 distinct PARP enzymes have been discovered and numbered accordingly. Recently a new nomenclature has been suggested that is based on their transferase activity (ARTD nomenclature [4]): PARP1 (ARTD1), PARP2 (ARTD2), PARP3 (ARTD3), PARP4 (ARTD4, vault-PARP), PARP5A (ARTD5, Tankyrase 1), PARP5B (ARTD6, Tankyrase 2), PARP6 (ARTD17), PARP7 (ARTD14, TIPARP, RM1), PARP8 (ARTD16), PARP9 (ARTD9, BAL1), PARP10 (ARTD10), PARP11 (ARTD11), PARP12 (ARTD12, ZH3HDC1), PARP13 (ARTD13, ZC3HAV1, ZAP1), PARP14 (ARTD8, BAL2 COAST6), PARP15 (ARTD7, BAL3), and PARP16 (ARTD15). However, not all PARP enzymes are proven poly(ADP-ribose) polymerases. Several of them seem to belong to the class of mono(ADP-ribosyl) transferases [5]. The nuclear PARP1 and PARP2 are the best characterized PARPs in mammals and are true poly(ADP-ribose) polymerases. PARP1 has a modular structure and starts its catalytic activity after binding to DNA nicks and breaks with the double zinc finger domain [2]. Recently, a third zinc finger-like structure was discovered. It is required for transmitting DNA-induced conformational changes to the catalytic domain [6,7].

The C-terminal catalytic domain sequentially transfers ADPR units from NAD^+ to protein acceptors generating PAR. Third, in the automodification domain, specific glutamatic acid and lysine residues serve as acceptors of ADPR allowing the enzyme to poly(ADP-ribosyl)ate itself. While the level of PAR is very low under physiological condition, it rises 200-fold under genotoxic stress conditions. After DNA damage, PARP1 is rapidly recruited and its catalytic activity increases 10- to 500-fold to synthesize long and branched PAR chains (Figure 1) [3,8]. PARP1 is the main enzyme-generating PAR and it is the main acceptor for covalent PAR modification. More than 90% of PAR originates from PARP1 during genotoxic stress [8,9]. PARP-bound PAR can recruit many other proteins involved in distinct cellular functions, including enzymes involved in DNA repair [10–16]. The system of PAR accepting and/or interacting proteins is very complex and not completely understood. Moreover, novel proteome-wide analyses allow the identification of putative PAR binding proteins [10]. Automodification of PARP1 diminishes the affinity of the enzyme for DNA breaks. This provides a mechanism for removing PARP1 from damaged DNA and for local modulation of chromatin compaction and transcriptional regulation [11,17–19].

Apart from PARP1, PARP2 is activated by DNA strand breaks, as well [20,21]. The DNA-binding domain (DBD) of PARP2 is structurally different from that of PARP1 [20,22]. PARP2 binds less efficiently to DNA single-strand breaks (SSB) than PARP1 but it detects gap and flap structures [23]. PARP2 consumes less NAD^+ compared to PARP1 due to its less efficient PAR synthesis [20]. It contributes only 5%–10% of the total PARP activity in response to DNA interruptions [20,24].

The enzymatic product of PARP1 and PARP2 activity PAR comprises a heterogeneous pool of negatively charged molecules that differ in length and branching. It is noteworthy that the hyperactivation of PARP1 and PARP2, due to genotoxic stress, consumes most of the NAD^+ in a cell [25]. Free or protein-associated PAR, synthesized after a genotoxic insult, are rapidly degraded *in vivo* with a half-life of less than 40 sec, while the residual fraction is catabolized with a half-life of 6 min [26–28]. The degradation of PAR is catalyzed by PARG (PAR glycohydrolase), an enzyme with both endo- and exoglycosidic activities that hydrolyze the glycosidic bond between ADPR units.

PARG produces primarily monomeric ADPR *via* exoglycosidic activity, albeit a few free PAR polymers may arise from endoglycosic cleavage (Figure 1) [29,30]. In addition, PARG displays less activity towards branched or short PAR ($K_M \approx 10$ μM) compared with long and linear PAR molecules ($K_M \approx 0.1$–0.4 μM) [29–34]. PARG is encoded by a single gene in mammals, and several splicing products are formed after transcription. They are translated into proteins of different molecular size, subcellular localization and the ability to cleave PAR. The nuclear mPARG-110/hPARG-111 isoform represents the full-length PARG protein in mice and humans and accounts for most of the PARG activity [35]. Recently another PAR-degrading enzyme has been described. The ADPR hydrolase 3 (ARH3) is structurally not related to PARG and is less efficient. Nevertheless, it provides PAR- degrading activity *in vitro* and in PAR-enriched mitochondria, as it has been demonstrated in a PARP overexpression system [36–38].

The two products of PARP/PARG interplay have been identified to exhibit different cellular signaling functions. The manipulation of either PARP or PARG activity modifies the occurrence of PAR and ADPR after genotoxic stress (Figure 1). This allows the study of distinct PAR and ADPR functions. Here, we discuss approaches to interfere with PAR metabolism to clarify the biological role of this nonprotein posttranslational modification (PAR) and its degradation product (ADPR).

EXPERIMENTAL TOOLS TO INVESTIGATE PAR AND ADPR

CHEMICAL INHIBITION OF PAR METABOLIZING ENZYMES

Within the last few decades, the concept of interfering with proteins involved in DNA repair and stress signaling has attracted a lot of attention in both basic and clinical research. To date, chemical inhibitors against PARP enzymes have reached the first level of clinical application. Almost

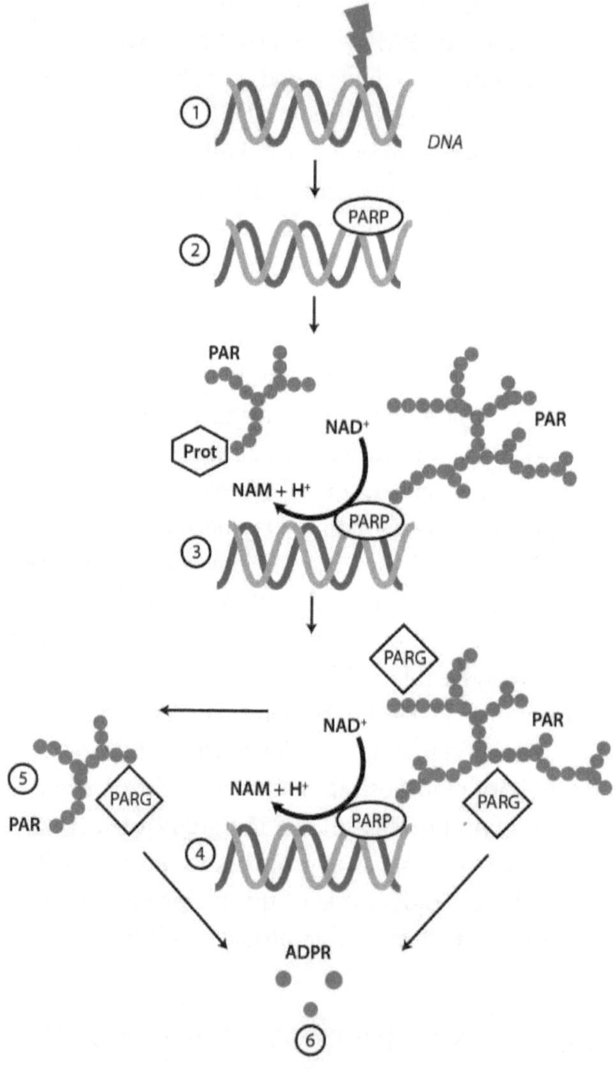

FIGURE 1. Poly(ADP-ribose) metabolism. Single-strand and double-strand breaks (SSB and DSB) in the DNA are induced by genotoxic stress (1). The nuclear PARPs bind to the damaged site and are thereby activated and produce ramified and highly negatively charged poly(ADPribose) (PAR) (2). They hydrolyze NAD^+, releasing nicotinamide (Nam) and H^+, catalyzing the successive transfer of ADPR to acceptors (Prot) (3). PARG is the catabolic enzyme with endo- and exoglycosidic activity (4). Its hydrolyzing activity leads to either free PAR chains (5) and/or monomeric ADPR (6).

all PARP inhibitors in preclinical and clinical studies compete with the substrate NAD^+ for the catalytic domain leading to a reversible inhibition of enzyme activity. The third generation PARP inhibitors veliparib (ABT-888) and olaparib (AZD2281/KU-0059436) are the most clinically investigated, and their half-maximal inhibitory concentration (EC_{50} or K_i) is in the nanomolar range for PARP1 and PARP2 (Figure 2) [39,40]. Both inhibitors are sufficiently bioavailable when administered orally and they are primarily used as anticancer drugs alone or in combination with other treatments (discussed in [41]). Moreover, the potential of PARP inhibitors to participate in the concept of synthetic lethality is under investigation [41,42].

The PARP inhibitors 3-Aminobenzamide (3-AB) [25,43–45], 1,5,7,8-Tetrahydro-2-methyl-4*H*- thiopyrano[4,3-*d*]pyrimidin-4-one (DR2313) [25,46], *N*-(6-Oxo-5,6-dihydrophenanthridin-2-yl)- (*N*,*N*-dimethylamino)acetamide (PJ-34) [25,47–50], 8-Hydroxy-2-methylquinazoline-4-one (NU1025) [51–53], and 3,4-Dihydro-5-[4-(1-piperidinyl)butoxyl]-1(2*H*)-isoquinolinone (DPQ) [54–57], were extensively used to suppress PARP activity *in vitro*, in cells and animals (Figure 2). They share the potential to interfere with both PAR-synthesizing enzymes activated by DNA damage due to the homology in the catalytic domain of PARP1 and PARP2, albeit with different kinetics. To overcome this problem, Moroni and co-workers developed a set of specific PARP2 inhibitors [58]. Beside synthetic PARP inhibitors, a panel of naturally occurring molecules was discovered that have PARP- suppressing activity. These compounds cover such different chemical structures as tryptophan derivatives, purines, xanthins, vitamins, hormones and metals [59].

Much less is known about the inhibitors of PARG, as was recently reviewed [60]. Whereas a number of natural and synthetic molecules have been described to exhibit PARG-suppressing activity, most of them are restricted in terms of bioavailability and/or specificity. Nevertheless, the use of Adenosine 5'-diphosphate (hydroxymethyl) pyrrolidinediol (ADP-HPD) [61,62], 3-Galloyl-α-β-D-glucose [63], (*Z*)-3-(5-(5-Bromo-1-(2,6-dichlorobenzyl)-2-oxoindolin-3-ylidene)-4-oxo-2- thioxo-thiazolidin-3-yl) propanoic acid [64], and 3,5-Dichloro-*N*-(3-chloro-4-(naphtalen-2-yloxy)

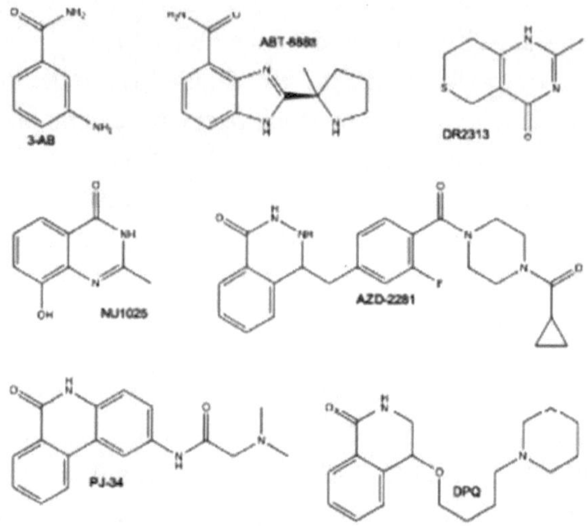

Drug name	IUPAC name	IC_{50}/EC_{50}	Reference
3-AB	3-Aminobenzamide	3.3 μM (IC_{50})	[43]
ABT-888 (Veliparib)	2-((*R*)-2-Methylpyrrolidin-2-yl)-1*H*-benzimidazole-4-carboxamide	2 nM (EC_{50})	[40]
AZD-2281 (Olaparib)	4-[(3-[(4-Cyclopropylcarbonyl)piperazin-4-yl] carbonyl)-4-fluorophenyl]methyl(2H)phthalazin-1-one	1–5 nM (IC_{50})	[39]
DPQ	3,4-Dihydro-5-[4-(1-piperidinyl)butoxyl]-1(2*H*)-isoquinolinone	40 nM (IC_{50})	[54]
DR2313	1,5,7,8-Tetrahydro-2-methyl-4*H*-thiopyrano[4,3-*d*] pyrimidin-4-one	0.2 μM (IC_{50})	[46]
NU1025	8-Hydroxy-2-methylquinazoline-4-one	0.4 μM (IC_{50})	[51]
PJ-34	*N*-(6-Oxo-5,6-dihydrophenanthridin-2-yl)-(*N,N*-dimethylamino)acetamide	20 nM (EC_{50})	[49]
ADP-HPD	Adenosine 5'-diphosphate (hydroxymethyl) pyrrolidinediol	0.12	[61]
Mono-galloyl glucose	3-galloyl-α,β-D-glucose	0.95 ± 0.02	[63]
RBPI-2	(*Z*)-3-(5-(5-Bromo-1-(2,6-dichlorobenzyl)-2-oxo-indolin-3-ylidene)-4-oxo-2-thioxothiazo-lidin-3-yl) propanoic acid	2.9 ± 1.8	[64]
Salicylanilide	3,5-Dichloro-*N*-(3-chloro-4-(naphtalen-2-yloxy) phenyl)-2-hydroxybenzamide	12 ± 2	[65]

FIGURE 2. PARP inhibitors.

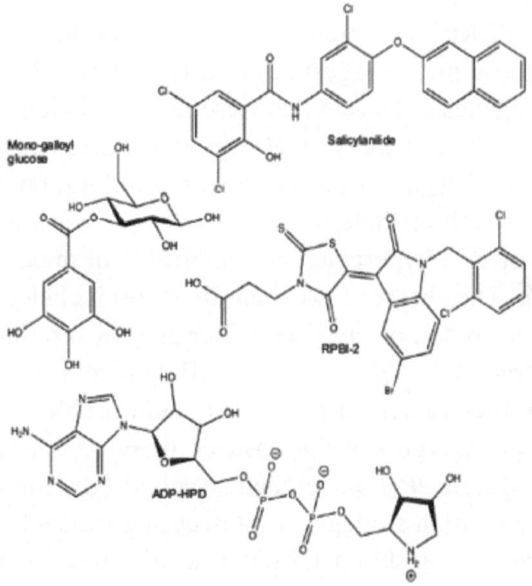

Common Name	IUPAC Name	IC_{50} [μM]	Reference
ADP-HPD	Adenosine 5'-diphosphate (hydroxymethyl) pyrrolidinediol	0.12	[61]
Mono-galloyl glucose	3-galloyl-α,β-D-glucose	0.95± 0.02	[63]
RBPI-2	(Z)-3-(5-(5-Bromo-1-(2,6-dichlorobenzyl)-2-oxo-indolin-3-ylidene)-4-oxo-2-thioxothiazo-lidin-3-yl) propanoic acid	2.9 ± 1.8	[64]
Salicylanilide	3,5-Dichloro-N-(3-chloro-4-(naphtalen-2-yloxy) phenyl)-2-hydroxybenzamide	12 ± 2	[65]

FIGURE 3. PARG inhibitors.

phenyl)-2- hydroxy benzamide [65] may inhibit PARG specifically and help understand its role within a wider biological context (Figure 3).

GENETIC DISRUPTION OF PARPS AND PARG

In the mid-1990s, the first PARP knock out (k. o.) mice were created to provide more detailed and specific information about the physiological role of PARP enzymes in different cellular responses. Wang *et al.* generated in 1995 a PARP1 k. o. mouse line by interrupting exon 2 [66]. The existence of these mice suggested initially that PARP1 is dispensable during embryogenesis. These mice were fertile, obviously healthy and revealed no dramatic phenotype. However, isolated cells derived from *Parp1*$^{-/-}$ mice had a marginally lower proliferation rate unrelated to DNA damage. Nevertheless, there was an unusual and unexpected development of skin hyperplasia in around 1/3 of mice related to progressive aging. The observed skin abnormalities included thickening of all layers of the epidermis, active proliferation of keratinocytes, development of intracellular edema and an inflammatory response [66]. As 70% of these mice remained free of any skin problems, a correlation between this phenotype and the absence of PARP1 seemed unlikely. Ménissier de Murcia *et al.* developed a second *Parp1* mouse strain using a different ablation strategy by interrupting exon 4 [67], which was completely free of any skin pathologies as discussed in more detail by Rhun *et al.* [68]. These PARP1 k. o. mice exhibited an extreme sensitivity to whole body J-radiation, indicating a functional link between PARP1 and DNA stress. This first report was later confirmed by Wang *et al.* investigating their first *Parp1*$^{-/-}$ mouse type [69]. Additionally, Ménissier de Murcia and colleagues showed that the whole body exposure to 8 Gy resulted in a very short half-life of four days with a complete lethality after eight days, compared to 50% lethality observed in control mice (14–20 days). Dramatic lethality with extensive necrosis occurred in the epithelial lining of the small intestine [67]. Furthermore, high sensitivity to intra-peritoneal administration of 75 mg/kg of body weight of the alkylating agent *N*-methyl-*N*-nitrosourea (MNU) was observed. They determined 80% lethality within the first week in *Parp1*$^{-/-}$ mice compared to 40% for control mice until week eight.

Results obtained in these two different *Parp1*$^{-/-}$ mice lines demonstrated the involvement of PARP1 in processing genotoxic attacks. Both authors concluded that the lack of the *Parp1* gene led to an impaired DNA repair capacity, an accumulation of DNA strand breaks and, subsequently, high genomic instability. The results of the k. o. experiments performed by Wang *et al.* and Ménissier de Murcia *et al.* are summarized in Table 1. None of the approaches to diminish the *Parp1* gene resulted in the expression of fully functional truncates. However, Wang *et al.* reported a truncated 1. 2 kb transcript generating a PARP1 fragment but without any enzymatic activity.

In 1999, a partial functional redundancy between PARP1 and PARP2 was observed by Amé *et al.* [20]. As a logical consequence, mice lacking PARP2 were needed. Therefore, Ménissier de Murcia and coworkers generated the first *Parp2*$^{-/-}$ mice line by disrupting exon 9 [70]. These manipulated mice were fertile and viable like the *Parp1*$^{-/-}$ mice. Interestingly, they showed an even higher sensitivity to genotoxic challenges than their *Parp1*$^{-/-}$ counterparts. Moreover, *Parp2*$^{-/-}$ mice displayed abnormalities in spermiogenesis caused by delayed nuclear elongation. They developed lipodystrophy characterized by adipodegeneration, as well as derailed differentiation of preadipocytes to adipocytes. It was also reported that genetic depletion of *Parp2*$^{-/-}$ resulted in reduced $CD4^{+}CD8^{+}$ thymocyte cellularity, the overexpression of the proapoptotic proteins Noxa and Puma as well as a reduced expression of the T-cell receptor (TCR) D [21,71–74]. The importance of the PAR metabolizing enzymes PARP1 and PARP2 is further illustrated by the fact that a double k. o. results in embryonic death. This can be observed already in the early stage of gastrulation leading to a complete loss of the offspring [70,75].

After the first demonstration of PARP1 and PARP2 functions using k. o. mice, another breakthrough was made by the disruption of the PAR degrading enzyme PARG. First, Koh and colleagues targeted exon 4 within the *parg* gene [76]. This led to early peri-implantation lethality resulting in the failure of the embryos to progress and the subsequent degeneration of the blastocyst. However, embryonic trophoblast stem cell lines established from early PARG null embryos could be kept viable in culture, but only when co-cultured with the traditional PARP inhibitor benzamide (0. 5 mM). Moreover, the *Parg*$^{-/-}$ cells were characterized by reduced growth, an accumulation of PAR and an increased sensitivity to the alkylating agent *N*-methyl-*N'*-nitro-*N*-nitrosoguanidine (MNNG, 20 and 100 PM). An improved strategy to impair *Parg* gene in mice to overcome lethality problems was reported by Cortes et al. [77]. They targeted exons 2 and 3 resulting in the depletion of only PARG110 protein isoform normally found in the cell nucleus. Due to this elegant approach the mice were viable, fertile and phenotypically normal but hypersensitive to alkylating agents (MNU, 150 mg/kg of body weight) and ionizing radiation (10 Gy of whole body J-radiation). However, Meyer-Ficca *et al.* reported a reduced number of litters as well as a decreased number of pups per litter in $Parg^{\Delta 2-3}$ mice [78]. Moreover, the $Parg^{\Delta 2-3}$ mice were prone to septic shock induced by lipopolysaccharides (LPS, 30 mg/kg of body weight). Both PARG manipulated mouse models are presented in Table 1.

RNA INTERFERENCE

Sense RNAs have been typically introduced as negative specificity controls using RNAs synthesized *in vitro* in antisense studies in the 1990s [79]. Interestingly, Guo and Kemphues found that control sense as well as antisense RNA molecules resulted in similar phenocopies when administered to the worm *Caenorhabditis elegans* [80]. Further studies in *C. elegans* revealed that the interfering RNA could be transported from the site of initial delivery to most cells and tissues in the worm and this systemic response was termed RNA interference or RNAi [81]. Subsequently, it was discovered that double-stranded RNA (dsRNA) rather than single-stranded antisense RNA (ssRNA) was responsible for the sequence specific degradation of targeted endogenous mRNA in *C. elegans* [82]. This form of posttranscriptional gene silencing in response to transgene sequences was previously demonstrated in plants and also in fungi representing an evolutionary well-conserved mechanism. Napoli *et al.* introduced a transgene designed to overexpress chalcone synthase putatively leading to an increase in purple pigment production in petunia flowers. Surprisingly, more than 40% of the transgenic plants developed white or variegated flowers rather than purple [83]. A similar phenomenon of target gene repression in a different model organism was discovered in experiments using the fungus *Neurosporus crassa* [84].

TABLE 1. Murine phenotypes of genetic PARP or PARG disruption.

Deletion	Phenotype	Ref.
Parp1	• Accumulation of DNA strand breaks and impaired DNA repair • High genomic instability Hypersensitivity to J-irradiation and alkylating agents • No defects in viability, fertility, development or tissue differentiation	[66–68]
Parp2	• High genomic instability • Hypersensitivity to J-irradiation and alkylating agents • Impaired adipogenesis, thymopoiesis, and spermatogenesis	[70,73]
Parp1, Parp2	• Embryonic lethality at onset of gastrulation	[70,75]
Parg	• Peri-implantation lethality	[76]
PargΔ2-3	• Increased responses to genotoxic treatment and septic shock • Phenotypically normal and viable	[77,78]

A first mechanistic model resolving this phenomenon originated from comprehensive biochemical and genetic studies on flies, worms, fungi, plants, and mammalian cells (Figure 4, reviewed by Hutvagner et al. [85]). The RNA-silencing response starts by processing the trigger dsRNA molecules into smaller fragments with 21 to 25 bp in size that are characterized by 3' dinucleotide overhangs. The activity of an enzyme called Dicer is required for this process and it depends on adenosine triphosphate (ATP). This multidomain protein contains an ATP-dependent RNA helicase, a Piwi Argonaut and Zwille (PAZ) domain, two tandem RNase III domains, and a dsRNA-binding domain [86]. The final products of the Dicer activity were termed short interfering RNAs (siRNAi) [87]. Each siRNA binds subsequently to the RNA-induced silencing complex (RISC) that is composed of different protein subunits. In a next step, the siRNAs become unwound due to a helicase component of RISC. This allows base pairing between the antisense strand and the target mRNA [88]. Endonucleolytic cleavage of the target mRNA *via* the endonuclease Ago2 leads to fragments missing polyA tail or missing 5' 7-Methylguanosine cap leading subsequently to degradation [79].

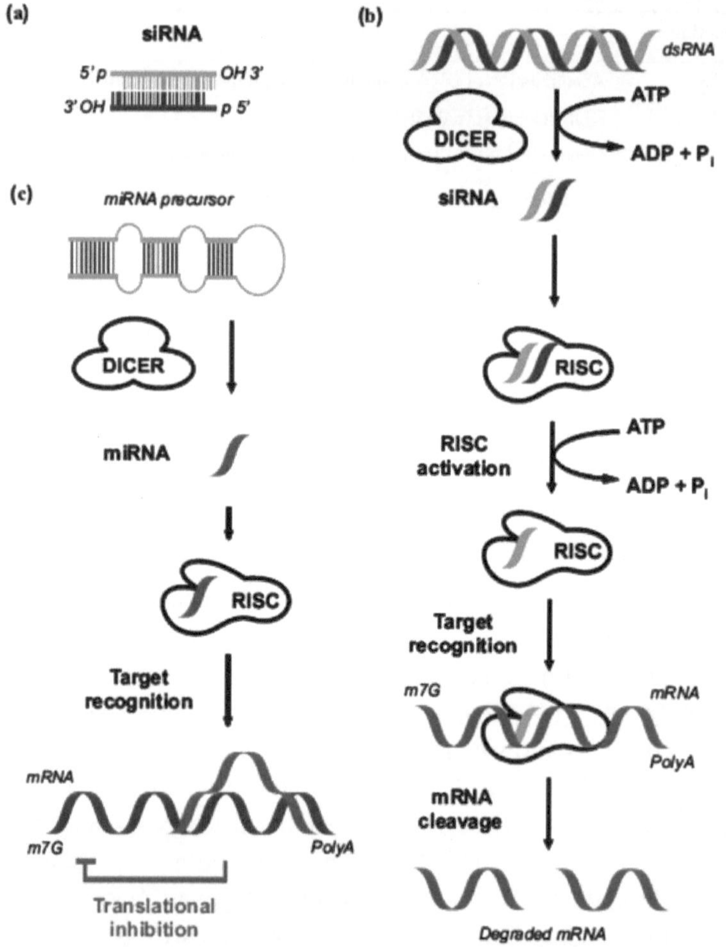

FIGURE 4. The pathway of RNA interference. **(a)** Highly specific siRNA which are 21–23 nt double-stranded RNA (dsRNA) duplexes in size with symmetric 2-3 nucleotide 3' overhangs and 5' phosphate (P) and 3' hydroxyl (OH) groups **(b)** The dsRNA is cleaved by Dicer in an ATP-dependent reaction. The siRNAs are subsequently incorporated into a multicomponent RNA-inducing silencing complex (RISC). Activated RISC unwinds the siRNA in an ATP-dependent fashion. The resulting antisense strand guides the RISC to its complementary mRNA before endonucleolytic cleavage of target mRNA. The free ends of the mRNA fragments are rapidly degraded by cytoplasmic nucleases. This ultimately results in the loss of protein expression. **(c)** Dicer cleaves the miRNA precursor to produce 22 nt miRNA. The single-stranded miRNAs are incorporated into RISC, followed by target recognition and final translational inhibition [104].

The mechanistic overview is shown in Figure 4. To date, four different types of small RNA molecules are described to possess RNA interference.

SHORT INTERFERING RNA (siRNA)

Synthetic siRNA was the first RNAi technology used in mammalian cells for sequence specific gene silencing [89]. The siRNA directly incorporates into RISC, where its guide strands binds and cleaves the target mRNA. After the release of cleaved mRNA, the guide-strand-bound RISC binds to another mRNA starting a new cleavage cycle. siRNAs are capable to cleave and subsequently suppress cytoplasmic or nuclear target mRNA [90].

SHORT HAIRPIN RNA (shRNA)

Short hairpin RNA (shRNA) has been developed as an alternative strategy to siRNA and was aimed for long-term gene silencing [91,92]. The shRNA is transcribed in the nucleus out of an expression vector bearing a short dsDNA sequence with a hairpin loop. The resulting transcript becomes further processed after coupling to the cytoplasmic RISC, following the identical fate as described above for siRNA. In practice, several aspects of shRNA differ intrinsically from siRNA reviewed by Rao *et al.* [93].

MICRORNA (miRNA)

MicroRNAs (miRNAs) are highly conserved small noncoding RNAs, which play a role in regulating physiological and pathological cell functions. They are initially transcribed in the nucleus as primary transcripts from precursors generally located within either intergenic regions or introns of protein coding sequences [94]. After processing by an RNAse III endonuclease, the pre-miRNAs are exported to the cytoplasm and further cleaved by the Dicer producing 20–23 bp mature miRNAs. The miRNAs are subsequently loaded into RISC eliciting transcriptional inhibition with target mRNA degradation or sequestration in cytoplasmic P-bodies [95]. Compared to siRNAs and shRNAs, which require a perfect match with the target mRNA, miRNAs exert typically translational inhibition by binding to a partially complementary mRNA (Figure 4). Therefore, the change in the expression of a single miRNA may affect hundreds of different genes [96,97].

BI-FUNCTIONAL shRNA (bi-shRNA)

Bi-functional shRNAs (bi-shRNAs) were developed to exploit both the cleavage and translational inhibitions mechanisms of RNAi [93]. They consist of two stem-loop shRNAs structures. The cleavage-dependent unit perfectly matches passenger and guide-strand, whereas the cleavage-independent unit is composed of a mismatched double-strand. Both units are embedded in a miR-30 scaffold encoded in a plasmid vector. The mature transcript of the cleavage-dependent part is loaded into RISC in complex with the endonuclease Ago2. The processed transcript of the cleavage-independent unit functions as a miRNA by binding to RISC inducing mRNA degradation/P-body sequestration or transcriptional inhibition. In general, the mechanism of bi-shRNAs leads to higher efficacy and greater durability compared with siRNA and miRNA [98].

The current understanding of therapeutic implementation of all these different small RNA approaches has been extensively reviewed: Seyhan discussed the progress and arising challenges in developing RNAi therapeutics against genetic diseases [99]. The current status of RNAi-based anticancer therapy and their potential clinical application is discussed by Wang *et al.* [98] and Bora *et al.* [100].

THE USE OF RNA INTERFERENCE AGAINST PARP ENZYMES

The possibility to suppress the expression of a specific protein in cell cultures has attracted a lot of interest. Moreover, the PAR research was extended to cell systems other than that of murine origin, due to RNAi techniques. Historically the first experiments using RNAi as a tool targeting PAR metabolism were performed aiming the knock down (k. d.) of PARP1 (Table 2).

Already in 2002, Gan *et al.* reported the silencing of PARP1 protein caused by the transfection of PARP1-specific dsRNA molecules in murine neuroblastoma cells [101]. They targeted the N- as well as the C-terminus of the *Parp1* gene, and both approaches diminished the expression of the corresponding protein. Further, this study confirmed the resistance of PARP1 abrogated cells to oxygen–glucose deprivation as a functional consequence. This phenotype was previously demonstrated in cells lacking PARP1 due to genetic ablation. Therefore they suggested the application of RNAi as a powerful tool to study gene functions in cells. In 2004, Kameoka *et al.* introduced RNAi against PARP1 in human cell cultures. In this report, the authors studied the HIV-1 replication within HeLa and J111 cells that have an impaired PARP1 expression [102]. One of the obvious advantages to prefer RNAi rather than chemicals for enzyme inhibition purposes is its specificity. This was demonstrated in an elegant study by Shah *et al.* in 2005 [103]. They developed and compared RNAi approaches targeting the C-terminus of the *parp1* gene in human, murine and hamster cells and investigated the impact of PARP1 k. d. on the expression level of PARP2. As discussed above PARP2 exhibits a significant amino acid and nucleotide sequence similarity with PARP1 in the C-terminal domain. The RNAi target sequences for PARP1 determined in this study have only three (for the murine) or greater than five (hamster, human) mismatches with the *parp2* gene. Shah *et al.* found no impact on PARP2 expression in the tested PARP1 k. d. models as tested by Western blot analyses. They concluded that RNAi against PARP1 does not interfere with its most related protein PARP2 and since other PARP-homologs have even greater disparity to PARP2, it is not likely to affect them by RNAi against PARP1 [103]. To date, RNAi against PARP1 has been applied to at least 20 different human cell types and to cells derived from other mammalian species (Table 2). These include immortalized, primary and cultures from embryonic stem cells.

TABLE 2. RNAi against PARP1 in mammalian cells. Studies labeled wit h * provide detailed targeting sequence information.

Cell line	Cell type	Species	References
293	Embryonic kidney cells	Human	[105]*, [106,107]
3T3-L1	Preadipocytes	Mouse	[108]*
A20	B-cell lymphoma	Mouse	[109]
A549	Lung adenocarcinoma epithelial	Human	[110,111]*, [107,112–114]
AGYNB010	Neuroblastoma	Mouse	[101]*
bEnd. 3	Cerebral vascular endothelial	Mouse	[114]
CHO	Ovarian cells	Chinese hamster	[103]*
EW7	Ewing sarcoma	Human	[115]*
GM00637	SV-40 transformed skin fibroblasts	Human	[116]*
H9 hESCs	Embryonic stem cells	Human	[117]*
HaCaT	Keratinocytes	Human	[118,119]
HCT-116	Colon adenocarcinoma	Human	[120,121]
HeLa	Cervix carcinoma	Human	[102,122–128]*, [129,130]
HepG2	Hepatocytes	Human	[131]
HUVEC	Endothelial cells	Human	[132]
J111	Acute monocytic leukemia	Human	[102]*
Jurkat	T-cell lymphoma (Type II)	Human	[133]*
MCF-7	Breast cancer	Human	[134]*
MEF	Fibroblasts	Mouse	[25,103,135,136]*
NB4	Acute promyelocytic leukemia	Human	[115]*
NIT-1	Insulinoma cells	Mouse	[137]*
PC12	Prostate cancer	Human	[138]
Primary	Fibroblasts	Human	[139]*
Primary	Cerebral cortex neurons	Rat	[140]
Primary	Rheumatoid arthritis synovial cells	Human	[141,142]
Primary	Vascular smooth muscle	Rat	[143]
Ramos	Burkitt's lymphoma	Human	[109]
SHSY5Y	Neuroblastoma	Human	[144]
SW480	Colorectal adenocarcinoma	Human	[145]
WRL-68	Liver cells	Human	[146]

RNAi molecules targeting PARP2 were next developed to study distinct and overlapping functions of PARP1 and PARP2 in a variety of mammalian cells (Table 3). As described above, PARP2 k. o. has created lethality in conjunction with PARP1 k. o. in mice [70]. Here the application of RNAi techniques helped to overcome this problem. Within the last years we have investigated the combination of either double RNAi (PARP1 and PARP2) in human cells [122,127] or the combination of genetic ablation of *parp1* gene together with RNAi against PARP2 in murine cell cultures [135]. None of the tested systems revealed a lethal phenotype and the cells kept the ability to attach and proliferate under normal culture conditions. Woodhouse *et al.* silenced PARP1 and PARP2 alone or in combination in cell cultures derived from primary human fibroblasts and no toxicity was reported due to double silencing [139]. This putative discrepancy compared to the *Parp1*$^{-/-}$ *Parp2*$^{-/-}$ mouse models is most likely related to the noncomplete gene disruption mediated by RNAi in all studies [122,135,139,145]. However, here the implication of RNAi provides advantages to study distinct functions of closely related proteins and overcomes lethality problems and the lack of specific inhibitors. Whereas PARP1 and PARP2 are the most studied members of the PARP enzyme family, RNAi was used as a tool to suppress other PARP enzymes including PARP3 [126,147,148], PARP4 (vault-PARP) [126], PARP5A (Tankyrase 1) [126,149], PARP5B (Tankyrase 2) [126], PARP7 (TIPARP) [150] and PARP16 [151].

TABLE 3. RNAi against PARP2 in mammalian cells. Studies labeled with * provide detailed targeting sequence information.

Cell line	Cell type	Species	References
293	Embryonic kidney cells	Human	[72]*, [152]
A549	Lung adenocarcinoma epithelial	Human	[111]
BHK	Baby hamster kidney fibroblast	Chinese hamster	[152]
C2C12	Myoblasts	Mouse	[153]*
HeLa	Cervix carcinoma	Human	[122,126,127]*
MEF	Fibroblasts	Mouse	[135]*
MOVAS	Aortic smooth muscle	Mouse	[154]*
SW480	Colorectal adenocarcinoma	Human	[145]

Moreover, RNAi approaches aiming at the organismic level of invertebrates, mice and plants have been performed to clarify PAR metabolism *in vivo* and/or to investigate a possible clinical application of RNAi itself. Gravel *et al.* published in 2004 RNAi experiments targeting the human tankyrase related gene *pme-5* in the invertebrate *C. elegans* [155]. The same group demonstrated later the effective application of RNAi against the *C. elegans* PARGs pme-3 and pme-4 [156]. The group of Tulin studied PAR metabolism in flies (*Drosophila melanogaster*). They established and investigated RNAi approaches against PARP1 and its counterpart PARG and clarified their involvement in gene expression [157,158].

Popoff *et al.* investigated antisense oligonucleotides targeting PARP2 in a colitis mouse model [159]. Recently, Goldberg and co-workers established whole body silencing of PARP1 studying nanoparticle-mediated siRNA delivery in mice [160]. In this analysis, they investigated the tumor dissemination in a Brca1-deficient genetic background after application of siRNAs targeting PARP1. The stress tolerance of PARP1 and PARP2 impaired *Arabidopsis thaliana* as well as oilseed rape (*Brassica napus*) plants have been determined using RNAi tools [161,162].

THE USE OF RNA INTERFERENCE AGAINST PARG

The lack of specific chemical inhibitors and the drastic lethal phenotype of $Parg^{-/-}$ cells as discussed above evolved into a set of studies using RNAi to suppress PARG in cells (Table 4). We reported previously that silencing of the *parg* gene in murine and human cells lead to an accumulation of PAR after oxidative and alkylating DNA insults [122,163]. The delayed degradation of PAR molecules had a protective effect in particular when cells were challenged with the oxidant hydrogen peroxide (H_2O_2). Recently, Feng *et al.* confirmed this finding in human breast cancer cells treated with alkylating agents [164]. Interestingly, the investigation of PARG silencing in mammalian cells after genotoxic challenge has resulted in a hypersensitivity phenotype, as well [111,165,166]. In these studies, cells treated with siRNA targeting PARG forced cell death due to PAR accumulation, similar to results obtained in cells with a genetic PARG impairment [76]. Andrabi et al. showed that artificially introduced PAR molecules trigger an apoptosis inducing factor (AIF) mediated cell death pathway [165]. By contrast, an increase of cytoplasmic Ca^{2+} independent from PARP activity has been implicated in the release of AIF from mitochondria subsequently leading to cell death [167]. We followed this hypothesis and found that the inhibition of PARG using RNAi diminishes the occurrence of free ADPR molecules that can trigger a rise in cytosolic Ca^{2+} via the transient receptor potential mediated channels 2 (TRPM2) [25]. As a consequence, AIF translocation—indicative of cell death—was abrogated. To date, this Ca^{2+} channel is the only one activated by monomeric ADPR, and it is dependent on PARP/PARG activity [25,168–171]. This finding resolves a long- lasting controversy about the roles of PAR and ADPR as cell death signals.

TABLE 4. RNAi against PARG in mammalian cells. Studies labeled with * provide detailed targeting sequence information.

Cell line	Cell type	Species	References
16HBE	Bronchial epithelial	Human	[172]*
293	Embryonic kidney cells	Human	[107]
A549	Lung adenocarcinoma epithelial	Human	[111]*, [107]
HeLa	Cervix carcinoma	Human	[122,163,165,166,173]*
LoVo	Colon carcinoma	Human	[174,175]*
MCF-7	Breast cancer	Human	[134,163,164]*, [176]
MEF	Fibroblasts	Mouse	[25,38,163]*
Primary	Rheumatoid arthritis synovial cells	Human	[142]
Primary	Glioblastoma	Human	[177]*
RAW 264. 7	Macrophages	Mouse	[178]*

CONCLUSIONS

The implementation of RNAi has markedly expanded our knowledge of the PAR metabolizing system. It has helped to demonstrate distinct functions attributable to PARP members and PARG in a nonmurine background. Moreover RNAi helped overcome lethality problems in mice that were discovered when PARP1 and PARP2 were ablated simultaneously or when the *parg* gene was depleted. The targeting specificity of the RNAi approach is considerably better than chemical inhibitors, which often do not distinguish between individual PARP members or fail to inhibit at all in a cellular context because of bioavailability problems (PARG).

REFERENCES

1. Ame, JC, Spenlehauer, C, de Murcia, G. The parp superfamily. *Bioessays* **2004**, *26*, 882–893.
2. Schreiber, V, Dantzer, F, Ame, JC, de Murcia, GPoly(adp-ribose): Novel functions for an old molecule. *Nat. Rev. Mol. Cell. Biol.* **2006**, *7*, 517–528.
3. Heeres, JT, Hergenrother, PJ. Poly(adp-ribose) makes a date with death. *Current Opin. Chem. Biol.* **2007**, *11*, 644–653.
4. Hottiger, MO, Hassa, PO, Luscher, B, Schuler, H, Koch-Nolte, F. Toward a unified nomenclature for mammalian adp-ribosyltransferases. *Trends Biochem. Sci.* **2010**, *35*, 208–219.
5. Gibson, BA, Kraus, WL. New insights into the molecular and cellular functions of poly(adp-ribose) and parps. *Nat. Rev. Mol. Cell Biol.* **2012**, *13*, 411–424.
6. Langelier, MF, Servent, KM, Rogers, EE, Pascal, JM. A third zinc-binding domain of human poly(adp-ribose) polymerase-1 coordinates DNA-dependent enzyme activation. *JBiol. Chem.* **2008**, *283*, 4105–4114.
7. Tao, Z, Gao, P, Hoffman, DW, Liu, HW. Domain c of human poly(adp-ribose) polymerase-1 is important for enzyme activity and contains a novel zinc-ribbon motif. *Biochemistry* **2008**, *47*, 5804–5813.
8. Hassa, PO, Haenni, SS, Elser, M, Hottiger, MO. Nuclear adp-ribosylation reactions in mammalian cells: Where are we today and where are we going? *Microbiol. Mol. Biol. Rev. MMBR* **2006**, *70*, 789–829.
9. Sallmann, FR, Vodenicharov, MD, Wang, Z Q, Poirier, GG. Characterization of sparp-1. An alteRNAtive product of parp-1 gene with poly(adp-ribose) polymerase activity independent of DNA strand breaks. *JBiol. Chem* **2000**, *275*, 15504–15511.
10. Gagne, JP, Isabelle, M, Lo, KS, Bourassa, S, Hendzel, MJ, Dawson, VL, Dawson, TM, Poirier, GG. Proteome-wide identification of poly(adp-ribose) binding proteins and poly(adp- ribose)-associated protein complexes. *Nucleic Acids Res* **2008**, *36*, 6959–6976.
11. Timinszky, G, Till, S, Hassa, PO, Hothorn, M, Kustatscher, G, Nijmeijer, B, Colombelli, J, Altmeyer, M, Stelzer, EH, Scheffzek, K, *et al.* A macrodomain-containing histone rearranges chromatin upon sensing parp1 activation. *Nat. Struct. Mol. Biol.* **2009**, *16*, 923–929.
12. Gottschalk, AJ, Timinszky, G, Kong, SE, Jin, J, Cai, Y, Swanson, SK, Washburn, MP, Florens, L, Ladurner, AG, Conaway, JW, *et al.* Poly(adp-ribosyl)ation directs

recruitment and activation of an atp-dependent chromatin remodeler. *Proc. Natl. Acad. Sci. USA* **2009**, *106*, 13770–13774.

13. Ahel, D, Horejsi, Z, Wiechens, N, Polo, SE, Garcia-Wilson, E, Ahel, I, Flynn, H, Skehel, M, West, SC, Jackson, SP, *et al.* Poly(adp-ribose)-dependent regulation of DNA repair by the chromatin remodeling enzyme alc1. *Science* **2009**, *325*, 1240–1243.
14. Realini, CA, Althaus, FR. Histone shuttling by poly(adp-ribosylation). *J Biol. Chem.* 1992, *267*, 18858–18865.
15. Malanga, M, Pleschke, JM, Kleczkowska, HE, Althaus, FR. Poly(adp-ribose) binds to specific domains of p53 and alters its DNA binding functions. *J Biol. Chem.* 1998, *273*, 11839–11843.
16. Pleschke, JM, Kleczkowska, HE, Strohm, M, Althaus, FR. Poly(adp-ribose) binds to specific domains in DNA damage checkpoint proteins. *J Biol. Chem.* **2000**, *275*, 40974–40980.
17. Ogata, N, Ueda, K, Kagamiyama, H, Hayaishi, O. Adp-ribosylation of histone h1. Identification of glutamic acid residues 2, 14, and the cooh-terminal lysine residue as modification sites. *JBiol. Chem.* 1980, *255*, 7616–7620.
18. Tulin, A, Spradling, A. Chromatin loosening by poly(adp)-ribose polymerase (parp) at drosophila puff loci. *Science* **2003**, *299*, 560–562.
19. Beneke, S. Regulation of chromatin structure by poly(adp-ribosyl)ation. *Front. Genet.* **2012**, *3*, 169.
20. Ame, JC, Rolli, V, Schreiber, V, Niedergang, C, Apiou, F, Decker, P, Muller, S, Hoger, T, Menissier-de Murcia, J, de Murcia, G. Parp-2, a novel mammalian DNA damage-dependent poly(adp-ribose) polymerase. *J Biol. Chem.* **1999**, *274*, 17860–17868.
21. Yelamos, J, Schreiber, V, Dantzer, F. Toward specific functions of poly(adp-ribose) polymerase-2. *Trends Mol. Med.* **2008**, *14*, 169–178.
22. Pion, E, Ullmann, GM, Ame, JC, Gerard, D, de Murcia, G, Bombarda, E. DNA-induced dimerization of poly(adp-ribose) polymerase-1 triggers its activation. *Biochemistry* **2005**, *44*, 14670–14681.
23. Schreiber, VParp-2, structure-function relationship. In *Poly(adp-ribosyl)ation*, Burkle, A, Ed, Landes Bioscience: Austin, TX, USA, **2004**, pp. 13–31.
24. Schreiber, V, Ame, JC, Dolle, P, Schultz, I, Rinaldi, B, Fraulob, V, Menissier-de Murcia, J, de Murcia, G. Poly(adp-ribose) polymerase-2 (parp-2) is required for efficient base excision DNA repair in association with parp-1 and xrcc1. *J Biol. Chem.* **2002**, *277*, 23028–23036.
25. Blenn, C, Wyrsch, P, Bader, J, Bollhalder, M, Althaus, FR. Poly(adp-ribose)glycohydrolase is an upstream regulator of ca2+ fluxes in oxidative cell death. *Cell. Mol. Life Sci. CMLS* **2011**, *68*, 1455–1466.
26. Wielckens, K, Bredehorst, R, Adamietz, P, Hilz, H. Mono adp-ribosylation and poly adp- ribosylation of proteins in normal and malignant tissues. *Adv. Enzyme Regul.* 1982, *20*, 23–37.
27. Wielckens, K, Schmidt, A, George, E, Bredehorst, R, Hilz, H. DNA fragmentation and nad depletion. Their relation to the turnover of endogenous mono(adp-ribosyl) and poly(adp-ribosyl) proteins. *J Biol. Chem.* 1982, *257*, 12872–12877.
28. Alvarez-Gonzalez, R, Althaus, FR. Poly(adp-ribose) catabolism in mammalian cells exposed to DNA-damaging agents. *Mutat. Res.* 1989, *218*, 67–74.

29. Brochu, G, Shah, GM, Poirier, GG. Purification of poly(adp-ribose) glycohydrolase and detection of its isoforms by a zymogram following one- or two-dimensional electrophoresis. *Anal. Biochem.* **1994**, *218*, 265–272.
30. Davidovic, L, Vodenicharov, M, Affar, EB, Poirier, GG. Importance of poly(adp-ribose) glycohydrolase in the control of poly(adp-ribose) metabolism. *Exp. Cell. Res.* **2001**, *268*, 7–13.
31. Braun, SA, Panzeter, PL, Collinge, MA, Althaus, FR. Endoglycosidic cleavage of branched polymers by poly(adp-ribose) glycohydrolase. *Eur. JBiochem.* **1994**, *220*, 369–375.
32. Alvarez-Gonzalez, R, Jacobson, MK. Characterization of polymers of adenosine diphosphate ribose generated in vitro and in vivo. *Biochemistry* **1987**, *26*, 3218–3224.
33. Ame, JC, Apiou, F, Jacobson, EL, Jacobson, MK. Assignment of the poly(adp-ribose) glycohydrolase gene (parg) to human chromosome 10q11. 23 and mouse chromosome 14b by in situ hybridization. *Cytogenet Cell. Genet.* **1999**, *85*, 269–270.
34. Hatakeyama, K, Nemoto, Y, Ueda, K, Hayaishi, O. Purification and characterization of poly(adp-ribose) glycohydrolase. Different modes of action on large and small poly(adp-ribose). *JBiol. Chem.* **1986**, *261*, 14902–14911.
35. Meyer, RG, Meyer-Ficca, ML, Whatcott, CJ, Jacobson, EL, Jacobson, MK. Two small enzyme isoforms mediate mammalian mitochondrial poly(adp-ribose) glycohydrolase (parg) activity. *Exp. Cell. Res.* **2007**, *313*, 2920–2936.
36. Oka, S, Kato, J, Moss, J. Identification and characterization of a mammalian 39-kda poly(adp- ribose) glycohydrolase. *J Biol. Chem.* **2006**, *281*, 705–713.
37. Niere, M, Kernstock, S, Koch-Nolte, F, Ziegler, M. Functional localization of two poly(adp- ribose)-degrading enzymes to the mitochondrial matrix. *Mol. Cell. Biol.* **2008**, *28*, 814–824.
38. Niere, M, Mashimo, M, Agledal, L, Dolle, C, Kasamatsu, A, Kato, J, Moss, J, Ziegler, M. Adp-ribosylhydrolase 3 (arh3), not poly(adp-ribose) glycohydrolase (parg) isoforms, is responsible for degradation of mitochondrial matrix-associated poly(adp-ribose). *J Biol. Chem.* **2012**, *287*, 16088–16102.
39. Menear, KA, Adcock, C, Boulter, R, Cockcroft, XL, Copsey, L, Cranston, A, Dillon, KJ, Drzewiecki, J, Garman, S, Gomez, S, *et al.* 4-[3-(4-cyclopropanecarbonylpiperazine-1- carbonyl)-4-fluorobenzyl]-2h-phthalazin- 1-one: A novel bioavailable inhibitor of poly(adp- ribose) polymerase-1. *J Med. Chem.* **2008**, *51*, 6581–6591.
40. Penning, TD, Zhu, GD, Gandhi, VB, Gong, J, Liu, X, Shi, Y, Klinghofer, V, Johnson, EF, Donawho, CK, Frost, DJ, *et al.* Discovery of the poly(adp-ribose) polymerase (parp) inhibitor 2-[(r)-2-methylpyrrolidin-2-yl]-1h-benzimidazole-4-carboxamide (abt-888) for the treatment of cancer. *J Med. Chem.* **2009**, *52*, 514–523.
41. Park, SR, Chen, A. Poly(adenosine diphosphate-ribose) polymerase inhibitors in cancer treatment. *Hematol. Oncol. Clin. North. Am.* **2012**, *26*, 649–670, ix.
42. Mangerich, A, Burkle, A. How to kill tumor cells with inhibitors of poly(adp-ribosyl) ation. *Int. J Cancer* **2011**, *128*, 251–265.
43. Rankin, PW, Jacobson, EL, Benjamin, RC, Moss, J, Jacobson, MK. Quantitative studies of inhibitors of adp-ribosylation in vitro and in vivo. *J Biol. Chem.* **1989**, *264*, 4312–4317.
44. Monti, D, Cossarizza, A, Salvioli, S, Franceschi, C, Rainaldi, G, Straface, E, Rivabene, R, Malorni, W. Cell death protection by 3-aminobenzamide and other

poly(adp-ribose)polymerase inhibitors: Different effects on human natural killer and lymphokine activated killer cell activities. *Biochem. Biophys. Res. Commun.* **1994**, *199*, 525–530.

45. Heller, B, Wang, Z Q, Wagner, EF, Radons, J, Burkle, A, Fehsel, K, Burkart, V, Kolb, H. Inactivation of the poly(adp-ribose) polymerase gene affects oxygen radical and nitric oxide toxicity in islet cells. *J Biol. Chem.* 1995, *270*, 11176–11180.
46. Nakajima, H, Kakui, N, Ohkuma, K, Ishikawa, M, Hasegawa, T. A newly synthesized poly(adp-ribose) polymerase inhibitor, dr2313 [2-methyl-3,5,7,8-tetrahydrothiopyrano[4,3-d]- pyrimidine-4-one]: Pharmacological profiles, neuroprotective effects, and therapeutic time window in cerebral ischemia in rats. *J Pharmacol. Exp. Ther.* **2005**, *312*, 472–481.
47. Abdelkarim, GE, Gertz, K, Harms, C, Katchanov, J, DiRNAgl, U, Szabo, C, Endres, M. Protective effects of pj34, a novel, potent inhibitor of poly(adp-ribose) polymerase (parp) in in vitro and in vivo models of stroke. *Int. J Mol. Med.* **2001**, *7*, 255–260.
48. Mabley, JG, Jagtap, P, Perretti, M, Getting, SJ, Salzman, AL, Virag, L, Szabo, E, Soriano, FG, Liaudet, L, Abdelkarim, GE, *et al.* Anti-inflammatory effects of a novel, potent inhibitor of poly (adp-ribose) polymerase. *Inflamm. Res.* **2001**, *50*, 561–569.
49. Garcia Soriano, F, Virag, L, Jagtap, P, Szabo, E, Mabley, JG, Liaudet, L, Marton, A, Hoyt, DG, Murthy, KG, Salzman, AL, *et al.* Diabetic endothelial dysfunction: The role of poly(adp-ribose) polymerase activation. *Nat. Med.* **2001**, *7*, 108–113.
50. Faro, R, Toyoda, Y, McCully, JD, Jagtap, P, Szabo, E, Virag, L, Bianchi, C, Levitsky, S, Szabo, C, Sellke, FW. Myocardial protection by pj34, a novel potent poly (adp-ribose) synthetase inhibitor. *Ann. Thorac. Surg.* **2002**, *73*, 575–581.
51. Boulton, S, Pemberton, LC, Porteous, JK, Curtin, NJ, Griffin, RJ, Golding, BT, Durkacz, BW. Potentiation of temozolomide-induced cytotoxicity: A comparative study of the Biological effects of poly(adp-ribose) polymerase inhibitors. *Br. J Cancer* **1995**, *72*, 849–856.
52. Griffin, RJ, Srinivasan, S, Bowman, K, Calvert, AH, Curtin, NJ, Newell, DR, Pemberton, LC, Golding, BT. Resistance-modifying agents. 5. Synthesis and *Biol.* ogical properties of quinazolinone inhibitors of the DNA repair enzyme poly(adp-ribose) polymerase (parp). *J Med. Chem.* **1998**, *41*, 5247–5256.
53. Delaney, CA, Wang, LZ, Kyle, S, White, AW, Calvert, AH, Curtin, NJ, Durkacz, BW, Hostomsky, Z, Newell, DR. Potentiation of temozolomide and topotecan growth inhibition and cytotoxicity by novel poly(adenosine diphosphoribose) polymerase inhibitors in a panel of human tumor cell lines. *Clin. Cancer Res.* **2000**, *6*, 2860–2867.
54. Suto, MJ, Turner, WR, Arundel-Suto, CM, Werbel, LM, Sebolt-Leopold, JS.Dihydroisoquinolinones: The design and synthesis of a new series of potent inhibitors of poly(adp-ribose) polymerase. *Anticancer Drug Des.* **1991**, *6*, 107–117.
55. Moroni, F, Meli, E, Peruginelli, F, Chiarugi, A, Cozzi, A, Picca, R, Romagnoli, P, Pellicciari, R, Pellegrini-Giampietro, DE. Poly(adp-ribose) polymerase inhibitors attenuate necrotic but not apoptotic neuronal death in experimental models of cerebral ischemiA*Cell Death Differ.* **2001**, *8*, 921–932.
56. Eliasson, MJ, Sampei, K, Mandir, AS, Hurn, PD, Traystman, RJ, Bao, J, Pieper, A, Wang, Z Q, Dawson, TM, Snyder, SH, *et al.* Poly(adp-ribose) polymerase gene disruption renders mice resistant to cerebral ischemia. *Nat. Med.* **1997**, *3*, 1089–1095.

57. Espinoza, LA, Smulson, ME, Chen, Z. Prolonged poly(adp-ribose) polymerase-1 activity regulates jp-8-induced sustained cytokine expression in alveolar macrophages. *Free Radical Biol. Med.* **2007**, *42*, 1430–1440.
58. Moroni, F, Formentini, L, Gerace, E, Camaioni, E, Pellegrini-Giampietro, DE, Chiarugi, A, Pellicciari, R. Selective parp-2 inhibitors increase apoptosis in hippocampal slices but protect cortical cells in models of post-ischaemic brain damage. *Br. J Pharmacol* **2009**, *157*, 854–862.
59. Banasik, M, Stedeford, T, Strosznajder, RP. Natural inhibitors of poly(adp-ribose) polymerase-1. *Mol. NeuroBiol.* **2012,** 46(1), 55–63.
60. Blenn, C, Wyrsch, P, Althaus, FR. The ups and downs of tannins as inhibitors of poly(adp- ribose)glycohydrolase. *Molecules* **2011**, *16*, 1854–1877.
61. Slama, JT, Aboul-Ela, N, Goli, DM, Cheesman, BV, Simmons, AM, Jacobson, MK. Specificinhibitionofpoly(adp-ribose)glycohydrolasebyadenosinediphosphate (hydroxymethyl)pyrrolidinediol. *J Med. Chem.* **1995**, *38*, 389–393.
62. Slama, JT, Aboul-Ela, N, Jacobson, MK. Mechanism of inhibition of poly(adp-ribose) glycohydrolase by adenosine diphosphate (hydroxymethyl)pyrrolidinediol. *J Med. Chem.* **1995**, *38*, 4332–4336.
63. Formentini, L, Arapistas, P, Pittelli, M, Jacomelli, M, Pitozzi, V, Menichetti, S, Ro mani, A, Giovannelli, L, Moroni, F, Chiarugi, A. Mono-galloyl glucose derivatives are potent poly(adp- ribose) glycohydrolase (parg) inhibitors and partially reduce parp-1-dependent cell death. *Br. J Pharmacol.* **2008**, *155*, 1235–1249.
64. Finch, KE, Knezevic, CE, Nottbohm, AC, Partlow, KC, Hergenrother, PJ. Selective small molecule inhibition of poly(adp-ribose) glycohydrolase (parg). *ACS Chem Biol.* **2012**, *7*, 563–570.
65. Steffen, JD, Coyle, DL, Damodaran, K, Beroza, P, Jacobson, MK. Discovery and structure-activity relationships of modified salicylanilides as cell permeable inhibitors of poly(adp-ribose) glycohydrolase (parg). *JMed. Chem.* **2011**, *54*, 5403–5413.
66. Wang, Z Q, Auer, B, Stingl, L, Berghammer, H, Haidacher, D, Schweiger, M, Wagner, EF. Mice lacking adprt and poly(adp-ribosyl)ation develop normally but are susceptible to skin disease. *Genes Dev.* **1995**, *9*, 509–520.
67. de Murcia, JM, Niedergang, C, Trucco, C, Ricoul, M, Dutrillaux, B, Mark, M, Oliver, FJ,Masson, M, Dierich, A, LeMeur, M, *et al.* Requirement of poly(adp-ribose) polymerase in recovery from DNA damage in mice and in cells. *Proc. Natl. Acad. Sci. USA* **1997**, *94*, 7303–7307.
68. Le Rhun, Y, Kirkland, JB, Shah, GM. Cellular responses to DNA damage in the absence of poly(adp-ribose) polymerase. *Biochem. Biophys. Res. Commun.* **1998**, *245*, 1–10.
69. Wang, Z Q, Stingl, L, Morrison, C, Jantsch, M, Los, M, Schulze-Osthoff, K, Wagner, EF. Parp is important for genomic stability but dispensable in apoptosis. *Genes Dev.* **1997**, *11*, 2347–2358.
70. Menissier de Murcia, J, Ricoul, M, Tartier, L, Niedergang, C, Huber, A, Dantzer, F, Schreiber, V, Ame, JC, Dierich, A, LeMeur, M, *et al.* Functional interaction between parp-1 and parp-2 in chromosome stability and embryonic development in mouse. *EMBO J* **2003**, *22*, 2255–2263.
71. Dantzer, F, Mark, M, Quenet, D, Scherthan, H, Huber, A, Liebe, B, Monaco, L, Chicheportiche, A, Sassone-Corsi, P, de Murcia, G, *et al.* Poly(adp-ribose)

polymerase-2 contributes to the fidelity of male meiosis i and spermiogenesis. *Proc. Natl. Acad. Sci. USA* **2006**, *103*, 14854–14859.
72. Bai, P, Houten, SM, Huber, A, Schreiber, V, Watanabe, M, Kiss, B, de Murcia, G, Auwerx, J, Menissier-de Murcia, J. Poly(adp-ribose) polymerase-2 [corrected] controls adipocyte differentiation and adipose tissue function through the regulation of the activity of the retinoid x receptor/peroxisome proliferator-activated receptor-gamma [corrected] heterodimer. *J Biol. Chem.* **2007**, *282*, 37738–37746.
73. Yelamos, J, Monreal, Y, Saenz, L, Aguado, E, Schreiber, V, Mota, R, Fuente, T, Minguela, A, Parrilla, P, de Murcia, G, *et al.* Parp-2 deficiency affects the survival of cd4+cd8+ double-positive thymocytes. *EMBO J* **2006**, *25*, 4350–4360.
74. Xi, H, Kersh, GJ. Sustained early growth response gene 3 expression inhibits the survival of cd4/cd8 double-positive thymocytes. *J Immunol.* **2004**, *173*, 340–348.
75. Oei, SL, Keil, C, Ziegler, M. Poly(adp-ribosylation) and genomic stability. *Biochem. Cell Biol.* **2005**, *83*, 263–269.
76. Koh, DW, Lawler, AM, Poitras, MF, Sasaki, M, Wattler, S, Nehls, MC, Stoger, T, Poirier, GG, Dawson, VL, Dawson, TM. Failure to degrade poly(adp-ribose) causes increased sensitivity to cytotoxicity and early embryonic lethality. *Proc. Natl. Acad. Sci. USA* **2004**, *101*, 17699–17704.
77. Cortes, U, Tong, WM, Coyle, DL, Meyer-Ficca, ML, Meyer, RG, Petrilli, V, Herceg, Z, Jacobson, EL, Jacobson, MK, Wang, ZQ. Depletion of the 110-kilodalton isoform of poly(adp-ribose) glycohydrolase increases sensitivity to genotoxic and endotoxic stress in mice. *Mol. Cell. Biol.* **2004**, *24*, 7163–7178.
78. Meyer-Ficca, ML, Lonchar, J, Credidio, C, Ihara, M, Li, Y, Wang, ZQ, Meyer, RG. Disruption of poly(adp-ribose) homeostasis affects spermiogenesis and sperm chromatin integrity in mice. *Biol. Reprod.* **2009**, *81*, 46–55.
79. Montgomery, MK. RNA interference: Historical overview and significance. *Meth. Mol. Biol.* **2004**, *265*, 3–21.
80. Guo, S, Kemphues, KJ. Par-1, a gene required for establishing polarity in c. Elegans embryos, encodes a putative ser/thr kinase that is asymmetrically distributed. *Cell* **1995**, *81*, 611–620.
81. Rocheleau, CE, Downs, WD, Lin, R, Wittmann, C, Bei, Y, Cha, Y H, Ali, M, Priess, JR, Mello, CC. Wnt signaling and an apc-related gene specify endoderm in early c. Elegans embryos. *Cell* **1997**, *90*, 707–716.
82. Fire, A, Xu, S, Montgomery, MK, Kostas, SA, Driver, SE, Mello, CC. Potent and specific genetic interference by double-stranded RNA in caenorhabditis elegans. *Nature* **1998**, *391*, 806–811.
83. Napoli, C, Lemieux, C, Jorgensen, R. Introduction of a chimeric chalcone synthase gene into petunia results in reversible co-suppression of homologous genes in trans. *Plant. Cell.* **1990**, *2*, 279–289.
84. Romano, N, Macino, G. Quelling: Transient inactivation of gene expression in neurospora crassa by transformation with homologous sequences. *Mol. Microbiol.* 1992, *6*, 3343–3353.
85. Hutvagner, G, Zamore, PD. RNAi: Nature abhors a double-strand. *Curr. Opin. Genet. Dev.* **2002**, *12*, 225–232.
86. Bernstein, E, Caudy, AA, Hammond, SM, Hannon, GJ. Role for a bidentate ribonuclease in the initiation step of RNA interference. *Nature* **2001**, *409*, 363–366.

87. Hammond, SM, Bernstein, E, Beach, D, Hannon, GJ. An RNA-directed nuclease mediates post- transcriptional gene silencing in drosophila cells. *Nature* **2000**, *404*, 293–296.
88. Zamore, PD, Tuschl, T, Sharp, PA, Bartel, DP. RNAi: Double-stranded RNA directs the atp- dependent cleavage of mRNA at 21 to 23 nucleotide intervals. *Cell* **2000**, *101*, 25–33.
89. Elbashir, SM, Harborth, J, Lendeckel, W, Yalcin, A, Weber, K, Tuschl, T. Duplexes of 21-nucleotide RNAs mediate RNA interference in cultured mammalian cells. *Nature* **2001**, *411*, 494–498.
90. Robb, GB, Brown, KM, Khurana, J, Rana, TM. Specific and potent RNAi in the nucleus of human cells. *Nat. Struct. Mol. Biol.* **2005**, *12*, 133–137.
91. Yu, JY, DeRuiter, SL, Turner, DL. RNA interference by expression of short-interfering RNAs and hairpin RNAs in mammalian cells. *Proc. Natl. Acad. Sci. USA* **2002**, *99*, 6047–6052.
92. Brummelkamp, TR, BeRNArds, R, Agami, R. A system for stable expression of short interfering RNAs in mammalian cells. *Science* **2002**, *296*, 550–553.
93. Rao, DD, Vorhies, JS, Senzer, N, Nemunaitis, J. SiRNA vs. ShRNA: Similarities and differences. *Adv. Drug Deliv. Rev.* **2009**, *61*, 746–759.
94. Bartel, DP. MicroRNA. s: Genomics, biogenesis, mechanism, and function. *Cell* **2004**, *116*, 281–297.
95. Liu, J, Rivas, FV, Wohlschlegel, J, Yates, JR, 3rd, Parker, R, Hannon, GJA role for the p-body component gw182 in microRNA function. *Nat. Cell. Biol.* **2005**, *7*, 1261–1266.
96. John, B, Enright, AJ, Aravin, A, Tuschl, T, Sander, C, Marks, DS. Human microRNA targets. *PLoS Biol.* **2004**, *2*, e363.
97. Chang, TC, Yu, D, Lee, Y S, Wentzel, EA, Arking, DE, West, KM, Dang, CV, Thomas-Tikhonenko, A, Mendell, JT, Widespread microRNA repression by myc contributes to tumorigenesis. *Nat. Genet.* **2008**, *40*, 43–50.
98. Wang, Z, Rao, DD, Senzer, N, Nemunaitis, J. RNA interference and cancer therapy. *Pharm. Res.* **2011**, *28*, 2983–2995.
99. Seyhan, AA. RNAi: A potential new class of therapeutic for human genetic disease. *Hum. Genet.* **2011**, *130*, 583–605.
100. Bora, RS, Gupta, D, Mukkur, TK, Saini, KS. RNA interference therapeutics for cancer: Challenges and opportunities (review). *Mol. Med. Report.* **2012**, *6*, 9–15.
101. Gan, L, Anton, KE, Masterson, BA, Vincent, VA, Ye, S, Gonzalez-Zulueta, M. Specific interference with gene expression and gene function mediated by long dsRNA in neural cells. *J Neurosci. Meth.* **2002**, *121*, 151–157.
102. Kameoka, M, Nukuzuma, S, Itaya, A, Tanaka, Y, Ota, K, Ikuta, K, Yoshihara, K. Rna interference directed against poly(adp-ribose) polymerase 1 efficiently suppresses human immunodeficiency virus type 1 replication in human cells. *J Virol.* **2004**, *78*, 8931–8934.
103. Shah, RG, Ghodgaonkar, MM, Affar el, B, Shah, GM. DNA vector-based RNAi approach for stable depletion of poly(adp-ribose) polymerase-1. *Biochem. Biophys. Res. Commun.* **2005**, *331*, 167–174.
104. Dykxhoorn, DM, Novina, CD, Sharp, PA. Killing the messenger: Short RNAs that silence gene expression. *Nature reviews. Mol. Cell Biol.* **2003**, *4*, 457–467.

105. Bai, P, Canto, C, Oudart, H, Brunyanszki, A, Cen, Y, Thomas, C, Yamamoto, H, Huber, A, Kiss, B, Houtkooper, RH, *et al.* Parp-1 inhibition increases mitochondrial metabolism through sirt1 activation. *Cell. Metab.* **2011**, *13*, 461–468.
106. Tempera, I, Deng, Z, Atanasiu, C, Chen, CJ, D'Erme, M, Lieberman, PM. Regulation of epstein-barr virus orip replication by poly(adp-ribose) polymerase 1. *JVirol.* **2010**, *84*, 4988–4997.
107. Erdelyi, K, Bai, P, Kovacs, I, Szabo, E, Mocsar, G, Kakuk, A, Szabo, C, Gergely, P, Virag, L. Dual role of poly(adp-ribose) glycohydrolase in the regulation of cell death in oxidatively stressed a549 cells. *Faseb J* **2009**.
108. Erener, S, Hesse, M, Kostadinova, R, Hottiger, MO. Poly(adp-ribose)polymerase-1 (parp1) controls adipogenic gene expression and adipocyte function. *Mol. Endocrinol.* **2012**, *26*, 79–86.
109. Ambrose, HE, Papadopoulou, V, Beswick, RW, Wagner, SD. Poly-(adp-ribose) polymerase-1 (parp-1) binds in a sequence-specific manner at the bcl-6 locus and contributes to the regulation of bcl-6 transcription. *Oncogene* **2007**, *26*, 6244–6252.
110. Huang, X, Dong, Y, Bey, EA, Kilgore, JA, Bair, JS, Li, LS, Patel, M, Parkinson, EI, Wang, Y, Williams, NS, *et al.* An nqo1 substrate with potent antitumor activity that selectively kills by parp1-induced programmed necrosis. *Cancer Res.* **2012**, *72*, 3038–3047.
111. Fisher, AE, Hochegger, H, Takeda, S, Caldecott, KW. Poly(adp-ribose) polymerase 1 accelerates single-strand break repair in concert with poly(adp-ribose) glycohydrolase. *Mol. Cell. Biol.* **2007**, *27*, 5597–5605.
112. Strom, CE, Johansson, F, Uhlen, M, Szigyarto, CA, Erixon, K, Helleday, T. Poly (adp-ribose) polymerase (parp) is not involved in base excision repair but parp inhibition traps a single-strand intermediate. *Nucleic Acids Res.* **2011**, *39*, 3166–3175.
113. Hegan, DC, Lu, Y, Stachelek, GC, Crosby, ME, Bindra, RS, Glazer, PM. Inhibition of poly(adp-ribose) polymerase down-regulates brca1 and rad51 in a pathway mediated by e2f4 and p130. *Proc. Natl. Acad. Sci. USA* **2010**, *107*, 2201–2206.
114. Modis, K, Gero, D, Erdelyi, K, Szoleczky, P, DeWitt, D, Szabo, C. Cellular bioenergetics is regulated by parp1 under resting conditions and during oxidative stress. *Biochem. Pharmacol.* **2012**, *83*, 633–643.
115. Mathieu, J, Flexor, M, Lanotte, M, Besancon, FA. parp-1/jnk1 cascade participates in the synergistic apoptotic effect of tnfalpha and all-trans retinoic acid in apl cells. *Oncogene* **2008**, *27*, 3361–3370.
116. Kandan-Kulangara, F, Shah, RG, Affar el, B, Shah, GM. Persistence of different forms of transient RNAi during apoptosis in mammalian cells. *PloS one* **2010**, *5*, e12263.
117. Cimadamore, F, Curchoe, CL, Alderson, N, Scott, F, Salvesen, G, Terskikh, AV. Nicotinamide rescues human embryonic stem cell-derived neuroectoderm from parthanatic cell death. *Stem Cells* **2009**, *27*, 1772–1781.
118. Ding, W, Liu, W, Cooper, KL, Qin, XJ, de Souza Bergo, PL, Hudson, LG, Liu, KJ. Inhibition of poly(adp-ribose) polymerase-1 by arsenite interferes with repair of oxidative DNA damage. *J Biol. Chem.* **2009**, *284*, 6809–6817.
119. Qin, XJ, Hudson, LG, Liu, W, Timmins, GS, Liu, KJ. Low concentration of arsenite exacerbates uvr-induced DNA strand breaks by inhibiting parp-1 activity. *Toxicol. Appl. Pharmacol* **2008**, *232*, 41–50.

120. Zhang, N, Chen, Y, Jiang, R, Li, E, Chen, X, Xi, Z, Guo, Y, Liu, X, Zhou, Y, Che, Y, *et al.* Parp and rip 1 are required for autophagy induced by 11'-deoxyverticillin a, which precedes caspase-dependent apoptosis. *Autophagy* **2011**, *7*, 598–612.
121. Vernole, P, Muzi, A, Volpi, A, Terrinoni, A, Dorio, AS, Tentori, L, Shah, GM, Graziani, G. Common fragile sites in colon cancer cell lines: Role of mismatch repair, rad51 and poly(adp- ribose) polymerase-1. *Mutation Res.* **2011**, *712*, 40–48.
122. Cohausz, O, Blenn, C, Malanga, M, Althaus, FR. The roles of poly(adp-ribose)-metabolizing enzymes in alkylation-induced cell death. *Cell. Mol. Life Sci.* **2008**, *65*, 644–655.
123. Ouararhni, K, Hadj-Slimane, R, Ait-Si-Ali, S, Robin, P, Mietton, F, Harel-Bellan, A, Dimitrov, S, Hamiche, A. The histone variant mh2a1. 1 interferes with transcription by down-regulating parp-1 enzymatic activity. *Genes Dev.* **2006**, *20*, 3324–3336.
124. Ghosh, U, Das, N, Bhattacharyya, NP. Inhibition of telomerase activity by reduction of poly(adp-ribosyl)ation of tert and tep1/tp1 expression in hela cells with knocked down poly(adp- ribose) polymerase-1 (parp-1) gene. *Mutation Res.* **2007**, *615*, 66–74.
125. Muthumani, K, Choo, AY, Zong, WX, Madesh, M, Hwang, DS, Premkumar, A, Thieu, KP, Emmanuel, J, Kumar, S, Thompson, CB, *et al.* The hiv-1 vpr and glucocorticoid receptor complex is a gain-of-function interaction that prevents the nuclear localization of parp-1. *Nat. Cell. Biol.* **2006**, *8*, 170–179.
126. Chang, P, Coughlin, M, Mitchison, TJ. Tankyrase-1 polymerization of poly(adp-ribose) is required for spindle structure and function. *Nat. Cell. Biol.* **2005**, *7*, 1133–1139.
127. Cohausz, O, Althaus, FR. Role of parp-1 and parp-2 in the expression of apoptosis-regulating genes in hela cells. *Cell. Biol. Toxicol.* **2009**, *25*, 379–391.
128. Beneke, S, Cohausz, O, Malanga, M, Boukamp, P, Althaus, F, Burkle, A. Rapid regulation of telomere length is mediated by poly(adp-ribose) polymerase-1. *Nucleic Acids Res.* **2008**, *36*, 6309–6317.
129. Szanto, A, Hellebrand, EE, Bognar, Z, Tucsek, Z, Szabo, A, Gallyas, FJr, Sumegi, B, Varbiro, G. Parp-1 inhibition-induced activation of pi-3-kinase-akt pathway promotes resistance to taxol. *Biochem. Pharmacol.* **2009**, *77*, 1348–1357.
130. Byun, JY, Kim, MJ, Eum, DY, Yoon, CH, Seo, WD, Park, KH, Hyun, JW, Lee, Y S, Lee, JS, Yoon, MY, *et al.* Reactive oxygen species-dependent activation of bax and poly(adp- ribose) polymerase-1 is required for mitochondrial cell death induced by triterpenoid pristimerin in human cervical cancer cells. *Mol. Pharmacol* **2009**, *76*, 734–744.
131. Pang, J, Gong, H, Xi, C, Fan, W, Dai, Y, Zhang, TM. Poly(adp-ribose) polymerase 1 is involved in glucose toxicity through sirt1 modulation in hepg2 hepatocytes. *J Cell. Biochem* **2011**, *112*, 299–306.
132. Mathews, MT, Berk, BC. Parp-1 inhibition prevents oxidative and nitrosative stress-induced endothelial cell death via transactivation of the vegf receptor 2. *Arterioscler Thromb Vasc Biol.* **2008**, *28*, 711–717.
133. Park, MT, Kim, MJ, Kang, Y H, Choi, SY, Lee, JH, Choi, JA, Kang, CM, Cho, CK, Kang, S, Bae, S, *et al.* Phytosphingosine in combination with ionizing radiation enhances apoptotic cell death in radiation-resistant cancer cells through ros-dependent and -independent aif release. *Blood* **2005**, *105*, 1724–1733.

134. Frizzell, KM, Gamble, MJ, Berrocal, JG, Zhang, T, Krishnakumar, R, Cen, Y, Sauve, AA, Kraus, WL. Global analysis of transcriptional regulation by poly(adp-ribose) polymerase-1 and poly(adp-ribose) glycohydrolase in mcf-7 human breast cancer cells. *JBiol. Chem.* **2009**, *284*, 33926–33938.
135. Wyrsch, P, Blenn, C, Bader, J, Althaus, FR. Cell death and autophagy under oxidative stress: Roles of poly(adp-ribose)polymerases and ca2+. *Mol. Cell. Biol.* **2012**.
136. Ariumi, Y, Turelli, P, Masutani, M, Trono, D. DNA damage sensors atm, atr, DNA-pkcs, and parp-1 are dispensable for human immunodeficiency virus type 1 integration. *J Virol.* **2005**, *79*, 2973–2978.
137. Lin, Y, Tang, X, Zhu, Y, Shu, T, Han, X. Identification of parp-1 as one of the transcription factors binding to the repressor element in the promoter region of cox-2. *Arch. Biochem. Biophys.* **2011**, *505*, 123–129.
138. Kondo, K, Obitsu, S, Ohta, S, Matsunami, K, Otsuka, H, Teshima, R. Poly(adp-ribose) polymerase (parp)-1-independent apoptosis-inducing factor (aif) release and cell death are induced by eleostearic acid and blocked by alpha-tocopherol and mek inhibition. *J Biol. Chem.* **2010**, *285*, 13079–13091.
139. Woodhouse, BC, Dianova, II, Parsons, JL, Dianov, GL. Poly(adp-ribose) polymerase-1 modulates DNA repair capacity and prevents formation of DNA double strand breaks. *DNA Repair* **2008**, *7*, 932–940.
140. Diaz-HeRNAndez, JI, Moncada, S, Bolanos, JP, Almeida, A. Poly(adp-ribose) polymerase-1 protects neurons against apoptosis induced by oxidative stress. *Cell. Death Differ.* **2007**, *14*, 1211–1221.
141. Kitamura, T, Sekimata, M, Kikuchi, S, Homma, Y. Involvement of poly(adp-ribose) polymerase 1 in erbb2 expression in rheumatoid synovial cells. *Am. JPhysiol. Cell. Physiol.* **2005**, *289*, C82–88.
142. Garcia, S, Bodano, A, Pablos, JL, Gomez-Reino, JJ, Conde, C. Poly(adp-ribose) polymerase inhibition reduces tumor necrosis factor-induced inflammatory response in rheumatoid synovial fibroblasts. *Ann. Rheum. Dis.* **2008**, *67*, 631–637.
143. Huang, D, Wang, Y, Wang, L, Zhang, F, Deng, S, Wang, R, Zhang, Y, Huang, K. Poly(adp- ribose) polymerase 1 is indispensable for transforming growth factor-beta induced smad3 activation in vascular smooth muscle cell. *PloS One* **2011**, *6*, e27123.
144. Lapucci, A, Pittelli, M, Rapizzi, E, Felici, R, Moroni, F, Chiarugi, A. Poly(adp-ribose) polymerase-1 is a nuclear epigenetic regulator of mitochondrial DNA repair and transcription. *Mol. Pharmacol.* **2011**, *79*, 932–940.
145. Bryant, HE, Schultz, N, Thomas, HD, Parker, KM, Flower, D, Lopez, E, Kyle, S, Meuth, M, Curtin, NJ, Helleday, T. Specific killing of brca2-deficient tumours with inhibitors of poly(adp-ribose) polymerase. *Nature* **2005**, *434*, 913–917.
146. Racz, B, Hanto, K, Tapodi, A, Solti, I, Kalman, N, Jakus, P, Kovacs, K, Debreceni, B, Gallyas, FJr, Sumegi, B. Regulation of mkp-1 expression and mapk activation by parp-1 in oxidative stress: A new mechanism for the cytoplasmic effect of parp-1 activation. *Free Radical Biol. Med.* **2010**, *49*, 1978–1988.
147. Boehler, C, Gauthier, LR, Mortusewicz, O, Biard, DS, Saliou, JM, Bresson, A, Sanglier- Cianferani, S, Smith, S, Schreiber, V, Boussin, F, *et al.* Poly(adp-ribose) polymerase 3 (parp3), a newcomer in cellular response to DNA damage and mitotic progression. *Proc. Natl. Acad. Sci. USA* **2011**, *108*, 2783–2788.

148. Loseva, O, Jemth, AS, Bryant, HE, Schuler, H, Lehtio, L, Karlberg, T, Helleday, T. Parp-3 is a mono-adp-ribosylase that activates parp-1 in the absence of DNA. *J Biol. Chem.* **2010**, *285*, 8054–8060.
149. Chang, W, Dynek, JN, Smith, S. Numa is a major acceptor of poly(adp-ribosyl) ation by tankyrase 1 in mitosis. *Biochem. J* **2005**, *391*, 177–184.
150. Diani-Moore, S, Ram, P, Li, X, Mondal, P, Youn, DY, Sauve, AA, Rifkind, AB. Identification of the aryl hydrocarbon receptor target gene tiparp as a mediator of suppression of hepatic gluconeogenesis by 2,3,7,8-tetrachlorodibenzo-p-dioxin and of nicotinamide as a corrective agent for this effect. *J Biol. Chem.* **2010**, *285*, 38801–38810.
151. Di Paola, S, Micaroni, M, Di Tullio, G, Buccione, R, Di Girolamo, M. Parp16/artd15 is a novel endoplasmic-reticulum-associated mono-adp-ribosyltransferase that interacts with, and modifies karyopherin-ss1. *PloS one* **2012**, *7*, e37352.
152. Hung, CF, Cheng, TL, Wu, RH, Teng, CF, Chang, WT. A novel bidirectional expression system for simultaneous expression of both the protein-coding genes and short hairpin RNAs in mammalian cells. *Biochem. Biophys. Res. Commun.* **2006**, *339*, 1035–1042.
153. Bai, P, Canto, C, Brunyanszki, A, Huber, A, Szanto, M, Cen, Y, Yamamoto, H, Houten, SM, Kiss, B, Oudart, H, *et al.* Parp-2 regulates sirt1 expression and whole-body energy expenditure. *Cell Metab.* **2011**, *13*, 450–460.
154. Szanto, M, Rutkai, I, Hegedus, C, Czikora, A, Rozsahegyi, M, Kiss, B, Virag, L, Gergely, P, Toth, A, Bai, P. Poly(adp-ribose) polymerase-2 depletion reduces doxorubicin-induced damage through sirt1 induction. *Cardiovasc. Res.* **2011**, *92*, 430–438.
155. Gravel, C, Stergiou, L, Gagnon, SN, Desnoyers, S. The c. Elegans gene pme-5: Molecular cloning and role in the DNA-damage response of a tankyrase orthologue. *DNA Repair* **2004**, *3*, 171–182.
156. St-Laurent, JF, Gagnon, SN, Dequen, F, Hardy, I, Desnoyers, S. Altered DNA damage response in caenorhabditis elegans with impaired poly(adp-ribose) glycohydrolases genes expression. *DNA Repair* **2007**, *6*, 329–343.
157. Boamah, EK, Kotova, E, Garabedian, M, Jarnik, M, Tulin, AV. Poly(adp-ribose) polymerase 1 (parp-1) regulates ribosomal biogenesis in drosophila nucleoli. *PLoS Genet.* **2012**, *8*, e1002442.
158. Tulin, A, Naumova, NM, Menon, AK, Spradling, AC. Drosophila poly(adp-ribose) glycohydrolase mediates chromatin structure and sir2-dependent silencing. *Genetics* **2006**, *172*, 363–371.
159. Popoff, I, Jijon, H, Monia, B, Tavernini, M, Ma, M, McKay, R, Madsen, K. Antisense oligonucleotides to poly(adp-ribose) polymerase-2 ameliorate colitis in interleukin-10-deficient mice. *JPharmacol. Exp. Ther.* **2002**, *303*, 1145–1154.
160. Goldberg, MS, Xing, D, Ren, Y, Orsulic, S, Bhatia, SN, Sharp, PA. Nanoparticle-mediated delivery of siRNA targeting parp1 extends survival of mice bearing tumors derived from brca1- deficient ovarian cancer cells. *Proc. Natl. Acad. Sci. USA* **2011**, *108*, 745–750.
161. De Block, M, Verduyn, C, De Brouwer, D, Cornelissen, M. Poly(adp-ribose) polymerase in plants affects energy homeostasis, cell death and stress tolerance. *Plant. J Cell Mol. Biol.* **2005**, *41*, 95–106.

162. Vanderauwera, S, De Block, M, Van de Steene, N, van de Cotte, B, Metzlaff, M, Van Breusegem, F. Silencing of poly(adp-ribose) polymerase in plants alters abiotic stress signal transduction. *Proc. Natl. Acad. Sci. USA* **2007**, *104*, 15150–15155.
163. Blenn, C, Althaus, FR, Malanga, M. Poly(adp-ribose) glycohydrolase silencing protects against h2o2-induced cell death. *Biochem. J* **2006**, *396*, 419–429.
164. Feng, X, Zhou, Y, Proctor, AM, Hopkins, MM, Liu, M, Koh, DW. Silencing of apoptosis- inducing factor and poly(adp-ribose) glycohydrolase reveals novel roles in breast cancer cell death after chemotherapy. *Mol. Cancer* **2012**, *11*, 48.
165. Andrabi, SA, Kim, NS, Yu, SW, Wang, H, Koh, DW, Sasaki, M, Klaus, JA, Otsuka, T, Zhang, Z, Koehler, RC, *et al.* Poly(adp-ribose) (par) polymer is a death signal. *Proc. Natl Acad Sci. USA* **2006**, *103*, 18308–18313.
166. Ame, JC, Fouquerel, E, Gauthier, LR, Biard, D, Boussin, FD, Dantzer, F, de Murcia, G, Schreiber, V. Radiation-induced mitotic catastrophe in parg-deficient cells. *J Cell Sci.* **2009**, *122*, 1990–**2002**.
167. Norberg, E, Gogvadze, V, Ott, M, Horn, M, Uhlen, P, Orrenius, S, Zhivotovsky, B. An increase in intracellular ca2+ is required for the activation of mitochondrial calpain to release aif during cell death. *Cell. Death Differ.* **2008**, *15*, 1857–1864.
168. Naziroglu, MT. rpm2 cation channels, oxidative stress and neurological diseases: Where are we now? *Neurochemical. Res.* **2011**, *36*, 355–366.
169. Perraud, AL, Fleig, A, Dunn, CA, Bagley, LA, Launay, P, Schmitz, C, Stokes, AJ, Zhu, Q, Bessman, MJ, Penner, R, *et al.* Adp-ribose gating of the calcium-permeable ltrpc2 channel revealed by nudix motif homology. *Nature* **2001**, *411*, 595–599.
170. Toth, B, Csanady, LI. dentification of direct and indirect effectors of the transient receptor potential melastatin 2 (trpm2) cation channel. *JBiol. Chem.* **2010**, *285*, 30091–30102.
171. Buelow, B, Song, Y, Scharenberg, AM. The poly(adp-ribose) polymerase parp-1 is required for oxidative stress-induced trpm2 activation in lymphocytes. *J Biol. Chem.* **2008**, *283*, 24571–24583.
172. Huang, HY, Cai, JF, Liu, QC, Hu, GH, Xia, B, Mao, JY, Wu, DS, Liu, JJ, Zhuang, ZX. Role of poly(adp-ribose) glycohydrolase in the regulation of cell fate in response to benzo(a)pyrene. *Exp. Cell Res.* **2012**, *318*, 682–690.
173. Biard, DS. Untangling the relationships between DNA repair pathways by silencing more than 20 DNA repair genes in human stable clones. *Nucleic Acids Res.* **2007**, *35*, 3535–3550.
174. Li, Q, Li, M, Wang, Y L, Fauzee, NJ, Yang, Y, Pan, J, Yang, L, Lazar, A. RNA interference of parg could inhibit the metastatic potency of colon carcinoma cells via pi3-kinase/akt pathway. *Cell. Physiol. Biochem.* **2012**, *29*, 361–372.
175. Pan, J, Fauzee, NJ, Wang, Y L, Sheng, YT, Tang, Y, Wang, JQ, Wu, WQ, Yan, JX, Xu, J. Effect of silencing parg in human colon carcinoma lovo cells on the ability of huvec migration and proliferation. *Cancer Gene Ther.* **2012**, 19(10), 715–22.
176. Fathers, C, Drayton, RM, Solovieva, S, Bryant, HE. Inhibition of poly(adp-ribose) glycohydrolase (parg) specifically kills brca2-deficient tumor cells. *Cell. Cycle* **2012**, *11*, 990–997.
177. Tang, JB, Svilar, D, Trivedi, RN, Wang, XH, Goellner, EM, Moore, B, Hamilton, RL, Banze, LA, Brown, AR, Sobol, RW. N-methylpurine DNA glycosylase and DNA

polymerase beta modulate ber inhibitor potentiation of glioma cells to temozolomide. *Neuro Oncol.* **2011**, *13*, 471–486.
178. Rapizzi, E, Fossati, S, Moroni, F, Chiarugi, A. Inhibition of poly(adp-ribose) glycohydrolase by gallotannin selectively up-regulates expression of proinflammatory genes. *Mol. Pharmacol.* **2004**, *66*, 890–898.

CHAPTER 9

EPIGENETIC DEREGULATION OF MICRORNAs IN RHABDOMYOSARCOMA AND NEUROBLASTOMA AND TRANSLATIONAL PERSPECTIVES

PAOLO ROMANIA, ALICE BERTAINA, GIORGIA BRACAGLIA, FRANCO LOCATELLI, DORIANA FRUCI, and ROSSELLA ROTA

INTRODUCTION

Epigenetic chromatin remodeling plays a pivotal role in normal mammalian development and post-natal tissue homeostasis. Indeed, lineage specification and cellular differentiation, which underlie embryo development and morphogenesis from a single pluripotent stem cell, are epigenetically regulated processes. The final result is the "plasticity" of an individual genotype that, through the activation of molecular cascades, timely and sequentially controlled, produces different phenotypes in response to different microenvironments. In the last 10 years, special attention has been paid to the non-protein coding portion of the genome such as non-coding small RNAs, among which are microRNAs (miRNAs), considered to be major regulators of developmental pathways [1–8]. Of note, chromatin remodeling and miRNA pathways have been shown to be interconnected and able to regulate each other. To date, it is recognized that the deregulation of the epigenetic- and miRNA-dependent control of gene expression underlies tumorigenesis.

Rhabdomyosarcoma (RMS) and neuroblastoma (NB) are pediatric cancers derived from cells sharing molecular features of skeletal muscle and neuronal progenitors, respectively, blocked at different stages of differentiation. It has been clearly demonstrated that deregulation of developmental pathways plays a major role in both tumors.

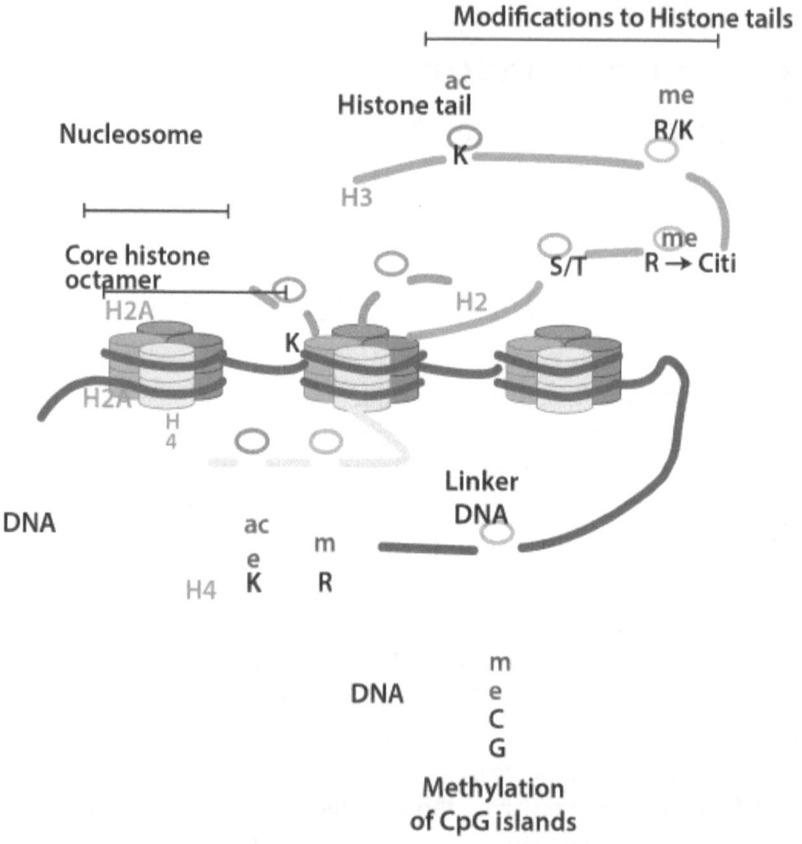

FIGURE 1. Schematic representation of chromatin structure. Eukaryotic DNA is wrapped around core histone proteins (histone octamer: H2A, H2B, H3 and H4) to form compact chromatin structures termed nucleosomes. Covalent modifications to histones (on histone tails) involve amino acidic residues (Lysine (K), Arginine (R) Serine (S) and Tyrosine (T)) that can be acetylated (ac), methylated (me), phosphorylated (ph) and/or ubiquitinilated (ub). CpG islands on DNA can be methylated. These post-translational modifications result in the epigenetic regulation of gene expression.

To fit within the nucleus of eukaryotic cells, short lengths of DNA are wrapped around an octamer of histone proteins (two copies of core histones H2A, H2B, H3 and H4) forming the basic units of chromatin termed "nucleosomes" (Figure 1).

Higher-order organizations are formed by further compaction of chromatin structure. The degree of packaging influences the chromatin accessibility to transcriptional-regulatory complexes switching a gene "on" or "off." Therefore, the "epigenome" encompasses genetic information that is the result of multiple processes involving chromatin modifications such as DNA methylation, histone proteins modification, histone variants replacement, nucleosome repositioning and mechanisms involving non-coding RNAs function. All these epigenetic modifications can be rapidly resumed and, therefore, fit well for the purpose of fine tuning the timely controlled developmental processes.

DNA methylation is catalyzed by DNA methyltransferases (DNMTs) and results in *de novo* methylation of unmethylated DNA and/or methylation maintenance of hemimethylated sequences. In normal tissues, DNA methylation is typically present and stable in the intergenic regions. In cancer cells, DNA methylation is exclusively found at the level of cytosines within CpG-rich regions of gene promoters leading to gene silencing. On the contrary, DNA hypomethylation of these CpG islands, also often aberrant in cancer, can increase gene expression [10]. The final result is either the silencing of tumor suppressor genes or the transcriptional activation of proto-oncogenes. In addition, DNA hypomethylation can lead to DNA helix breakpoints and, ultimately, to loss of heterozygosity (LOH) or aberrant chromosomal rearrangements. In strict conjunction with these mechanisms, the histone code generated by covalent modifications on histone tails, regulates chromatin remodeling for the accessibility of the transcription machinery to genes up to DNA repair, replication and segregation [11]. The major modifications of histone tails are controlled by histone acetyltransferases (HAT) and histone demethylases (HDM) in a competitive manner with histone deacetylases (HDAC) and histone methyltransferases (HMT), respectively. Among HMTs, the Polycomb group (PcG) protein EZH2, acomponent of the Polycomb repressor complex 2 (PRC2), is responsible for the trimethylation of Lysine 27 on histone H3 (H3K27me3) on gene promoters (Figure 2).

This modification allows the recruitment of the PRC1 complex, which inhibits gene transcription through histone H2A ubiquitination. HDACs participate to the PRC complexes reinforcing the inhibition of gene expression by deacetylating H3K27, thus favoring H3K27 trimethylation.

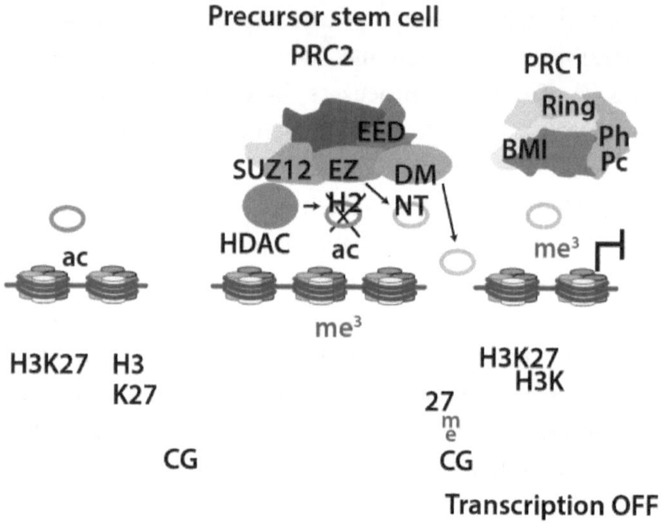

FIGURE 2. Schematic representation of transcriptional gene repression by PRC2. In an uncommitted stem cell, the core component EZH2 methylates histone H3 on K27, thus generating the epigenetic mark H3K27me3. The HDAC activity favors this EZH2 effect deacetylating H3K27. In this way, the PRC1 complex is recruited and binds to DNA, thus stabilizing the repressive state of the chromatin. The final result is the transcriptional repression of developmental genes.

Finally, PcG complexes recruit DNMTs to specific gene loci to induce transcriptional silencing through DNA methylation (Figure 2). The final result is the maintenance of cell stemness and the support of self-renewal and pluripotency. Consistently, components of the PRC complexes are often aberrantly over-expressed in tumors (reviewed in [12]). Of note, epigenetic modifications that are maintained through mitosis and inherited during cell development and differentiation can be reversed by treatment with appropriate drugs. Therefore, compounds acting on "epimutations" can be used in association with conventional chemotherapy to induce growth arrest, differentiation and tumor cell death.

THE SMALL NON-CODING RNAS, "MicroRNAS"

MiRNAs are a class of non-coding ~19–22 nucleotides (nt) single-strand RNAs transcribed in a developmental and tissue-specific manner during embryogenesis [13]. More than 1500 mature miRNAs have been identified in humans (http://www. mirbase. org), which are highly conserved across species. miRNAs are involved in post-translational gene silencing by binding complementary sequences in the 3'-untraslated regions (*UTRs*) of a target mRNA through their "seed" sequence leading to translational repression or mRNA degradation [14,15].

The expression of more than 60% of human protein-coding genes is controlled by miRNAs (reviewed in [13]). Several miRNAs can target the same mRNA and, conversely, each miRNA can target several mRNAs leading to additional layers of post-transcriptional control of gene expression. To further expand the scenario, specific miRNAs have been recently described to upregulate gene transcription also in quiescent cells [16] and others to target sequences within the 5'UTR [17], exonic regions [18] or gene promoter regions [19].

Mature miRNAs are the result of sequential processing of longer RNA precursors transcribed by RNA polymerase II and/or III and called pri-miRNA (Figure 3) [20,21]. This long transcript is processed by Drosha, a ribonuclease-III protein, in collaboration with DGCR8, a protein responsible for anchoring the pri-miRNA into the complex. The resulting 70-nt-long molecule termed pre-miRNA, is exported to the cytoplasm by the nuclear export protein Exportin-5 [13,22]. In the cytoplasm, the pre-miRNA is released and processed by Dicer to produce the functional ~19–22-nt double-strand RNAs. The less stable single strand is loaded by the Dicer-TRPB complex into the RNA-induced silencing complex (RISC) which carries it to complementary 3'UTR mRNA sequences.

MiRNAs regulate fundamental physiologic processes such as embryonic development, lineage/tissue identity specification and homeostasis [23–27]. Therefore, it is not surprising that miRNAs display an aberrant expression profile in a wide range of human diseases [13]. In tumor cells, as reported for protein-coding gene, miRNA promoters can be aberrantly modified by the deregulated epigenetic machinery leading to changes in their expression profile [28–31]. Increasing evidence suggesting tumor

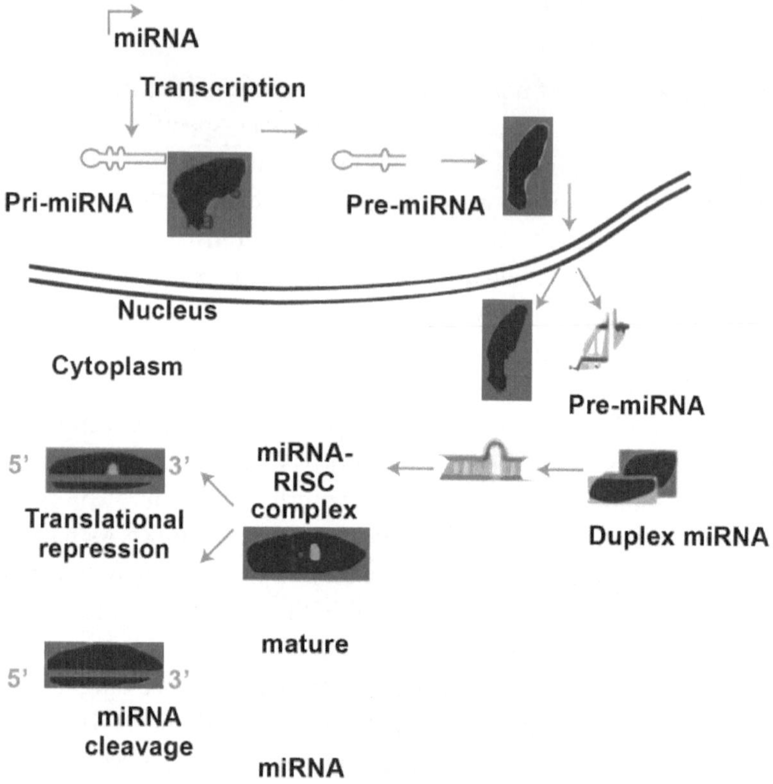

FIGURE 3. The biogenesis of miRNAs. MiRNAs are transcribed as longer RNA precursor (pri-miRNA) by RNA polymerase II and/or III and processed by the Drosha complex in pre-miRNAs. These fragments are exported from the nucleus by exportin 5 (XPO5). In the cytoplasm pre-miRNA are further processed by Dicer and TAR RNA-binding protein 2 (TARBD2) to generate mature miRNAs, which are loaded into the RNA-induced silencing complex (RISC). The translation of mRNAs targets into proteins is repressed by mRNA degradation or, alternatively, block of mRNA translation.

suppressor or oncogenic role for miRNAs has been obtained in adult cancers [1,13,32–34]. Consistently, the re-expression of tumor suppressor miRNAs in tumor cells has been suggested as a potential anti-cancer therapeutic approach [35–40].

Recently, miRNAs have been shown, by us and other groups, to play a role also in the pathogenesis of embryonal pediatric cancers such as

RMS and NB [41–43]. So far, the majority of tumor suppressor miRNAs involved in RMS and NB regulates the differentiation of skeletal muscle and neuronal compartments, respectively. These miRNAs are often silenced by aberrant DNA or histone methylation or located in unstable chromosomal regions often involved in LOH, ie miR-34a [13,44]. Consistently, their re-expression after treatment with HDAC inhibitors, de-methylating compounds or differentiating agents is sufficient to allow tumor cell differentiation and to impair tumorigenesis. Therefore, an epigenetic approach aimed at re-expressing these miRNAs could have a therapeutic value.

RHABDOMYOSARCOMA

RMS is a skeletal muscle-derived tumor and the most common soft-tissue sarcoma of childhood [45]. RMS cells express transcription factors specific to skeletal muscle progenitors, such as Myogenic Differentiation (MyoD) and myogenin [46,47], but their intrinsic myogenic program is disrupted at different stages of differentiation. In line with these observations, restoring terminal differentiation in RMS cells leads to inhibition of cell proliferation *in vitro* and tumor growth *in vivo*.

Pediatric RMS includes two major histological subtypes, namely alveolar (~25% of cases) and embryonal (~75% of cases) RMS [45,48]. The five-year overall survival rate of non-metastatic embryonal RMS patients is around 70%, while that of alveolar RMS and metastatic patients is still in the order of 25%, supporting the urgent need of novel therapeutic approaches.

The majority of alveolar RMS are characterized by chromosomal translocations, such as t (2; 13) or t (1; 13), leading to the expression of PAX3/FOXO1 or PAX7/FOXO1 fusion proteins, respectively [49–51]. These fusion products, especially PAX3/FOXO1, are a hallmark of high-risk tumors and correlate with poor prognosis [52]. Approximately 20% of alveolar RMS do not present known chromosomal translocations and are molecularly and clinically indistinguishable from embryonal ones [53–55]. However, novel chromosomal translocations involving PAX3 have been recently discovered in a subset of alveolar RMS previously diagnosed as fusion negative [51,56].

During skeletal muscle tissue development, PAX3 expression is detected in pluripotent progenitors and is followed by that of PAX7, which characterizes myogenic committed cells [57,58]. These transcription factors, on one hand, induce the expression of MyoD that marks myoblasts, and on the other hand, stimulate self-renewal and proliferation of myogenic cells. Thus, their expression is fundamental to assure the balance between cell proliferation and differentiation during embryogenesis leading to the right number of committed progenitors in order to form multinucleated myofibers. During differentiation, pro-myogenic miRNAs downregulate PAX3 and PAX7 to achieve complete myogenesis (reviewed in [59]).

An altered expression of miRNAs acting as pro-myogenic regulators has been reported in RMS, [43,59–61]. These miRNAs behave as tumor suppressors when re-expressed, halting tumorigenic processes working in negative feedback circuitries with epigenetic regulators. These findings demonstrate that aberrant epigenetic control of miRNA expression concurs to RMS pathogenesis.

EPIGENETICALLY DEREGULATED miRNAS IN RHABDOMYOSARCOMA

Muscle-specific miRNAs that regulate myogenesis are termed "myo-miRs." To this group belong three miRNA clusters, miR-1-1/miR-133a-2, miR-1-2/miR133a-1 and miR-206/miR-133b, encoded by three bicistronic miRNA genes on separate chromosomes (reviewed in [62]).

Although no direct epigenetic regulation of the expression of miR-1 and miR-206 clusters has been highlighted in RMS, a recent study provided several evidences linking these miRNAs to the epigenetic machinery in normal myoblasts [63]. In this work, YY1 was shown to regulate the expression of miR-1 and miR-206 clusters in murine myoblasts *in vitro* and *in vivo*. In particular, the authors found YY1 binding sites in previously identified muscle-specific enhancers [64] located (i) upstream of miR-1-2 (E1) and (ii) in an intron between miR-1-2 and miR-133a-1

(E2), and (iii) between miR-1-1 and miR-133a-2 gene loci (E3). Moreover, they discovered a previously unknown enhancer upstream of miR-206 and miR-133b coding regions (E4) showing YY1 binding site. They demonstrated that YY1 repressed the activity of all these four enhancers impairing the expression of miRNAs [63]. Interestingly, Ezh2 was present and active on the enhancers E1 and E3 but not on E2 and E4 suggesting that YY1 can function in myoblasts in a HMT-independent manner. Furthermore, the same authors uncovered a feedback loop between miR-1 and YY1 demonstrating that miR-1 directly targets the 3'UTR of YY1 mRNA. Having demonstrated that miR-1 is able to target PAX7, as already reported for miR-206 [65], which is consistently upregulated in YY1 over-expressing myoblasts, the authors depicted an anti-myogenic network in which YY1 plays a central role in repressing miR-1/miR-206 (Figure 4). This network is flexible and bi-univocal due to the feedback control of miR-1 on YY1 expression [63].

Additional links with epigenetic networks in myoblasts have been shown for both miR-1 and miR-206 clusters. The ectopic expression of these miRNAs inhibits HDAC4, which sustains cell proliferation by preventing the expression of the cyclin-dependent kinase inhibitor $p21^{Cip1}$, essential for terminal muscle differentiation [66,67]. This finding is of interest for the RMS context due to the potentiality of therapeutic approaches with HDACs inhibitors [68,69]. Interestingly, miR-29b directly targets HDAC4 during osteoblast differentiation [70], suggesting an intricate complex of miRNAs pathways acting similarly in different tissue contexts. Collectively, these observations suggest that miRNAs can influence gene transcription through the control of several types of epigenetic repressors. Of note, crucial components of molecular pathways related to RMS aggressiveness, such as IGF and RAS, can be post-transcriptionally downregulated by pro-myogenic miRNAs. Indeed, (i) miR-1 and miR-133 target IGF receptor 1, thus avoiding aberrant muscle hypertrophy [71,72]; (ii) miR-214, which is repressed by EZH2 in proliferating myoblasts, is able to favor myogenesis by downregulating the anti-myogenic N-RAS oncogene [73,74].

In addition to myomiRs, there are non-muscle-specific miRNAs that participate to skeletal muscle differentiation such as miR-181a/miR-181b, miR-27a, miR-27b, miR-26a and miR-29b2/miR-29c. Wang and colleagues provided the first evidence of an epigenetic deregulation of

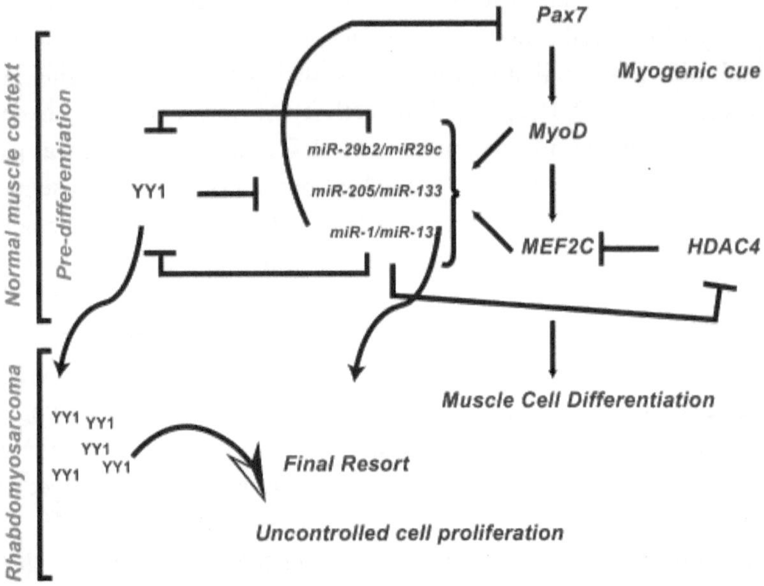

FIGURE 4. Circuits involving the Polycomb Group (PcG) protein YY1 and miRNAs in normal and Rhabdomyosarcoma contexts. During myoblasts expansion, YY1 represses miR-29, miR-1 and miR-206 cluster expression. Under a myogenic stimulus, Pax7 induces the muscle regulatory factor MyoD that allows MEF2C activation, leading to miR-29, miR-1 and miR-206 de-repression. MiR-1 and miR-29 clusters post-transcriptionally block YY1 expression. Moreover, downregulation of Pax7 caused by miR-1 and miR-206 and that of HDAC4 caused by miR-1 ultimately leads to differentiation of myogenic precursors. In the RMS context, YY1 expression is highly deregulated resulting in pro-myogenic miRNAs repression that causes uncontrolled cell proliferation and muscle differentiation impairment.

miRNAs in RMS discovering a regulatory circuitry between miR-29b2/miR-29c and the PcG protein YY1 [60]. The same authors had previously demonstrated that during the expansion of myogenic precursors, NF-kB inhibited terminal differentiation maintaining a high level of YY1, which, in turn, repressed the expression of myofibrillary genes favoring cell precursor proliferation [75]. Therefore, reasoning that miRNAs are regulators of myogenesis, the authors investigated whether the anti-differentiation effect of YY1 was at least in part related to the repression of pro-myogenic miRNAs [60]. Indeed, they identified miR-29b2/miR-29c cluster on chromosome 1 being repressed by YY1 in proliferating myoblasts (Figure 4).

This repression involved the recruitment of both HDAC1 and the EZH2 PcG protein, which deacetylated and trimethylated, respectively, the Lys 27 on histone H3 in a highly conserved region 20 kb upstream of the miR-29b2/miR-29c gene locus. Under myogenic cues, miR-29b2/miR-29c gene cluster is expressed and directly targeted YY1 reinforcing myogenesis. Of note, in myogenic precursors undergoing early-stage differentiation, YY1 and EZH2 have been shown to work in concert to repress the transcription of late-stage muscle-specific genes such as Muscle Creatine Kinase (MCK) and Myosin Heavy Chain (MHC) to avoid premature differentiation [76]. Therefore, a fine tuning between anti- and pro-myogenic molecular stimuli underlies a proper myogenesis during which both epigenetic and miRNAs pathways appeared highly interconnected.

The miR-29b2/miR-29c-YY1 epigenetic negative feedback circuitry [60] was shown by the authors to be disrupted in RMS cells, in which YY1 and EZH2 over-expression resulted in persistent activation of stemness maintenance program. The tumor suppressor role of miR-29b2/miR-29c was clearly evidenced by gain-of-function experiments demonstrating that forced over-expression of this miRNA cluster in RMS cells is sufficient to impair the tumorigenic properties both *in vitro* and *in vivo* by repressing YY1 expression. Consistent with these observations, tumor tissues from RMS patients showed upregulation of YY1 and EZH2 [60]. In line with this report, preliminary results from our laboratory showed that EZH2 downregulation in RMS cells impairs tumorigenesis, thereby allowing the de-repression of several tumor suppressor miRNAs, including the miR-29b2/miR-29c cluster.

Interestingly, even if YY1 binding sites were not found in the promoter of miR-29a/miR-29b1, this miRNA cluster appeared downregulated in primary sarcoma samples as compared to skeletal muscle tissues [77]. Interestingly, the miR-29 family has been shown to directly target DNMT3A and DNMT3B in several types of cancer, thus suggesting a link between their reduced expression and pathological gene hyper-methylation [78,79].

Altogether, these results highlight the importance of bi-univocal regulation of epigenetic molecules and miRNAs in skeletal muscle differentiation that should be considered also in the context of RMS in which these pathways are deregulated.

NEUROBLASTOMA

NB is a neuroectodermal tumor that originates from precursor cells of the sympathetic nervous system and represents the third leading cause of cancer-related death in childhood [80]. The heterogeneous clinical behavior, ranging from spontaneous regression to rapid progression, is attributable to biological and genetic characteristics of the tumor.

NB diagnosed in patients under one year of age usually has a favorable prognosis since tumor cells undergo spontaneous differentiation or regression, whereas NB occurring in patients over one year of age tend to grow aggressively resulting in a fatal outcome. The prognosis of stage I-III NB, with a tumor confined to the originating organ or surrounding tissue, is quite favorable, whereas that of stage IV NB, where the tumor is metastatic, is dismal. Stage IV-S NB is a metastatic disease seen exclusively in infants, which is associated with high survival rate due to the spontaneous maturation and regression of tumor cells. Nevertheless, since disease and risk staging are not comprehensive and fully precise, they should be considered as surrogate markers of the underlying tumor biology.

NB cells correspond to adrenal neuroblasts arrested at different stages of sympathoadrenal development, thus representing multipotent progenitor cells with specific tumorigenic potential [81,82]. The degree of differentiation of NB is another important factor for establishing prognosis. Based on this, NB tumors are classified in different risk categories spanning from more mature and benign ganglioneuromas (GN), to intermediate and potentially malignant ganglioneuroblastomas (GNB), to undifferentiated NB, always malignant with worst prognosis. The factors responsible for malignant transformation from sympathetic neuroblasts into neuroblastoma cells are not well understood. In several NB cell lines, all-*trans* retinoic acid (ATRA)-induced differentiation is associated with increased expression of neurotrophic receptors, including Trk family receptors and glial cell-derived neurotrophic factor receptor, Ret. Of note, ATRA is a compound related to 13-cis-retinoic acid that is used as differentiating agent in children with high-risk NB.

From a clinical point of view, the presence of any segmental chromosomal imbalances correlates with a more aggressive phenotype, whereas tumors containing mostly whole chromosome gains or losses are associated with a benign phenotype and high propensity for spontaneous regression or differentiation. The most widely characterized cytogenetic alterations in NB tumors include amplification of the *MYCN* oncogenic transcription factor at chromosome band 2p24, LOH or rearrangements of the distal portion of the short arm of chromosome 1 (1p31-term), chromosome 3 (3p22) and chromosome 11 (11q23), or gains of chromosome arm 1q or 17q. Besides these chromosomal/molecular abnormalities, gains of chromosomes 4q, 6p, 7q, 11q and 18q, amplification of *MDM2* and *MYC* genes and *LOH* at 14q, 10q and 19q13 have also been described [83]. In particular, loss of chromosome 1p region, occurring mainly through an unbalanced translocation that results in the gain of 17q [84], arises preferentially in tumors with *MYCN* amplification [85]. Some of these genetic alterations were proven to be independently associated with a poor clinical outcome. The region of 1p, lost in NB tumors, is quite large and contains multiple genes that have been shown to contribute to NB pathogenesis [86,87]. Similarly, the minimal common region of gain at distal 17q23 has been shown to contain at least 15 genes that remain to be investigated [88]. Of note, *MYCN* is one of the major players in normal neural crest differentiation, suggesting that its amplification/deregulation is the result of a developmental defect that probably has occurred during embryonic development [89]. However, although *MYCN* amplification is considered the most important prognostic marker of highly aggressive tumors; it is present in less than 30% of NB tumors [90].

More recently, activating mutations in the anaplastic lymphoma kinase (*ALK*) gene have been associated with the majority of hereditary NB and a 10% of sporadic NB cases [91–94]. ALK is a receptor tyrosine kinase involved in several other human cancers, including anaplastic large-cell lymphoma and non-small cell lung cancer [95–97]. ALK signaling drives cell transformation *in vitro* and *in vivo* through several pathways such as cell-cycle, survival and cell migration [98,99]. Nevertheless, additional molecular alterations concur to NB pathogenesis and their identification is essential to improve the prognostic classification of NB patients.

In the attempt to improve the risk stratification of NB subtypes, several groups have characterized and correlated methylation profiling of biologically and clinically different subgroups of NB patients with clinical risk factors and survival (for a detailed review see [100]). The first DNA methylation studies in NB have revealed that silencing of caspase 8 and RAS-association domain family 1 isoform A (RASSF1A) are important in the development and progression of disease [101,102]. Both genes are often methylated in primary NB cells and their methylation status negatively correlates with survival.

Recent advances in genome-wide methylation screening methodologies, such as re-expression analysis after treatment with 5-aza-2'-deoxycytidine (DAC), DNA methylation promoter assay after affinity-based capture, methylation microarray after bisulfate treatment and next-generation sequencing techniques, have allowed the identification of 75 different DNA methylation biomarkers involved in fundamental biological processes operative in NB ([103–105] and reviewed in [100]).

Epigenetic inactivation of miRNAs with tumor-suppressor activities is also recognized as a major hallmark of NB tumors. A recent study from Chavali group [106] showed that the majority of the miRNA genes are flanked by scaffold/matrix-attachment region (S/MAR) elements that regulate their tissue and cell type-specific expression by binding to epigenetic regulators MAR binding proteins. In this study, the authors showed that the miR-17-92 cluster, a well-known example of oncomiR that is over-expressed in high-risk NB [41], has two conserved MARs elements, one upstream and one downstream the cluster, being the former strongly bound by HMG I/Y protein, a chromatin modulator that promotes an open conformation facilitating mRNA transcription. Consistent with this finding, the miR-17-92 cluster expression was down-regulated in cells interfered for HMG I/Y family expression [106].

EPIGENETICALLY DEREGULATED miRNAS IN NEUROBLASTOMA

Recent studies from Stallings' group revealed extensive epigenomic changes in NB cells upon ATRA exposure, a phenomenon that is associated with a complex series of molecular events, including modifications of both chromatin compaction and DNA methylation status [107].

Intriguingly, in a previous paper, Das and colleagues showed that several miRNAs that are upmodulated as a consequence of ATRA treatment target DNA methyltransferases involved in the process of ATRA-induced differentiation of NB cell lines [108]. More recently, the same authors revealed that DNA methylation changes occurred during ATRA treatment include several regulatory regions of tumor suppressor miRNAs [109]. The authors identified 67 miRNAs sensitive to the effects of DNA methylation in 18 primary NB tumors and four NB cell lines by the combined use of methylated DNA immunoprecipitation, miRNA and mRNA target analysis. Of note, a relative high proportion of these epigenetically silenced miRNAs (42%) were significantly associated with poor patient survival when under-expressed in tumors, whereas other 10 miRNA (15%) were found hyper-methylated in favorable tumors and over-expressed in tumors from patients with poor survival. Remarkably, not only miRNAs, but also their predicted gene targets, were significantly associated with poor patient survival when over-expressed in tumors. In particular, miRNA-340 was functionally inactivated by DNA methylation, re-expressed in NB cell lines following DAC treatment and significantly associated with poor survival when under-expressed in primary NB tumors. According to the NB cell line used, over-expression of miR-340 induced either apoptosis or differentiation by targeting SOX2, an SRY box containing a transcription factor family member that acts in regulating the maintenance and differentiation of neural stem cells [109].

Several of the epigenetically regulated miRNAs identified in this integrated analysis were previously found to negatively impact NB growth both *in vivo* and *in vitro* (Figure 5). The main findings regarding their implications in NB pathogenesis are described as follows.

MiR-10a and -10b were found to induce NB differentiation through multiple gene targeting (for a detailed review see [110]). Meseguer *et al.* demonstrated that these microRNAs directly target the SR-family splicing factor (SFRS1), a regulator of alternative splicing that plays a role also in enhancing translation initiation of mRNAs containing specific SFRS1 binding sequences [111]. Of note, this factor seems to be a proto-oncogene deregulated in many forms of tumor [112]. Foley and colleagues identified the nuclear receptor co-repressor 2 (NCOR2) as the primary target of miR-10a and -10b responsible for causing differentiation

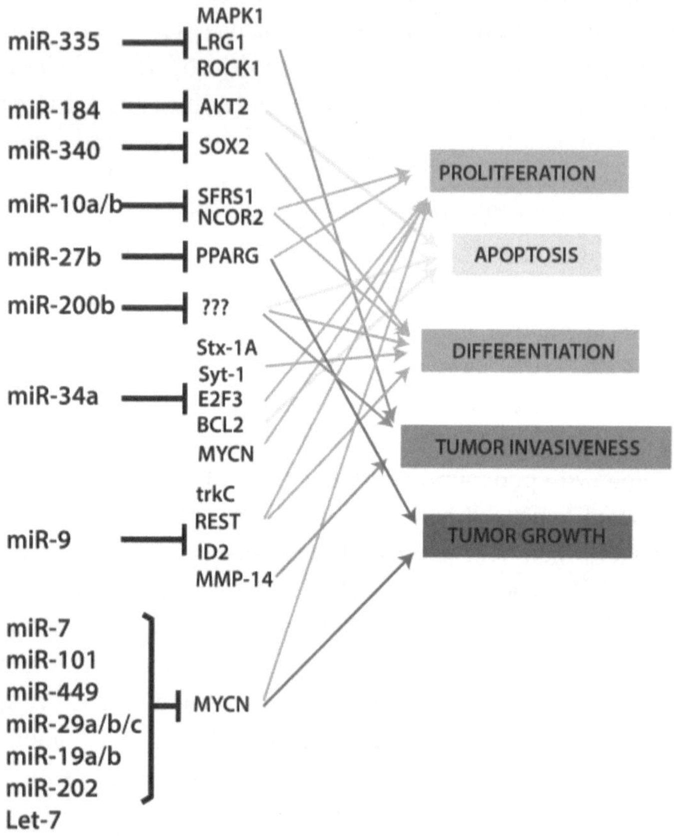

FIGURE 5. Scheme of cellular pathways affected by methylation-sensitive miRNAs in Neuroblastoma. Experimental validated gene targets and pathways affected by altered expression of methylation-sensitive miRNAs are indicated.

[113]. NCOR2 is a transcriptional co-repressor that recruits a complex of proteins that include SIN3A/B and histone deacetylases HDAC1, HDAC2 and HDAC3 to DNA promoter regions. Interestingly, neural stem cells from mice lacking NCOR2 exhibit extensive neurite outgrowth and reduction in cell proliferation [114]. More intriguingly, siRNA-mediated inhibition of NCOR2 resulted in a significant increase in miR-10a level that suggests a possible regulatory feedback loop in the differentiation process [113].

A number of these epigenetically regulated miRNAs were previously found to be involved in a complex gene network involving MYCN. A study from Buechner and colleagues demonstrated that the tumor suppressor miRNAs let-7 and miR-101 inhibit proliferation and clonogenic growth of Kelly NB cell line by targeting MYCN [115]. Furthermore, they experimentally validate miR-449, miR-19a/b, miR-29a/b/c, and miR-202 as direct MYCN-targeting miRNAs. A very recent paper by Molenaar and colleagues, elegantly demonstrated the existence of a LIN28B-let-7-MYCN axis in NB [116]. Using a mouse model with neural crest-specific over-expression of LIN28B, these authors showed that LIN28B drives NB upregulating MYCN protein through let-7 repression. Notably, MYCN has been recently discovered to bind and transcriptionally downregulate another epigenetically controlled miRNA, miR-335, which in turn regulates genes in the TGF-β non-canonical pathway, such as the Rho-associated coiled-coil containing protein (ROCK1), MAPK1 and putative member LRG1, leading to inhibition of invasiveness and migratory potential of NB cells [117]. The same group had previously discovered a pro-apoptotic effect of miR-184 both *in vitro* and *in vivo*, through the direct targeting of AKT2, a serine/threonine kinase active downstream of the PI3K pathway. Interestingly, MYCN could exert its tumorigenic effect also by suppressing either directly or indirectly the same miRNA [118,119].

Several groups identified miR-542-5p as a putative tumor suppressor, whose expression has a prognostic value and might be important in NB tumor biology [120–122]. Of note, miR-542-5p is predicted to target TBR1, a gene essential for proper development of neurons. Functional studies that address this or other putative targets should be performed in the future.

The brain specific miR-9 was demonstrated by Laneve *et al.* to play an important role in controlling NB cell proliferation through a regulatory circuit involving TRKc and other two miRs, miR-125a and -125b [123]. The same authors discovered a regulatory feedback mechanism involving miR-9 and the neuro-restrictive silencer factor REST, suggesting a fine interplay for the maintenance of the neuronal differentiation program [124]. More recently, Nasi and colleagues found that the transcription factor ID2, which is involved in the self-renewal and proliferation of neural precursor cells, is a target of miR-9 [125]. At the same time, another group demonstrated that miR-9 reduced matrix metalloproteinase (MMP)-14 level by inhibiting the invasion, metastasis, and angiogenesis of SH- SY5Y and SK-N-SH NB cell lines both *in vitro* and *in vivo* [126].

Lee and colleagues showed that the epigenetically regulated miR-27b acts as a tumor suppressor in NB by inhibiting the tumor-promoting function of peroxisome proliferators-activated receptor (PPAR)y and blocking cell growth *in vitro* and tumor growth in mouse xenografts [127]. In addition, miR-27b indirectly regulates also NF-kB activity and transcription of inflammatory target genes, which triggers an increased inflammatory response.

As described above, LOH of 1p36 is commonly found in NB with *MYCN* amplification and is associated with poor outcome. Of note, five microRNAs (miR-200a, miR-200b, miR-429, miR-34a and miR-551a) map within the first 10Mb on chromosome 1 short arm (1p36. 22 to 1pter), suggesting that the loss of these miRNAs could contribute to the acquisition of an aggressive NB phenotype.

MiR-200b was originally identified through a positional gene candidate approach looking for miRNAs that map within chromosomal regions frequently altered in NB [128]. Conversely, Ragusa and colleagues discovered that the miR-200b region is unaffected by the frequent deletions involving 1p36 in the NB cell lines analyzed [129]. Moreover, they found a CpG island near the gene-encoding miR-200b and showed that, upon treatment with demethylating agents, miR-200b levels were upregulated in six NB cell lines. Finally, they provided several evidences supporting a role for miR-200b in NB cell invasiveness, differentiation and apoptosis.

MiR-34a expression is silenced in several types of cancer due to aberrant CpG methylation of its promoter [130]. In NB cell lines and primary NB tumors, miR-34a is absent or expressed at low levels, while it is upregulated in a dose-dependent manner following ATRA treatment [131]. Given that miR-34a has been implicated in NB growth inhibition, and mutations of the retained allele are not a common mechanism towards achieving biallelic inactivation, it is plausible that other mechanisms such as epigenetic regulation could contribute for loss of miR-34a expression in NB. Of note, miR-34a expression has been shown to increase after HDAC inhibition in a bladder cancer cell line [132]. It will be of interest to validate this finding also in the NB context since several studies demonstrated the involvement of this microRNA in NB pathogenesis. Welch and colleagues demonstrated a tumor suppressive role for miR-34a through the transcription factor E2F3 targeting [131]. Moreover, Wei *et al.* provided evidence that MYCN is a direct target of miR-34a, as further confirmed by a functional screen study by Cole *et al.*, which also demonstrated a BCL2 direct targeting [128,133]. Recently, synaptotagmin I (Syt-I) and syntaxin 1A (Stx-1A), two proteins involved in synaptogenesis and neuronal differentiation, were found to be targets of miR-34a [134].

All these findings reveal a network of epigenetic pathways and miRNA interactions whose dysregulation contributes to NB, and that has translational consequences for the management of the disease.

TRANSLATIONAL PERSPECTIVES AND CONCLUSIONS

The majority of miRNAs aberrantly downregulated in RMS and NB behave, when re-expressed, as tumor suppressors re-establishing a proper differentiation program. Therefore, they represent promising therapeutic targets. Moreover, miRNAs detection in cancer patients could represent a powerful prognostic and diagnostic tool. A miRNA signature has been recently reported for RMS primary samples by two groups [77,135]. Intriguingly, the first group suggested that an embryonal sample had been previously misdiagnosed allowing a re-evaluation of the tumor followed by a diagnosis as alveolar [77]. The second group showed that miRNA expression clusterings in RMS samples are correlated with PAX3/FOXO1, PAX7/FOXO1 and embryonal subgroups, and, thus, are potentially useful for risk stratification [135].

Recently, miRNA signatures have been shown to be related also to NB prognosis. Lin and colleagues found that, using a combination of 15 biomarkers that consist of 12 miRNAs' signature, expression levels of Dicer and Drosha and age at diagnosis, it was possible to segregate 66 NB patients into four distinct patterns that were highly predictive of clinical outcome [136]. Subsequently, a signature of 25 miRNAs was built and validated on an independent set of 304 frozen tumor samples and 75 archived formalin-fixed paraffin-embedded samples. This signature was significantly able to discriminate the patients with respect to progression-free and overall survival [137].

Furthermore, the possibility to detect and quantify miRNAs in body fluids whose profile mirrors physiological and pathological condition could also represent a reliable method to improve prognosis and diagnosis [52,138,139].

MiRNAs re-expression in tumors rises doubts regarding delivery, efficacy and safety of these molecules *in vivo*. Viral and non-viral vectors have shown pre-clinical feasibility and efficacy in animal models including primates [36,140–145]. However, their application on human patients could induce side effects linked to their immunogenicity or could be ineffective due to lack of specificity (reviewed in [146]). The use of nanoparticles represents an efficient system to deliver cDNA to cells and, therefore, it could be potentially useful for the delivery of miRNAs *in vivo*. Indeed, a first promising study using nanoparticles encapsulating miR-34a and conjugated to a GD2 antibody demonstrate the feasibility of a targeted delivery miRNA-based therapy for NB [147].

Based on these observations, therapeutic approaches able to modulate miRNA gene expression targeting epigenetic regulators present several advantages. Conversely to DNA mutations, epigenetic modifications are reversible and responsive to specific compounds, which gave promising results in clinical trials for adult tumors. Indeed, low doses of DNA methyltransferases inhibitors azacitidine (AZA) and DAC have shown therapeutic effects in myelodysplasia/leukemia as well as in lung cancer patients [148–153]. It is noteworthy that older trials with high doses of epigenetic inhibitors have highlighted extreme toxicities and no effect on epigenome [154,155]. Very recently, an elegant work of Tsai *et al.*, investigated the mechanisms involved in the efficacy of low nanomolar doses of DAC and AZA in preclinical models of acute myeloid leukemia, breast, colon and lung cancers [156]. The authors showed that high-dose treatments resulted in high cytotoxic effects on tumor cells. Nevertheless, residual cells engrafted in immunodeficient mice were still able to grow and give tumors. Conversely, low concentrations of AZA or DAC reduced self-renewal of tumor cells both *in vitro* and *in vivo,* sustaining global DNA demethylation and re-expression of anti-tumor master gene pathways [156]. These results strongly confirm the potential efficacy of an epigenetic therapy that could re-express tumor suppressor genes and miRNAs.

Moreover, epigenetic targeted therapy with inhibitors of both DNA methyltransferases and HDACs, has shown anti-tumor effectiveness combined with a restored expression of tumor suppressor genes in preclinical models [157–159]. More important, well-tolerated combination regimens for low-dose AZA and HDAC inhibitors have been found effective in patients with myelodysplasia/ leukemia [160,161]. Finally, very encouraging results have been reported in a phase I/II clinical trial on the use of a combined epigenetic low-dose therapy in recurrent metastatic non-small cell lung (NSCL) cancer refractory to conventional therapy [153]. In brief, AZA and entinostat, an inhibitor that blocks HDAC1-3 and HDAC9 function, were used on 45 NSCL patients, resulting in limited side effects while having efficacy on the aberrant epigenome of tumor cells. These findings suggest the potential use of these compounds also in solid cancers.

Novel agents that inhibit histone methyltransferases among which EZH2 are of particular interest for future applications since they do not require cell division to target cancer cells [162,163]. These compounds have been shown to synergize with other epigenetic agents in preclinical studies, representing promising drugs in order to de-repress EZH2 targeted miRNAs [164,165].

However, as reported for other targeted therapies, escape strategies for the miRNA re-expression approach could select resistant cells expressing mRNAs with mutated miRNA binding sites in their 3'UTRs. Moreover, PAX3/ or PAX7/FOXO1 mRNAs expressed in fusion-positive alveolar RMS, being devoid of 3'UTR regions of PAX genes, are resistant to miRNAs' post-transcriptional regulation. In addition, the de-repression of oncogenes could be a potential side effect of an epigenetic therapy [166]. Therefore, although the field of investigation in the epigenetic treatment of cancer appears very promising, further studies are needed to enhance the knowledge on the epigenome and its regulation both in normal and tumor contexts.

REFERENCES

1. Iorio, M. V., Croce, C. M. MicroRNA involvement in human cancer. *Carcinogenesis* **2012**, *33*, 1126–1133.
2. Crippa, S., Cassano, M., Sampaolesi, M. Role of miRNAs in muscle stem cell biology: proliferation, differentiation and death. *Curr. Pharmaceut. Des.* **2012**, *18*, 1718–1729.
3. Mondol, V., Pasquinelli, A. E. Let's make it happen: the role of let-7 microRNA in development. *Curr. Top. Dev. Biol.* **2012**, *99*, 1–30.
4. Mansfield, J. H., McGlinn, E. Evolution, expression, and developmental function of Hox-embedded miRNAs. *Curr. Top. Dev. Biol.* **2012**, *99*, 31–57.
5. Sokol, N. S. The role of microRNAs in muscle development. *Curr. Top. Dev. Biol.* **2012**, *99*, 59–78.
6. Cochella, L., Hobert, O. Diverse functions of microRNAs in nervous system development. *Curr. Top. Dev. Biol.* **2012**, *99*, 115–143.
7. O'Connell, R. M., Baltimore, D. MicroRNAs and hematopoietic cell development. *Curr. Top. Dev. Biol.* **2012**, *99*, 145–174.
8. Boettger, T., Braun, T. A new level of complexity: The role of microRNAs in cardiovascular development. *Circ. Res.* **2012**, *110*, 1000–1013.
9. Bell, J. T., Spector, T. D. A twin approach to unraveling epigenetics. *Trends Genet.* **2011**, *27*, 116–125.
10. Baylin, S. B. DNA methylation and gene silencing in cancer. *Nat. Clin. Pract. Oncol.* **2005**, *2*, S4–S11.
11. Martin-Subero, J. I. How epigenomics brings phenotype into being. *Pediatr. Endocrinol. Rev.* **2011**, *9*, 506–510.
12. Richly, H., Aloia, L., Di Croce, L. Roles of the Polycomb group proteins in stem cells and cancer. *Cell Death. Dis.* **2011**, *2*, e204.
13. Esteller, M. Non-coding RNAs in human disease. *Nat. Rev.* **2011**, *12*, 861–874.
14. Elbashir, S. M., Harborth, J., Lendeckel, W., Yalcin, A., Weber, K., Tuschl, T. Duplexes of 21-nucleotide RNAs mediate RNA interference in cultured mammalian cells. *Nature* **2001**, *411*, 494–498.
15. Lewis, B. P., Burge, C. B., Bartel, D. P. Conserved seed pairing, often flanked by adenosines, indicates that thousands of human genes are microRNA targets. *Cell* **2005**, *120*, 15–20.
16. Vasudevan, S., Tong, Y., Steitz, J. A. Switching from repression to activation: microRNAs can up-regulate translation. *Science* **2007**, *318*, 1931–1934.
17. Orom, U. A., Nielsen, F. C., Lund, A. H. MicroRNA-10a binds the 5'UTR of ribosomal protein mRNAs and enhances their translation. *Mol. Cell* **2008**, *30*, 460–471.
18. Duursma, A. M., Kedde, M., Schrier, M., le Sage, C., Agami, R. miR-148 targets human DNMT3b protein coding region. *RNA* **2008**, *14*, 872–877.

19. Place, R. F., Li, L. C., Pookot, D., Noonan, E. J., Dahiya, R. MicroRNA-373 induces expression of genes with complementary promoter sequences. *Proc. Natl. Acad. Sci. USA* **2008**, *105*, 1608–1613.
20. Lee, Y., Kim, M., Han, J., Yeom, K. H., Lee, S., Baek, S. H., Kim, V. N. MicroRNA genes are transcribed by RNA polymerase II. *EMBO J.* **2004**, *23*, 4051–4060.
21. Borchert, G. M., Lanier, W., Davidson, B. L. RNA polymerase III transcribes human microRNAs. *Nat. Struct. Mol. Biol.* **2006**, *13*, 1097–1101.
22. Winter, J., Jung, S., Keller, S., Gregory, R. I., Diederichs, S. Many roads to maturity: MicroRNA biogenesis pathways and their regulation. *Nat. Cell Biol.* **2009**, *11*, 228–334.
23. Butcher, J., Abdou, H., Morin, K., Liu, Y. Micromanaging oligodendrocyte differentiation by noncoding RNA: toward a better understanding of the lineage commitment process. *J. Neurosci.* **2009**, *29*, 5365–5366.
24. Gangaraju, V. K., Lin, H. MicroRNAs: Key regulators of stem cells. *Nat. Rev. Mol. Cell Biol.* **2009**, *10*, 116–125.
25. Wang, Y., Russell, I., Chen, C. MicroRNA and stem cell regulation. *Curr. Opin. Mol. Ther.* **2009**, *11*, 292–298.
26. Mallanna, S. K., Rizzino, A. Emerging roles of microRNAs in the control of embryonic stem cells and the generation of induced pluripotent stem cells. *Dev. Biol.* **2010**, *344*, 16–25.
27. Laurent, L. C. MicroRNAs in embryonic stem cells and early embryonic development. *J. Cell Mol. Med.* **2008**, *12*, 2181–2188.
28. Calin, G. A., Croce, C. M. MicroRNA signatures in human cancers. *Nat. Rev. Cancer* **2006**, *6*, 857–866.
29. Garzon, R., Fabbri, M., Cimmino, A., Calin, G. A., Croce, C. M. MicroRNA expression and function in cancer. *Trends Mol. Med.* **2006**, *12*, 580–587.
30. Iorio, M. V., Piovan, C., Croce, C. M. Interplay between microRNAs and the epigenetic machinery: An intricate network. *Biochim. Biophys. Acta* **2010**, *1799*, 694–701.
31. Fabbri, M., Calore, F., Paone, A., Galli, R., Calin, G. A. Epigenetic Regulation of miRNAs in Cancer. *Adv. Exp. Med. Bio.* **2013**, *754*, 137–148.
32. Chhabra, R., Dubey, R., Saini, N. Cooperative and individualistic functions of the microRNAs in the miR-23a~27a~24–2 cluster and its implication in human diseases. *Mol. Cancer* **2010**, *9*, 232.
33. Di Leva, G., Croce, C. M. Roles of small RNAs in tumor formation. *Trends Mol. Med.* **2010**, *16*, 257–267.
34. Perera, R. J., Ray, A. Epigenetic regulation of miRNA genes and their role in human melanomas. *Epigenomics* **2012**, *4*, 81–90.
35. Ferretti, E., De Smaele, E., Miele, E., Laneve, P., Po, A., Pelloni, M., Paganelli, A., Di Marcotullio, L., Caffarelli, E., Screpanti, I., *et al.* Concerted microRNA control of Hedgehog signalling in cerebellar neuronal progenitor and tumour cells. *EMBO J.* **2008**, *27*, 2616–2627.
36. Kota, J., Chivukula, R. R., O'Donnell, K. A., Wentzel, E. A., Montgomery, C. L., Hwang, H. W., Chang, T. C., Vivekanandan, P., Torbenson, M., Clark, K. R., *et al.* Therapeutic microRNA delivery suppresses tumorigenesis in a murine liver cancer model. *Cell* **2009**, *137*, 1005–1017.

37. Huang, T. H., Esteller, M. Chromatin remodeling in mammary gland differentiation and breast tumorigenesis. *Cold Spring Harb. Perspect. Biol.* **2010**, *2*, a004515.
38. Guessous, F., Zhang, Y., Kofman, A., Catania, A., Li, Y., Schiff, D., Purow, B., Abounader, R. microRNA-34a is tumor suppressive in brain tumors and glioma stem cells. *Cell Cycle* **2010**, *9*, 6.
39. Taulli, R., Bersani, F., Ponzetto, C. Micro-orchestrating differentiation in cancer. *Cell Cycle* **2010**, *9*, 918–922.
40. Wiggins, J. F., Ruffino, L., Kelnar, K., Omotola, M., Patrawala, L., Brown, D., Bader, A. G. Development of a lung cancer therapeutic based on the tumor suppressor microRNA-34. *Cancer Res.* **2010**, *70*, 5923–5930.
41. Fontana, L., Fiori, M. E., Albini, S., Cifaldi, L., Giovinazzi, S., Forloni, M., Boldrini, R., Donfrancesco, A., Federici, V., Giacomini, P., *et al.* Antagomir-17–5p abolishes the growth of therapy-resistant neuroblastoma through p21 and BIM. *PloS One* **2008**, *3*, e2236.
42. Ciarapica, R., Russo, G., Verginelli, F., Raimondi, L., Donfrancesco, A., Rota, R., Giordano, A. Deregulated expression of miR-26a and Ezh2 in rhabdomyosarcoma. *Cell Cycle* **2009**, *8*, 172–175.
43. Taulli, R., Bersani, F., Foglizzo, V., Linari, A., Vigna, E., Ladanyi, M., Tuschl, T., Ponzetto, C. The muscle-specific microRNA miR-206 blocks human rhabdomyosarcoma growth in xenotransplanted mice by promoting myogenic differentiation. *J. Clin. Invest.* **2009**, *119*, 8, 2366–2378.
44. Calin, G. A., Sevignani, C., Dumitru, C. D., Hyslop, T., Noch, E., Yendamuri, S., Shimizu, M., Rattan, S., Bullrich, F., Negrini, M., *et al.* Human microRNA genes are frequently located at fragile sites and genomic regions involved in cancers. *Proc. Natl. Acad. Sci. USA* **2004**, *101*, 2999–3004.
45. Loeb, D. M., Thornton, K., Shokek, O. Pediatric soft tissue sarcomas. *Surg. Clin. N. Am.* **2008**, *88*, 615–627.
46. Tapscott, S. J., Thayer, M. J., Weintraub, H. Deficiency in rhabdomyosarcomas of a factor required for MyoD activity and myogenesis. *Science* **1993**, *259*, 1450–1453.
47. De Giovanni, C., Landuzzi, L., Nicoletti, G., Lollini, P. L., Nanni, P. Molecular and cellular biology of rhabdomyosarcoma. *Future Oncol.* **2009**, *5*, 1449–1475.
48. 48. Kohashi, K., Oda, Y., Yamamoto, H., Tamiya, S., Takahira, T., Takahashi, Y., Tajiri, T., Taguchi, T., Suita, S., Tsuneyoshi, M. Alterations of RB1 gene in embryonal and alveolar rhabdomyosarcoma: special reference to utility of pRB immunoreactivity in differential diagnosis of rhabdomyosarcoma subtype. *J. Cancer Res. Clin. Oncol.* **2008**, *134*, 1097–1103.
49. Crist, W. M., Anderson, J. R., Meza, J. L., Fryer, C., Raney, R. B., Ruymann, F. B., Breneman, J., Qualman, S. J., Wiener, E., Wharam, M., *et al.* Intergroup rhabdomyosarcoma study-IV: Results for patients with nonmetastatic disease. *J. Clin. Oncol.* **2001**, *19*, 3091–3102.
50. Sorensen, P. H., Lynch, J. C., Qualman, S. J., Tirabosco, R., Lim, J. F., Maurer, H. M., Bridge, J. A., Crist, W. M., Triche, T. J., Barr, F. G. PAX3-FKHR and PAX7-FKHR gene fusions are prognostic indicators in alveolar rhabdomyosarcoma: a report from the children's oncology group. *J. Clin. Oncol.* **2002**, *20*, 2672–2679.
51. Wachtel, M., Dettling, M., Koscielniak, E., Stegmaier, S., Treuner, J., Simon-Klingenstein, K., Buhlmann, P., Niggli, F. K., Schafer,B. W. Gene expression signatures

identify rhabdomyosarcoma subtypes and detect a novel t(2,2) (q35,p23) translocation fusing PAX3 to NCOA1. *Cancer Res.* **2004**, *64*, 5539–5545.

52. Missiaglia, E., Shepherd, C. J., Patel, S., Thway, K., Pierron, G., Pritchard-Jones, K., Renard, M., Sciot, R., Rao, P., Oberlin, O., *et al.* MicroRNA-206 expression levels correlate with clinical behaviour of rhabdomyosarcomas. *Br. J. Cancer* **2010**, *102*, 1769–1777.
53. Williamson, D., Missiaglia, E., de Reynies, A., Pierron, G., Thuille, B., Palenzuela, G., Thway, K., Orbach, D., Lae, M., Freneaux, P., *et al.* Fusion gene-negative alveolar rhabdomyosarcoma is clinically and molecularly indistinguishable from embryonal rhabdomyosarcoma. *J. Clin. Oncol.* **2010**, *28*, 2151–2158.
54. Davicioni, E., Anderson, M. J., Finckenstein, F. G., Lynch, J. C., Qualman, S. J., Shimada, H., Schofield, D. E., Buckley, J. D., Meyer, W. H., Sorensen, P. H., *et al.* Molecular classification of rhabdomyosarcoma-genotypic and phenotypic determinants of diagnosis: A report from the Children's Oncology Group. *Am. J. Pathol.* **2009**, *174*, 550–564.
55. Lae, M., Ahn, E. H., Mercado, G. E., Chuai, S., Edgar, M., Pawel, B. R., Olshen, A., Barr, F. G., Ladanyi, M. Global gene expression profiling of PAX-FKHR fusion-positive alveolar and PAX-FKHR fusion-negative embryonal rhabdomyosarcomas. *J. Pathol.* **2007**, *212*, 143–151.
56. Sumegi, J., Streblow, R., Frayer, R. W., Dal Cin, P., Rosenberg, A., Meloni-Ehrig, A., Bridge, J. A. Recurrent t(2,2) and t(2,8) translocations in rhabdomyosarcoma without the canonical PAX-FOXO1 fuse PAX3 to members of the nuclear receptor transcriptional coactivator family. *Genes Chromosomes Cancer* **2010**, *49*, 224–236.
57. Lagha, M., Sato, T., Regnault, B., Cumano, A., Zuniga, A., Licht, J., Relaix, F., Buckingham, M. Transcriptome analyses based on genetic screens for Pax3 myogenic targets in the mouse embryo. *BMC Genom.* **2010**, *11*, 696.
58. Seale, P., Sabourin, L. A., Girgis-Gabardo, A., Mansouri, A., Gruss, P., Rudnicki, M. A. Pax7 is required for the specification of myogenic satellite cells. *Cell* **2000**, *102*, 777–786.
59. Rota, R., Ciarapica,R.,Giordano,A.,Miele,L.,Locatelli,F. MicroRNAsin rhabdomyosarcoma: pathogenetic implications and translational potentiality. *Mol. Cancer* **2011**, *10*, 120.
60. Wang, H., Garzon, R., Sun, H., Ladner, K. J., Singh, R., Dahlman, J., Cheng, A., Hall, B. M., Qualman, S. J., Chandler, D. S., *et al.* NF-kappaB-YY1-miR-29 regulatory circuitry in skeletal myogenesis and rhabdomyosarcoma. *Cancer Cell* **2008**, *14*, 369–381.
61. Subramanian, S., Kartha, R. V. MicroRNA-mediated gene regulations in human sarcomas. *Cell Mol. Life Sci.* **2012**, *69*, 3571–3585.
62. Gagan, J., Dey, B. K., Dutta, A. MicroRNAs regulate and provide robustness to the myogenic transcriptional network. *Curr. Opin. Pharmacol.* **2012**, *12*, 383–388.
63. Lu, L., Zhou, L., Chen, E. Z., Sun, K., Jiang, P., Wang, L., Su, X., Sun, H., Wang, H. A Novel YY1-miR-1 regulatory circuit in skeletal myogenesis revealed by genome-wide prediction of YY1-miRNA network. *PLoS One* **2012**, *7*, e27596.
64. Liu, N., Williams, A. H., Kim, Y., McAnally, J., Bezprozvannaya, S., Sutherland, L. B., Richardson, J. A., Bassel-Duby, R., Olson, E. N. An intragenic MEF2-dependent

enhancer directs muscle-specific expression of microRNAs 1 and 133. *Proc. Natl. Acad. Sci. USA* **2007**, *104*, 20844–20849.
65. Dey, B. K., Gagan, J., Dutta, A. miR-206 and -486 induce myoblast differentiation by downregulating Pax7. *Mol. Cell. Biol.* **2010**, *31*, 203–214.
66. Chen, J. F., Mandel, E. M., Thomson, J. M., Wu, Q., Callis, T. E., Hammond, S. M., Conlon, F. L., Wang, D. Z. The role of microRNA-1 and microRNA-133 in skeletal muscle proliferation and differentiation. *Nat. Genet.* **2006**, *38*, 228–233.
67. Williams, A. H., Valdez, G., Moresi, V., Qi, X., McAnally, J., Elliott, J. L., Bassel-Duby, R., Sanes, J. R., Olson, E. N. MicroRNA-206 delays ALS progression and promotes regeneration of neuromuscular synapses in mice. *Science* **2009**, *326*, 1549–1554.
68. Hecker, R. M., Amstutz, R. A., Wachtel, M., Walter, D., Niggli, F. K., Schafer, B. W. p21 Downregulation is an important component of PAX3/FKHR oncogenicity and its reactivation by HDAC inhibitors enhances combination treatment. *Oncogene* **2010**, *29*, 3942–3952.
69. Ecke, I., Petry, F., Rosenberger, A., Tauber, S., Monkemeyer, S., Hess, I., Dullin, C., Kimmina, S., Pirngruber, J., Johnsen, S. A., *et al.* Antitumor effects of a combined 5-aza-2'deoxycytidine and valproic acid treatment on rhabdomyosarcoma and medulloblastoma in Ptch mutant mice. *Cancer Res.* **2009**, *69*, 887–895.
70. Li, Z., Hassan, M. Q., Jafferji, M., Aqeilan, R. I., Garzon, R., Croce, C. M., van Wijnen, A. J., Stein, J. L., Stein, G. S., Lian, J. B. Biological functions of miR-29b contribute to positive regulation of osteoblast differentiation. *J. Biol. Chem.* **2009**, *284*, 15676–15684.
71. 71. Elia, L., Contu, R., Quintavalle, M., Varrone, F., Chimenti, C., Russo, M. A., Cimino, V., De Marinis, L., Frustaci, A., Catalucci, D., *et al.* Reciprocal regulation of microRNA-1 and insulin-like growth factor-1 signal transduction cascade in cardiac and skeletal muscle in physiological and pathological conditions. *Circulation* **2009**, *120*, 2377–2385.
72. Huang, M. B., Xu, H., Xie, S. J., Zhou, H., Qu, L. H. Insulin-like growth factor-1 receptor is regulated by microRNA-133 during skeletal myogenesis. *PLoS One* **2011**, *6*, e29173.
73. Olson, E. N., Spizz, G., Tainsky, M. A. The oncogenic forms of N-ras or H-ras prevent skeletal myoblast differentiation. *Mol. Cell. Biol.* **1987**, *7*, 2104–2111.
74. Liu, J., Luo, X. J., Xiong, A. W., Zhang, Z. D., Yue, S., Zhu, M. S., Cheng, S. Y. MicroRNA-214 promotes myogenic differentiation by facilitating exit from mitosis via down-regulation of proto-oncogene N-ras. *J. Biol. Chem.* **2010**, *285*, 26599–26607.
75. Wang, H., Hertlein, E., Bakkar, N., Sun, H., Acharyya, S., Wang, J., Carathers, M., Davuluri, R., Guttridge,D. C. NF-kappaBregulationofYY1inhibitsskeletalmyogenesisthrough transcriptional silencing of myofibrillar genes. *Mol. Cell. Biol.* **2007**, *27*, 4374–4387.
76. Caretti, G., Di Padova, M., Micales, B., Lyons, G. E., Sartorelli, V. The Polycomb Ezh2 methyltransferase regulates muscle gene expression and skeletal muscle differentiation. *Genes Dev.* **2004**, *18*, 2627–2638.
77. Subramanian, S., Lui, W. O., Lee, C. H., Espinosa, I., Nielsen, T. O., Heinrich, M. C., Corless, C. L., Fire, A. Z., van de Rijn, M. MicroRNA expression signature of human sarcomas. *Oncogene* **2008**, *27*, 2015–2026.

78. Fabbri, M., Garzon, R., Cimmino, A., Liu, Z., Zanesi, N., Callegari, E., Liu, S., Alder, H., Costinean, S., Fernandez-Cymering, C., *et al.* MicroRNA-29 family reverts aberrant methylation in lung cancer by targeting DNA methyltransferases 3A and 3B. *Proc. Natl. Acad. Sci. USA* **2007**, *104*, 15805–15810.
79. Amodio, N., Leotta, M., Bellizzi, D., Di Martino, M. T., D'Aquila, P., Lionetti, M., Fabiani, F., Leone, E., Gulla, A. M., Passarino, G., *et al.* DNA-demethylating and anti-tumor activity of synthetic miR-29b mimics in multiple myeloma. *Oncotarget* **2012**, *3*, 1246–1258.
80. Hoehner,J. C.,Gestblom,C.,Hedborg,F.,Sandstedt,B.,Olsen,L.,Pahlman,S. A developmental model of neuroblastoma: differentiating stroma-poor tumors' progress along an extra-adrenal chromaffin lineage. *Lab. Investig.* **1996**, *75*, 659–675.
81. Cooper, M. J., Hutchins, G. M., Cohen, P. S., Helman, L. J., Mennie, R. J., Israel, M. A. Human neuroblastoma tumor cell lines correspond to the arrested differentiation of chromaffin adrenal medullary neuroblasts. *Cell Growth Differ.* **1990**, *1*, 149–159.
82. Gaetano, C., Matsumoto, K., Thiele, C. J. *In vitro* activation of distinct molecular and cellular phenotypes after induction of differentiation in a human neuroblastoma cell line. *Cancer Res.* **1992**, *52*, 4402–4407.
83. Brodeur, G. M. Neuroblastoma: biological insights into a clinical enigma. *Nat. Rev. Cancer* **2003**, *3*, 203–216.
84. Van Roy, N., Laureys, G., Cheng, N. C., Willem, P., Opdenakker, G., Versteeg, R., Speleman, F. 1,17 translocations and other chromosome 17 rearrangements in human primary neuroblastoma tumors and cell lines. *Genes. Chromosomes Cancer* **1994**, *10*, 103–114.
85. Fong, C. T., Dracopoli, N. C., White, P. S., Merrill, P. T., Griffith, R. C., Housman, D. E., Brodeur, G. M. Loss of heterozygosity for the short arm of chromosome 1 in human neuroblastomas: correlation with *N*-myc amplification. *Proc. Natl. Acad. Sci. USA* **1989**, *86*, 3753–3757.
86. Janoueix-Lerosey, I., Novikov, E., Monteiro, M., Gruel, N., Schleiermacher, G., Loriod, B., Nguyen, C., Delattre, O. Gene expression profiling of 1p35–36 genes in neuroblastoma. *Oncogene* **2004**, *23*, 5912–5922.
87. Fujita, T., Igarashi, J., Okawa, E. R., Gotoh, T., Manne, J., Kolla, V., Kim, J., Zhao, H., Pawel, B. R., London, W. B., *et al. CHD5*, a tumor suppressor gene deleted from 1p36. 31 in neuroblastomas. *J. Natl. Cancer Inst.* **2008**, *100*, 940–949.
88. Saito-Ohara, F., Imoto, I., Inoue, J., Hosoi, H., Nakagawara, A., Sugimoto, T., Inazawa, J. PPM1D is a potential target for 17q gain in neuroblastoma. *Cancer Res.* **2003**, *63*, 1876–1883.
89. Van Noesel, M. M., Versteeg, R. Pediatric neuroblastomas: Genetic and epigenetic "danse macabre". *Gene* **2004**, *325*, 1–15.
90. Oppenheimer, O., Alaminos, M., Gerald, W. L. Genomic medicine and neuroblastoma. *Expert. Rev. Mol. Diagn.* **2003**, *3*, 39–54.
91. Mosse, Y. P., Laudenslager, M., Longo, L., Cole, K. A., Wood, A., Attiyeh, E. F., Laquaglia, M. J., Sennett, R., Lynch, J. E., Perri, P., *et al.* Identification of ALK as a major familial neuroblastoma predisposition gene. *Nature* **2008**, *455*, 930–935.
92. Janoueix-Lerosey, I., Lequin, D., Brugieres, L., Ribeiro, A., de Pontual, L., Combaret, V., Raynal, V., Puisieux, A., Schleiermacher, G., Pierron, G., *et al.* Somatic and

germline activating mutations of the ALK kinase receptor in neuroblastoma. *Nature* **2008**, *455*, 967–970.
93. Chen, Y., Takita, J., Choi, Y. L., Kato, M., Ohira, M., Sanada, M., Wang, L., Soda, M., Kikuchi, A., Igarashi, T., *et al.* Oncogenic mutations of ALK kinase in neuroblastoma. *Nature* **2008**, *455*, 971–974.
94. George, R. E., Sanda, T., Hanna, M., Frohling, S., Luther, W., II, Zhang, J., Ahn, Y., Zhou, W., London, W. B., McGrady, P., *et al.* Activating mutations in ALK provide a therapeutic target in neuroblastoma. *Nature* **2008**, *455*, 975–978.
95. Morris, S. W., Kirstein, M. N., Valentine, M. B., Dittmer, K. G., Shapiro, D. N., Saltman, D. L., Look, A. T. Fusion of a kinase gene, ALK, to a nucleolar protein gene, NPM, in non-Hodgkin's lymphoma. *Science* **1994**, *263*, 1281–1284.
96. Rikova, K., Guo, A., Zeng, Q., Possemato, A., Yu, J., Haack, H., Nardone, J., Lee, K., Reeves, C., Li, Y., *et al.* Global survey of phosphotyrosine signaling identifies oncogenic kinases in lung cancer. *Cell* **2007**, *131*, 1190–1203.
97. Soda, M., Choi, Y. L., Enomoto, M., Takada, S., Yamashita, Y., Ishikawa, S., Fujiwara, S., Watanabe, H., Kurashina, K., Hatanaka, H., *et al.* Identification of the transforming EML4-ALK fusion gene in non-small-cell lung cancer. *Nature* **2007**, *448*, 561–566.
98. Osajima-Hakomori, Y., Miyake, I., Ohira, M., Nakagawara, A., Nakagawa, A., Sakai, R. Biological role of anaplastic lymphoma kinase in neuroblastoma. *Am. J. Pathol.* **2005**, *167*, 213–222.
99. Lim, M. S., Carlson, M. L., Crockett, D. K., Fillmore, G. C., Abbott, D. R., Elenitoba-Johnson, O. F., Tripp, S. R., Rassidakis, G. Z., Medeiros, L. J., Szankasi, P., *et al.* The proteomic signature of NPM/ALK reveals deregulation of multiple cellular pathways. *Blood* **2009**, *114*, 1585–1595.
100. Decock, A., Ongenaert, M., Vandesompele, J., Speleman, F. Neuroblastoma epigenetics: From candidate gene approaches to genome-wide screenings. *Epigenetics* **2011**, *6*, 962–790.
101. Teitz, T., Wei, T., Valentine, M. B., Vanin, E. F., Grenet, J., Valentine, V. A., Behm, F. G., Look, A. T., Lahti, J. M., Kidd, V. J. Caspase 8 is deleted or silenced preferentially in childhood neuroblastomas with amplification of MYCN. *Nat. Med.* **2000**, *6*, 529–535.
102. Astuti, D., Agathanggelou, A., Honorio, S., Dallol, A., Martinsson, T., Kogner, P., Cummins, C., Neumann, H. P., Voutilainen, R., Dahia, P., *et al.* RASSF1A promoter region CpG island hypermethylation in phaeochromocytomas and neuroblastoma tumours. *Oncogene* **2001**, *20*, 7573–7577.
103. Murphy, D. M., Buckley, P. G., Bryan, K., Das, S., Alcock, L., Foley, N. H., Prenter, S., Bray, I., Watters, K. M., Higgins, D., *et al.* Global MYCN transcription factor binding analysis in neuroblastomarevealsassociationwithdistinctE-boxmotifsandregionsofDNA hypermethylation. *PLoS One* **2009**, *4*, e8154.
104. Buckley, P. G., Alcock, L., Bryan, K., Bray, I., Schulte, J. H., Schramm, A., Eggert, A., Mestdagh, P., De Preter, K., Vandesompele, J., *et al.* Chromosomal and microRNA expression patterns reveal biologically distinct subgroups of 11q-neuroblastoma. *Clin. Cancer Res.* **2010**, *16*, 2971–2978.
105. Caren, H., Djos, A., Nethander, M., Sjoberg, R. M., Kogner, P., Enstrom, C., Nilsson, S., Martinsson, T. Identification of epigenetically regulated genes that predict patient outcome in neuroblastoma. *BMC Cancer* **2011**, *11*, 66.

106. Chavali, P. L., Funa, K., Chavali, S. Cis-regulation of microRNA expression by scaffold/matrix-attachment regions. *Nucleic Acids Res.* **2011**, *39*, 6908–6918.
107. Angrisano, T., Sacchetti, S., Natale, F., Cerrato, A., Pero, R., Keller, S., Peluso, S., Perillo, B., Avvedimento, V. E., Fusco, A., *et al.* Chromatin and DNA methylation dynamics during retinoic acid-induced RET gene transcriptional activation in neuroblastoma cells. *Nucleic Acids Res.* **2011**, *39*, 1993–2006.
108. Das, S., Foley, N., Bryan, K., Watters, K. M., Bray, I., Murphy, D. M., Buckley, P. G., Stallings, R. L. MicroRNA mediates DNA demethylation events triggered by retinoic acid during neuroblastoma cell differentiation. *Cancer Res.* **2010**, *70*, 7874–7881.
109. Das, S., Bryan, K., Buckley, P. G., Piskareva, O., Bray, I. M., Foley, N., Ryan, J., Lynch, J., Creevey, L., Fay, J., *et al.* Modulation of neuroblastoma disease pathogenesis by an extensive network of epigenetically regulated microRNAs. *Oncogene* **2012**, doi:10. 1038/onc. 2012. 311.
110. Stallings, R. L., Foley, N. H., Bray, I. M., Das, S., Buckley, P. G. MicroRNA and DNA methylation alterations mediating retinoic acid induced neuroblastoma cell differentiation. *Semin. Cancer Biol.* **2011**, *21*, 283–290.
111. Meseguer, S., Mudduluru, G., Escamilla, J. M., Allgayer, H., Barettino, D. MicroRNAs-10a and -10b contribute to retinoic acid-induced differentiation of neuroblastoma cells and target the alternative splicing regulatory factor SFRS1 (SF2/ASF). *J. Biol. Chem.* **2011**, *286*, 4150–4164.
112. Karni, R., de Stanchina, E., Lowe, S. W., Sinha, R., Mu, D., Krainer, A. R. The gene encoding the splicing factor SF2/ASF is a proto-oncogene. *Nat. Struct. Mol. Biol.* **2007**, *14*, 185–193.
113. Foley, N. H., Bray, I., Watters, K. M., Das, S., Bryan, K., Bernas, T., Prehn, J. H., Stallings, R. L. MicroRNAs 10a and 10b are potent inducers of neuroblastoma cell differentiation through targeting of nuclear receptor corepressor 2. *Cell Death Differ.* **2011**, *18*, 1089–1098.
114. Jepsen, K., Solum, D., Zhou, T., McEvilly, R. J., Kim, H. J., Glass, C. K., Hermanson, O., Rosenfeld, M. G. SMRT-mediated repression of an H3K27 demethylase in progression from neural stem cell to neuron. *Nature* **2007**, *450*, 415–419.
115. Buechner, J., Tomte, E., Haug, B. H., Henriksen, J. R., Lokke, C., Flaegstad, T., Einvik, C. Tumour-suppressor microRNAs let-7 and mir-101 target the proto-oncogene *MYCN* and inhibit cell proliferation in *MYCN*-amplified neuroblastoma. *Br. J. Cancer* **2011**, *105*, 296–303.
116. Molenaar, J. J., Domingo-Fernandez, R., Ebus, M. E., Lindner, S., Koster, J., Drabek, K., Mestdagh, P., van Sluis, P., Valentijn, L. J., van Nes, J., *et al.* LIN28B induces neuroblastoma and enhances MYCN levels via let-7 suppression. *Nat. Genet.* **2012**, *44*, 1199–1206.
117. Lynch, J., Fay, J., Meehan, M., Bryan, K., Watters, K. M., Murphy, D. M., Stallings, R. L. MiRNA-335 suppresses neuroblastoma cell invasiveness by direct targeting of multiple genes from the non-canonical TGF-beta signalling pathway. *Carcinogenesis* **2012**, *33*, 976–985.
118. Foley, N. H., Bray, I. M., Tivnan, A., Bryan, K., Murphy, D. M., Buckley, P. G., Ryan, J., O'Meara, A., O'Sullivan, M., Stallings, R. L. MicroRNA-184 inhibits neuroblastoma cell survival through targeting the serine/threonine kinase AKT2. *Mol. Cancer* **2010**, *9*, 83.

119. Tivnan, A., Foley, N. H., Tracey, L., Davidoff, A. M., Stallings, R. L. MicroRNA-184-mediated inhibition of tumour growth in an orthotopic murine model of neuroblastoma. *Anticancer Res.* **2010**, *30*, 4391–4395.
120. Bray, I., Tivnan, A., Bryan, K., Foley, N. H., Watters, K. M., Tracey, L., Davidoff, A. M., Stallings, R. L. MicroRNA-542–5p as a novel tumor suppressor in neuroblastoma. *Cancer Lett.* **2011**, *303*, 56–64.
121. Schulte, J. H., Marschall, T., Martin, M., Rosenstiel, P., Mestdagh, P., Schlierf, S., Thor, T., Vandesompele, J., Eggert, A., Schreiber, S., *et al.* Deep sequencing reveals differential expression of microRNAs in favorable *versus* unfavorable neuroblastoma. *Nucleic Acids Res.* **2010**, *38*, 5919–5928.
122. Schulte, J. H., Schowe, B., Mestdagh, P., Kaderali, L., Kalaghatgi, P., Schlierf, S., Vermeulen, J., Brockmeyer, B., Pajtler, K., Thor, T., *et al.* Accurate prediction of neuroblastoma outcome based on miRNA expression profiles. *Int. J. Cancer* **2010**, *127*, 2374–2385.
123. Laneve, P., Di Marcotullio, L., Gioia, U., Fiori, M. E., Ferretti, E., Gulino, A., Bozzoni, I., Caffarelli,E. TheinterplaybetweenmicroRNAsandtheneurotrophinreceptor tropomyosin-related kinase C controls proliferation of human neuroblastoma cells. *Proc. Natl. Acad. Sci. USA* **2007**, *104*, 7957–7962.
124. Laneve, P., Gioia, U., Andriotto, A., Moretti, F., Bozzoni, I., Caffarelli, E. A minicircuitry involving REST and CREB controls miR-9–2 expression during human neuronal differentiation. *Nucleic Acids Res.* **2010**, *38*, 6895–6905.
125. Annibali, D., Gioia, U., Savino, M., Laneve, P., Caffarelli, E., Nasi, S. A new module in neural differentiation control: two microRNAs upregulated by retinoic acid, miR-9 and -103, target the differentiation inhibitor ID2. *PLoS One* **2012**, *7*, e40269.
126. Zhang, H., Qi, M., Li, S., Qi, T., Mei, H., Huang, K., Zheng, L., Tong, Q. MicroRNA-9 targets matrix metalloproteinase 14 to inhibit invasion, metastasis, and angiogenesis of neuroblastoma cells. *Mol. Cancer Therapeut.* **2012**, *11*, 1454–1466.
127. Lee, J. J., Drakaki, A., Iliopoulos, D., Struhl, K. MiR-27b targets PPARgamma to inhibit growth, tumor progression and the inflammatory response in neuroblastoma cells. *Oncogene* **2012**, *31*, 3818–3825.
128. Cole, K. A., Attiyeh, E. F., Mosse, Y. P., Laquaglia, M. J., Diskin, S. J., Brodeur, G. M., Maris, J. M. A functional screen identifies miR-34a as a candidate neuroblastoma tumor suppressor gene. *Mol. Cancer Res.* **2008**, *6*, 735–742.
129. Ragusa, M., Majorana, A., Banelli, B., Barbagallo, D., Statello, L., Casciano, I., Guglielmino, M. R., Duro, L. R., Scalia, M., Magro, G., *et al.* MIR152, MIR200B, and MIR338, human positional and functional neuroblastoma candidates, are involved in neuroblast differentiation and apoptosis. *J. Mol. Med.* **2010**, *88*, 1041–1053.
130. Lodygin, D., Tarasov, V., Epanchintsev, A., Berking, C., Knyazeva, T., Korner, H., Knyazev, P., Diebold, J., Hermeking, H. Inactivation of miR-34a by aberrant CpG methylation in multiple types of cancer. *Cell Cycle* **2008**, *7*, 2591–2600.
131. Welch, C., Chen, Y., Stallings, R. L. MicroRNA-34a functions as a potential tumor suppressor by inducing apoptosis in neuroblastoma cells. *Oncogene* **2007**, *26*, 5017–5022.
132. Saito, Y., Liang, G., Egger, G., Friedman, J. M., Chuang, J. C., Coetzee, G. A., Jones, P. A. Specific activation of microRNA-127 with downregulation of the

proto-oncogene BCL6 by chromatin-modifying drugs in human cancer cells. *Cancer Cell* **2006**, *9*, 435–443.
133. Wei, J. S., Song, Y. K., Durinck, S., Chen, Q. R., Cheuk, A. T., Tsang, P., Zhang, Q., Thiele, C. J., Slack, A., Shohet, J., *et al.* The *MYCN* oncogene is a direct target of miR-34a. *Oncogene* **2008**, *27*, 5204–5213.
134. Agostini, M., Tucci, P., Steinert, J. R., Shalom-Feuerstein, R., Rouleau, M., Aberdam, D., Forsythe, I. D., Young, K. W., Ventura, A., Concepcion, C. P., *et al.* MicroRNA-34a regulates neurite outgrowth, spinal morphology, and function. *Proc. Natl. Acad. Sci. USA* **2011**, *108*, 21099–21104.
135. Gougelet, A., Perez, J., Pissaloux, D., Besse, A., Duc, A., Decouvelaere, A. V., Ranchere-Vince, D., Blay, J. Y., Alberti, L. miRNA Profiling: How to Bypass the Current Difficulties in the Diagnosis and Treatment of Sarcomas. *Sarcoma* **2011**, *2011*, 460650.
136. Lin, R. J., Lin, Y. C., Chen, J., Kuo, H. H., Chen, Y. Y., Diccianni, M. B., London, W. B., Chang, C. H., Yu, A. L. microRNA signature and expression of Dicer and Drosha can predict prognosis and delineate risk groups in neuroblastoma. *Cancer Res.* **2010**, *70*, 7841–7850.
137. De Preter, K., Mestdagh, P., Vermeulen, J., Zeka, F., Naranjo, A., Bray, I., Castel, V., Chen, C., Drozynska, E., Eggert, A., *et al.* MiRNA expression profiling enables risk stratification in archived and fresh neuroblastoma tumor samples. *Clin. Cancer Res.* **2011**, *17*, 7684–7692.
138. Cortez, M. A., Calin, G. A. MicroRNA identification in plasma and serum: A new tool to diagnose and monitor diseases. *Expert Opin. Biol. Ther.* **2009**, *9*, 703–711.
139. Miyachi, M., Tsuchiya, K., Yoshida, H., Yagyu, S., Kikuchi, K., Misawa, A., Iehara, T., Hosoi, H. Circulating muscle-specific microRNA, miR-206, as a potential diagnostic marker for rhabdomyosarcoma. *Biochem. Biophys. Res. Commun.* **2010**, *400*, 89–93.
140. Hogrefe, R. I., Lebedev, A. V., Zon, G., Pirollo, K. F., Rait, A., Zhou, Q., Yu, W., Chang, E. H. Chemically modified short interfering hybrids (siHYBRIDS): nanoimmunoliposome delivery *in vitro* and *in vivo* for RNAi of HER-2. *Nucleos. Nucleot. Nucleic Acids* **2006**, *25*, 889–907.
141. Pirollo, K. F., Rait, A., Zhou, Q., Hwang, S. H., Dagata, J. A., Zon, G., Hogrefe, R. I., Palchik, G., Chang, E. H. Materializing the potential of small interfering RNA via a tumor-targeting nanodelivery system. *Cancer Res.* **2007**, *67*, 2938–2943.
142. Elmen, J., Lindow, M., Schutz, S., Lawrence, M., Petri, A., Obad, S., Lindholm, M., Hedtjarn, M., Hansen, H. F., Berger, U., *et al.* LNA-mediated microRNA silencing in non-human primates. *Nature* **2008**, *452*, 896–899.
143. Lanford, R. E., Hildebrandt-Eriksen, E. S., Petri, A., Persson, R., Lindow, M., Munk, M. E., Kauppinen, S., Orum, H. Therapeutic silencing of microRNA-122 in primates with chronic hepatitis C virus infection. *Science* **2010**, *327*, 198–201.
144. Anand, S., Majeti, B. K., Acevedo, L. M., Murphy, E. A., Mukthavaram, R., Scheppke, L., Huang, M., Shields, D. J., Lindquist, J. N., Lapinski, P. E., *et al.* MicroRNA-132-mediated loss of p120RasGAP activates the endothelium to facilitate pathological angiogenesis. *Nat. Med.* **2010**, *16*, 909–914.
145. Su, J., Baigude, H., McCarroll, J., Rana, T. M. Silencing microRNA by interfering nanoparticles in mice. *Nucleic Acids Res.* **2011**, *39*, e38.
146. Almeida, M. I., Reis, R. M., Calin, G. A. MicroRNA history: discovery, recent applications, and next frontiers. *Mutat. Res.* **2011**, *717*, 1–8.

147. Tivnan, A., Orr, W. S., Gubala, V., Nooney, R., Williams, D. E., McDonagh, C., Prenter, S., Harvey, H., Domingo-Fernandez, R., Bray, I. M., *et al.* Inhibition of neuroblastoma tumor growth by targeted delivery of microRNA-34a using anti-disialoganglioside GD2 coated nanoparticles. *PLoS One* **2012**, *7*, e38129.
148. Issa, J. P., Garcia-Manero, G., Giles, F. J., Mannari, R., Thomas, D., Faderl, S., Bayar, E., Lyons, J., Rosenfeld, C. S., Cortes, J., *et al.* Phase 1 study of low-dose prolonged exposure schedules of the hypomethylating agent 5-aza-2'-deoxycytidine (decitabine) in hematopoietic malignancies. *Blood* **2004**, *103*, 1635–1640.
149. Kantarjian, H., Issa, J. P., Rosenfeld, C. S., Bennett, J. M., Albitar, M., DiPersio, J., Klimek, V., Slack, J., de Castro, C., Ravandi, F., *et al.* Decitabine improves patient outcomes in myelodysplastic syndromes: results of a phase III randomized study. *Cancer* **2006**, *106*, 1794–1803.
150. Vigil, C. E., Martin-Santos, T., Garcia-Manero, G. Safety and efficacy of azacitidine in myelodysplastic syndromes. *Drug Des. Devel. Ther.* **2010**, *4*, 221–229.
151. Candelaria, M., Herrera, A., Labardini, J., Gonzalez-Fierro, A., Trejo-Becerril, C., Taja-Chayeb, L., Perez-Cardenas, E., de la Cruz-Hernandez, E., Arias-Bofill, D., Vidal, S., *et al.* Hydralazine and magnesium valproate as epigenetic treatment for myelodysplastic syndrome. Preliminary results of a phase-II trial. *Ann. Hematol.* **2010**, *90*, 379–387.
152. Fu, S., Hu, W., Iyer, R., Kavanagh, J. J., Coleman, R. L., Levenback, C. F., Sood, A. K., Wolf, J. K., Gershenson, D. M., Markman, M., *et al.* Phase 1b-2a study to reverse platinum resistance through use of a hypomethylating agent, azacitidine, in patients with platinum-resistant or platinum-refractory epithelial ovarian cancer. *Cancer* **2010**, *117*, 1661–1669.
153. Juergens, R. A., Wrangle, J., Vendetti, F. P., Murphy, S. C., Zhao, M., Coleman, B., Sebree, R., Rodgers, K., Hooker, C. M., Franco, N., *et al.* Combination epigenetic therapy has efficacy in patients with refractory advanced non-small cell lung cancer. *Cancer Discov.* **2011**, *1*, 598–607.
154. Abele, R., Clavel, M., Dodion, P., Bruntsch, U., Gundersen, S., Smyth, J., Renard, J., van Glabbeke, M., Pinedo, H. M. The EORTC Early Clinical Trials Cooperative Group experience with 5-aza-2'-deoxycytidine (NSC 127716) in patients with colo-rectal, head and neck, renal carcinomas and malignant melanomas. *Eur. J. Cancer Clin. Oncol.* **1987**, *23*, 1921–1924.
155. Momparler, R. L., Bouffard, D. Y., Momparler, L. F., Dionne, J., Belanger, K., Ayoub, J. Pilot phase I-II study on 5-aza-2'-deoxycytidine (Decitabine) in patients with metastatic lung cancer. *Anti-Cancer Drugs* **1997**, *8*, 358–368.
156. Tsai, H. C., Li, H., Van Neste, L., Cai, Y., Robert, C., Rassool, F. V., Shin, J. J., Harbom, K. M., Beaty, R., Pappou, E., *et al.* Transient low doses of DNA-demethylating agents exert durable antitumor effects on hematological and epithelial tumor cells. *Cancer Cell* **2012**, *21*, 430–446.
157. Shoemaker, A. R., Mitten, M. J., Adickes, J., Ackler, S., Refici, M., Ferguson, D., Oleksijew, A., O'Connor, J. M., Wang, B., Frost, D. J., *et al.* Activity of the Bcl-2 family inhibitor ABT-263 in a panel of small cell lung cancer xenograft models. *Clin. Cancer Res.* **2008**, *14*, 3268–3277.

158. Jain, H. V., Meyer-Hermann, M. The molecular basis of synergism between carboplatin and ABT-737 therapy targeting ovarian carcinomas. *Cancer Res.* **2011**, *71*, 705–715.
159. Belinsky, S. A., Klinge, D. M., Stidley, C. A., Issa, J. P., Herman, J. G., March, T. H., Baylin, S. B. Inhibition of DNA methylation and histone deacetylation prevents murine lung cancer. *Cancer Res.* **2003**, *63*, 7089–7093.
160. Gore, S. D., Baylin, S., Sugar, E., Carraway, H., Miller, C. B., Carducci, M., Grever, M., Galm, O., Dauses, T., Karp, J. E., *et al.* Combined DNA methyltransferase and histone deacetylase inhibition in the treatment of myeloid neoplasms. *Cancer Res.* **2006**, *66*, 6361–6369.
161. Fandy, T. E., Herman, J. G., Kerns, P., Jiemjit, A., Sugar, E. A., Choi, S. H., Yang, A. S., Aucott, T., Dauses, T., Odchimar-Reissig, R., *et al.* Early epigenetic changes and DNA damage do not predict clinical response in an overlapping schedule of 5-azacytidine and entinostat in patients with myeloid malignancies. *Blood* **2009**, *114*, 2764–2773.
162. Suva, M. L., Riggi, N., Janiszewska, M., Radovanovic, I., Provero, P., Stehle, J. C., Baumer, K., Le Bitoux, M. A., Marino, D., Cironi, L., *et al.* EZH2 is essential for glioblastoma cancer stem cell maintenance. *Cancer Res.* **2009**, *69*, 9211–9218.
163. Tan, J., Yang, X., Zhuang, L., Jiang, X., Chen, W., Lee, P. L., Karuturi, R. K., Tan, P. B., Liu, E. T., Yu, Q. Pharmacologic disruption of Polycomb-repressive complex 2-mediated gene repression selectively induces apoptosis in cancer cells. *Genes Dev.* **2007**, *21*, 1050–1063.
164. Fiskus, W., Wang, Y., Sreekumar, A., Buckley, K. M., Shi, H., Jillella, A., Ustun, C., Rao, R., Fernandez, P., Chen, J., *et al.* Combined epigenetic therapy with the histone methyltransferase EZH2 inhibitor 3-deazaneplanocin A and the histone deacetylase inhibitor panobinostat against human AML cells. *Blood* **2009**, *114*, 2733–2743.
165. 165. Hayden, A., Johnson, P. W., Packham, G., Crabb, S. J. *S*-adenosylhomocysteine hydrolase inhibition by 3-deazaneplanocin A analogues induces anti-cancer effects in breast cancer cell lines and synergy with both histone deacetylase and HER2 inhibition. *Breast Cancer Res. Treat.* **2010**, *102*, 109–119.
166. Yu, Y., Zeng, P., Xiong, J., Liu, Z., Berger, S. L., Merlino, G. Epigenetic drugs can stimulate metastasis through enhanced expression of the pro-metastatic Ezrin gene. *PLoS One* **2010**, *5*, e12710.

CHAPTER 10

EPIGENETIC EFFECTS OF ENVIRONMENTAL CHEMICALS BISPHENOL A AND PHTHALATES

SHER SINGH and STEVEN SHOEI-LUNG LI

INTRODUCTION

Plastics are widely used in modern life, and their unbound chemicals bisphenol A (BPA) and phthalates can leach out into the surrounding environment. BPA and phthalates have recently attracted the special attention of the scientific community, regulatory agencies and the general public because of their high production volume, widespread use of plastics, and adverse health effects [1]. BPA is now used in the production of polycarbonate plastic containers such as baby bottles and epoxy resins that line metal cans for food and beverages. BPA is also used as a plasticizer to soften and increase the flexibility of polyvinyl chloride (PVC) plastic products. BPA has another medical use in dental sealants and composites used for filling. It is thought that human exposure mainly occurs through food and drink. However, exposure may also occur through dermal contact with thermal paper, used widely in cash register receipts. Phthalates are a group of similar diesters of phthalic acid used as plasticizers to soften and increase the flexibility of PVC plastics [2]. Human exposure to phthalates mainly occurs through foods, because of their uses in wrapping materials and food processing [3]. When ingested through food contamination, diethylhexyl phthalate (DEHP) is converted by intestinal lipases to mono-(2-ethylhexyl) phthalate (MEHP), which is then preferentially absorbed. Dibutyl phthalate (DBP) is used as a component of latex adhesives. It is also used in cosmetics and other personal care products, as a plasticizer in cellulose plastics, and as a solvent for dyes [4]. Monobutyl phthalate (MBP) is the toxic metabolite of DBP and butylbenzyl phthalate (BBP).

Epigenetics is the study of heritable changes in gene expression occurring without changes in DNA sequence. Epigenetic mechanisms include DNA methylation, histone modifications (acetylation, methylation, phosphorylation, ubiquitination, sumoylation and ADP ribosylation), and expression of non-coding RNAs (including microRNAs). In mammals, DNA methylation patterns are established during embryogenesis through the coopearation of DNA methyltransferases (DNMTs) and associated proteins. DNMT1 is responsible for the maintenance of methylation patterns throughout DNA replication (*i.e.,* specific for hemi-methylated sequences). DNMT2 may be involved in embryonic stem cells and potential RNA methylation. DNMT3A and DNMT3B are involved in active *de novo* DNA methylation at CpG sites. The early developmental period is thought to be the most susceptible to epigenetic insults because the DNA synthesis rate is high, and the elaborate DNA methylation patterning and chromatin organization required for normal tissue development is established at this time [5].

Epigenetics can influence the gene expression profiles of most organs and cell types. Furthermore, epigenetics is an important mechanism in the ability of environmental chemicals to influence human health and disease [6]. Environmental chemicals such as BPA and phthalates may play some critical roles in the etiology of many human disease risks [5,7,8]. Multiple lines of evidences from *in vitro* and *in vivo* models have established that epigenetic modifications caused by *in utero* exposure to environmental toxicants can induce alterations in gene expression that may persist throughout life. Thus, the environmentally induced epigenetic changes become increasingly relevant to human health and disease [9–11].

In the last few years, many investigations have examined the relationships between exposure to environmental chemicals and epigenetic effects, and identified several toxicants that modify epigenetic marks. Most of these studies conducted so far have focused on DNA methylation, whereas only a few recent investigations have studied the effects of environmental chemicals on histone modifications and expression of microRNAs [12]. Here, we review the epigenetic effects, as well as toxicogenomics, toxicities and health effects, of environmental toxicants BPA and phthalates derived from *in vitro* models, animal and human studies.

TOXICITIES AND HEALTH EFFECTS OF BISPHENOL A AND PHTHALATES

BPA and phthalates have long been known to have weak estrogenic properties and act as endocrine-disruptors owing to their ability to compete with endogenous steroid hormones binding to receptors. BPA was originally discovered as an artificial estrogen, and its estrogenic effect was used to enhance the rapid growth of cattle and poultry. BPA was also used for a few years as estrogen replacement for women. Since BPA can bind weakly to estrogen receptors ESR1 and ESR2, it is likely to be an endocrine disruptor. The impacts of BPA exposure on human health has been extensively reviewed and reported by the National Toxicology Program-Center for the Evaluation of Risks to Human Reproduction [13]. There is extensive literature showing the adverse effects of acute exposure of low doses of BPA in experimental animals [14,15]. Epidemiological studies had found associations between blood levels of BPA in women and impaired health, including endometrial hyperplasia and obesity [16]. BPA had been shown to have adverse health effects, including secondary sexual developmentalchanges and neurobehavioral alterations, in fetal through early childhood development [17]. Elevated exposure of pregnant women and children is of particular concern because of known windows of vulnerability to BPA that put the developing fetus and children at higher risk, compared with adults exposed to the same levels of BPA [14,18].

The impacts of phthalate exposure on human health have also been extensively reviewed and reported by the National Toxicology Program-Center for the Evaluation of Risks to Human Reproduction [19]. There is sufficient evidence in rodents that phthalate exposure causes developmental and reproductive toxicities. In humans, dysmorphic disorders of the genital tract, observed in male infants, were significantly associated with prenatal exposure to phthalates [20]. DBP/BBP/MBP were shown to have profound effects on the male reproductive development if exposure occurred during the critical periods of sexual differentiation (*i.e.*, late in the gestation). The phenotypic alterations observed in male offspring rats exposed to DBP/BBP/MBP during the perinatal period had remarkable similarities with common humanreproductive disorders, including cryptorchidism, hypospadias and low sperm counts [21]. The antiandrogenic activities of phthalate mixtures and bisphenol A display additive interactions. They show a tendency to synergistic activities at high and antagonistic activities at low concentrations [22].

Biomonitoring of BPA through human blood and/or urine testing may underestimate the total body burden of this potential toxicant. Sweat analysis should be considered as an additional method for monitoring bioaccumulation of BPA in humans. Induced sweating appears to be a potential method for elimination of BPA [23].

TOXICOGENOMICS OF BISPHENOL A AND PHTHALATES

In the Comparative Toxicogenomics Database [24], BPA and the five most frequently curated phthalates (DEHP/MEHP and DBP/BBP/MBP) were found to have 1232 and 265 interactions with unique genes/proteins, respectively [25,26]. The GeneGo pathway maps, GeneGo processes, GeneGo toxicity networks and GeneGo diseases of the 1232 unique genes/proteins interacting with BPA were compared using MetaCore with those of the 265 unique genes/proteins interacting with five phthalates. BPA and phthalates were found to exhibit similar toxicogenomics, as well as adverse effects on human health, owing to their 89 common interacting genes/proteins. All of the top ten GeneGo pathway maps with highest probabilities were from the 89 common genes/proteins interacting with both BPA and phthalates, while those interacting with either BPA- or phthalate-specific genes/proteins had lower and little probabilities. All top 10 BPA- and phthalate-specific GeneGo processes were similar to those of the 89 common genes/proteins. It is of importance that five of the top 10 GeneGo toxicity networks predicted by the 89 common genes/proteins were involved in inflammation, because many chronic human diseases are due to immune and inflammatory dysfunctions [27]. It is also of interest that six of the top 10 GeneGo diseases were urogenital, prostatic, male genital, female genital, endometrial, and breast neoplasms. The diseases and disorders, as well as molecular and cellular functions, and physiological system development and functions, of the 89 common genes/proteins interacting with both BPA and phthalates were further analyzed using IPA, and cancer, developmental disorder and reproductive diseases were found to be the top three categories. Finally, these 89 genes/proteins may serve as biomarkers to assay the toxicities of environmental chemicals BPA and phthalates leached out from the widely used plastics.

EPIGENETIC EFFECTS OF BISPHENOL A AND PHTHALATES

Bisphenol A (BPA) and phthalates (DEHP/MEHP and DBP/BBP/MBP) are epigenetically toxic (Figure 1). The epigenetic effect of BPA was clearly demonstrated in viable yellow mice [28]. The maternal exposure to BPA shifted the coat color distribution of viable yellow mouse offspring toward yellow by decreasing CpG methylation in the IAP retrotransposable sequenceinserted upstream of the Agouti gene. Interestingly, this effect on DNA methylation and the associated change in coat color of the exposed animals were prevented by maternal dietary supplementation with a source of methyl group such as folic acid or the phytoestrogen genistein [29].

The *in utero* and neonatal exposure to low doses of bisphenol A (BPA) and/or phthalates (DEHP/MEHP and BBP/DBP/MBP) may cause DNA hypermethylation/hypomethylation at CpG islands near gene promoter regions, histone modifications (acetylation, methylation, phosphorylation, ubiquitination, sumoylation and ADP ribosylation), and expression of non-coding RNAs, including microRNAs. These epigenetic marks can induce up/down alterations in gene expression that may persist throughout a lifetime. These permanent changes will result in adverse health effects such as neural and immune disorders, infertility, and late-onset complex diseases (cancers and diabetes). The transient exposure

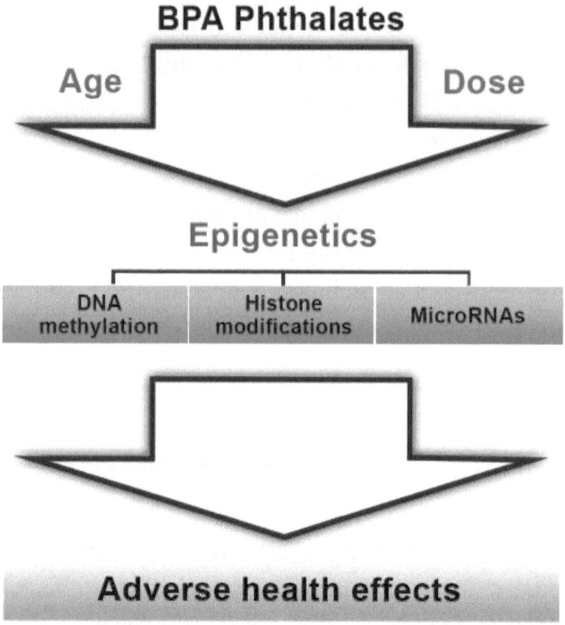

FIGURE 1. Epigenetic mechanisms of bisphenol A and phthalates.

to BPA and phthalates of gestating female rats was further shown to be a transgenerationally differential DNA methylation of the F3 generation.

Exposure to endocrine disrupting chemicals such as BPA and phthalates is of particular concern in the context of development. Neonatal exposure of rats to BPA resulted in an increased incidence of prostate intraepithelial neoplasia, and the prostate tissues showed consistent methylation changes. For example, the phosphodiesterase type 4 variant 4 (Pde4d4) gene of the rat was found to have hypomethylation in the regulatory CpG island and an elevated expression in the adult prostate [30,31]. Neonatal exposure of the rat to BPA was also reported to alter the promoter methylation and expression of nucleosome binding protein-1 (Nsbp1) and hippocalcin-like 1 (Hpcal1) genes [32]. The neonatal exposure to BPA was shown to induce hypermethylation of estrogen receptor promoter regions in rat testis, indicating methylation mediated epigenetic changes as one of the possible mechanisms of BPA induced adverse effects on spermatogenesis and fertility [33].

BPA has been shown to alter the methylation status of the *Hoxa10* gene in mouse *in utero* exposure model [34]. The *in utero* BPA treatment increased the expression of the developmental homeobox gene *Hoxa10* in the uterus of female offspring at two weeks of age. This change in gene expression was associated with significant demethylation of specific CpG sites in both promoter and intron of the *Hoxa10* gene. Genome-wide effects of BPA on DNA methylation in brain tissue have also been investigated. Maternal exposure to BPA was associated with either hypo- or hyper-methylation of the promoter-associated CpG islands in several loci in the fetal mouse brain [35]. Gene-specific changes were confirmed at 13 loci, and changes in DNA methylation state of two genes, encoding transport-related proteins, were associated with altered gene expression profiles. Exposure of human primary breast epithelial cells to low-dose BPA was reported to increase DNA methylation at CpG islands of lysosomal-associated membrane protein 3 (*LAMP3*) gene and repress the expression of *LAMP3* gene [36].

BPA effects on histone modifications were found to increase expression of the histone methyltransferase Enhancer of Zeste Homolog 2 (*EZH2*) level in human breast cancer MCF7 cells and mammary glands of six-week-old mice exposed to BPA *in utero* [37]. Both *in intro* and *in vivo*, these changes were accomplished by an increase in histone H3 trimethylation at lysine 27, which is the main histone modification catalyzed by EZH2 and is typically associated with gene expression [38].

Concerning microRNAs (miRNAs), BPA exposure of human placental cell lines has been shown to alter miRNA expression levels, and specifically, miR-146a was strongly induced by BPA treatment. This resulted in both slower proliferation rate and higher sensitivity to the DNA damaging agent bleomycin [39]. A mouse sertoli cell line TM4 exposed to BPA for 24 h was reported to have two-fold up or down-regulated 37 miRNAs, and most of miRNAs were down-regulated over the course of BPA treatment [40].

As to phthalates, treatment of human breast cancer MCF7 cells with BBP led to the demethylation of estrogen receptor (ESR1) promoter-associated CpG islands, indicating that altered *ESR1* mRNA expression by

BBP is related to aberrant DNA methylation in the promoter region of the receptor gene [41]. Maternal exposure to DEHP was shown to increase DNA methylation and expression levels of DNA methyltransferases in mouse testis. Fetal testis was a main target for DEHP as evidenced in testicular dysgenesis syndrome due to a reduction in insulin-like hormone 3 (*INSL3*) expression and testosterone production [42].

Molecular mechanisms that underlie the long-lasting effects of BPA and phthalates continue to be elucidated, and they likely involve disruption of epigenetic programming of gene expression during development. It will be important to determine whether epigenetic markers in more accessible tissues correlate with epigenetic markers in target tissues. Many studies strongly imply that exposures to endocrine-disrupting chemicals (EDCs) may have cumulative adverse effects on future generations, and that these effects could be mediated through epigenetic mechanisms [43].

Finally, the transient exposure to a plastic mixture (BPA and phthalates) of gestating female rats during the period of embryonic sex determination was shown to promote early-onset female puberty transgenerationally (F3 generation) and decrease the pool size of ovarian primordial follicles. Spermatogenic cell apoptosis was also affected transgenerationally, and differential DNA methylation of the F3 generation sperm promoter regions was found in all exposed lineage males [44].

CONCLUSION AND REMARKS

The hypomethylation of the mouse Agouti gene caused by exposure to BPA can be prevented by maternal dietary supplementation with a source of methyl group [29]. However, it remains to be investigated if any bioaccumulation of epigenetic impacts can be reversed/eliminated after exposure to BPA and phthalates is discontinued. The differential DNA methylation was reported to be transgenerational after exposure of gestating female rats to mixture of BPA and phthalate, but the synergistic impact of both BPA and phthalate remain to be determined.

The growing evidence indicates that epigenetics holds substantial potential for developing biological markers to predict which chemicals would put exposed subjects at risk and which individuals would be more susceptible to developing disease. It is still important to note that the mechanisms by which environmental toxicants modulate the epigenetic landscape of individual cells are yet to be elucidated in order to better understand the biology of epigenetic alterations and the health effects of toxic exposures on these disease-associated epigenetic alterations. Better defined mechanisms will lead to better prediction of the toxic potential of environmental chemicals such as BPA and phthalates and allow for more targeted and appropriate disease prevention strategies.

In human studies, the use of laboratory methods with enhanced precision, sensitivity and coverage will be required, so that epigenetic changes can be detected as early as possible and well ahead of disease diagnosis. New technologies available now allow for global analysis of epigenetic alterations and these may provide insight into the extent and patterns of alterations between human normal and diseased tissues. Appropriate *in vitro* models must be considered. In this context, human embryonic stem cells may be extremely useful in bettering the understanding of epigenetic effects on human development, health and disease, because the formation of embryoid bodies *in vitro* is very similar to the early stage of embryogenesis [45,46].

REFERENCES

1. Halden, R. U. Plastics and health risks. *Annu. Rev. Public Health* **2010**, *31*, 179–194.
2. Heudorf, U., Mersch-Sundermann, V., Angerer, J. Phthalates: Toxicology and exposure. *Int. J. Hyg. Environ. Health* **2007**, *210*, 623–634.
3. Wormuth, M., Scheringer, M., Vollenweider, M., Hungerbühler, K. What are the sources of exposure to eight frequently used phthalic acid esters in Europeans? *Risk Anal.* **2006**, *26*, 803–824.
4. Thomas, J. A., Thomas, M. J., Gangolli, S. D. Biological effects of di-(2-ethylhexyl) phthalate and other phthalic acid esters. *Crit. Rev. Toxicol.* **1984**, *13*, 283–317.
5. Dolinoy, D. C., Jirtle, R. L. Environmental epigenomics in human health and disease. *Environ. Mol. Mutagen.* **2008**, *49*, 4–8.
6. Anway, M. D., Rekow, S. S., Skinner, M. K. Transgenerational epigenetic programming of the embryonic testis transcriptome. *Genomics* **2008**, *91*, 30–40.
7. Schwartz, D., Collins, F. Environmental biology and human disease. *Science* **2007**, *316*, 695–696.
8. Perera, F., Herbstman, J. Prenatal environmental exposures, epigenetics, and disease. *Reprod. Toxicol.* **2011**, *31*, 363–373.

9. Liu, L., Li, Y., Tollefsbol, T. O. Gene-environment interactions and epigenetic basis of human diseases. *Curr. Issues Mol. Biol.* **2008**, *10*, 25–36.
10. Choudhuri, S., Cui, Y., Klaassen, C. D. Molecular targets of epigenetic regulation and effectors of environmental influences. *Toxicol. Appl. Pharmacol.* **2010**, *245*, 378–393.
11. Kundakovic, M., Champagne, F. A. Epigenetic perspective on the developmental effects of bisphenol A. *Brain Behav. Immun.* **2011**, *25*, 1084–1093.
12. Baccarelli, A., Bollati, V. Epigenetics and environmental chemicals. *Curr. Opin. Pediatr.* **2009**, *21*, 243–251.
13. Shelby, M. D. NTP-CERHR monograph on the potential human reproductive and developmental effects of bisphenol A. *NTP CERHR MON.* **2008**, *22*, v, vii–ix, 1–64 passim.
14. Talsness, C. E., Andrade, A. J. M., Kuriyama, S. N., Taylor, J. A., vom Saal, F. S. Components of plastic: Experimental studies in animals and relevance for human health. *Phil. Trans. Biol. Sci.* **2009**, *364*, 2079–2096.
15. Rubin, B. S. Bisphenol A: An endocrine disruptor with widespread exposure and multiple effects. *J. Steroid Biochem. Mol. Biol.* **2011**, *127*, 27–34.
16. Lang, I. A., Galloway, T. S., Scarlett, A., Henley, W. E., Depledge, M., Wallace, R. B., Melzer, D. Association of urinary bisphenol a concentration with medical disorders and laboratory abnormalities in adults. *J. Am. Med. Assoc.* **2008**, *300*, 1303–1310.
17. Welshons, W. V., Nagel, S. C., vom Saal, F. S. Large effects from small exposures. III. Endocrine mechanisms mediating effects of bisphenol A at levels of human exposure. *Endocrinology* **2006**, *147*, s56–s69.
18. Golub, M. S., Wu, K. L., Kaufman, F. L., Li, L. -H., Moran-Messen, F., Zeise, L., Alexeeff, G. V., Donald, J. M. Bisphenol A: Developmental toxicity from early prenatal exposurea. *Birth. Defects Res. B Dev. Reprod. Toxicol.* **2010**, *89*, 441–466.
19. Shelby, M. D. NTP-CERHR monograph on the potential human reproductive and developmental effects of di-(2-ethylhexyl) phthalate (DEHP). *NTP CERHR MON.* **2006**, *18*, v, vii-7, II-iii-xiii passim.
20. Swan, S. H., Main, K. M., Liu, F., Stewart, S. L., Kruse, R. L., Calafat, A. M., Mao, C. S., Redmon, J. B., Ternand, C. L., Sullivan, S., *et al.* Decrease in anogenital distance among male infants with prenatal phthalate exposure. *Environ. Health Perspect.* **2005**, *113*, 1056–1061.
21. Martino-Andrade, A. J., Chahoud, I. Reproductive toxicity of phthalate esters. *Mol. Nutr. Food Res.* **2010**, *54*, 148–157.
22. Christen, V., Crettaz, P., Oberli-Schrämmli, A., Fent, K. Antiandrogenic activity of phthalate mixtures: Validity of concentration addition. *Toxicol. Appl. Pharmacol.* **2012**, *259*, 169–176.
23. Genuis, S. J., Beesoon, S., Birkholz, D., Lobo, R. A. Human excretion of bisphenol A: Blood, urine, and sweat (BUS) study. *J. Environ. Public Health* **2012**, *2012*, 185731.
24. Davis, A. P., Murphy, C. G., Saraceni-Richards, C. A., Rosenstein, M. C., Wiegers, T. C., Mattingly, C. J. Comparative toxicogenomics database: A knowledgebase and discovery tool for chemical-gene-disease networks. *Nucleic Acids Res.* **2009**, *37*, D786–D792.
25. Singh, S., Li, S. S. -L. Phthalates: Toxicogenomics and inferred human diseases. *Genomics* **2011**, *97*, 148–157.

26. Singh, S., Li, S. S. -L. Bisphenol A and phthalates exhibit similar toxicogenomics and health effects. *Gene* **2012**, *494*, 85–91.
27. Dietert, R. R., DeWitt, J. C., Germolec, D. R., Zelikoff, J. T. Breaking patterns of environmentally influenced disease for health risk reduction: Immune perspectives. *Environ. Health Perspect.* **2010**, *118*, 1091–1099.
28. Morgan, H. D., Sutherland, H. G. E., Martin, D. I. K., Whitelaw, E. Epigenetic inheritance at the agouti locus in the mouse. *Nat. Genet.* **1999**, *23*, 314–318.
29. Dolinoy, D. C., Huang, D., Jirtle, R. L. Maternal nutrient supplementation counteracts bisphenol A-induced DNA hypomethylation in early development. *Proc. Natl. Acad. Sci. USA* **2007**, *104*, 13056–13061.
30. Ho, S. -M., Tang, W. -Y., Belmonte de Frausto, J., Prins, G. S. Developmental exposure to estradiol and bisphenol a increases susceptibility to prostate carcinogenesis and epigenetically regulates phosphodiesterase type 4 variant 4. *Cancer Res.* **2006**, *66*, 5624–5632.
31. Prins, G. S., Tang, W. -Y., Belmonte, J., Ho, S. -M. Developmental exposure to bisphenol a increases prostate cancer susceptibility in adult rats: Epigenetic mode of action is implicated. *Fertil. Steril.* **2008**, *89*, e41.
32. Tang, W. -Y., Morey, L. M., Cheung, Y. Y., Birch, L., Prins, G. S., Ho, S. -M. Neonatal exposure to estradiol/bisphenol A alters promoter methylation and expression of *Nsbp1* and *Hpcal1* genes and transcriptional programs of Dnmt3a/b and Mbd2/4 in the rat prostate gland throughout life. *Endocrinology* **2012**, *153*, 42–55.
33. Doshi, T., Mehta, S. S., Dighe, V., Balasinor, N., Vanage, G. Hypermethylation of estrogen receptor promoter region in adult testis of rats exposed neonatally to bisphenol A. *Toxicology* **2011**, *289*, 74–82.
34. Bromer, J. G., Zhou, Y., Taylor, M. B., Doherty, L., Taylor, H. S. Bisphenol—A exposure *in utero* leads to epigenetic alterations in the developmental programming of uterine estrogen response. *FASEB J.* **2010**, *24*, 2273–2280.
35. Yaoi, T., Itoh, K., Nakamura, K., Ogi, H., Fujiwara, Y., Fushiki, S. Genome-wide analysis of epigenomic alterations in fetal mouse forebrain after exposure to low doses of bisphenol A. *Biochem. Biophys. Res. Comm.* **2008**, *376*, 563–567.
36. Weng, Y. -I., Hsu, P. -Y., Liyanarachchi, S., Liu, J., Deatherage, D. E., Huang, Y. -W., Zuo, T., Rodriguez, B., Lin, C. -H., Cheng, A. -L., *et al.* Epigenetic influences of low-dose bisphenol A in primary human breast epithelial cells. *Toxicol. Appl. Pharmacol.* **2010**, *248*, 111–121.
37. Doherty, L., Bromer, J., Zhou, Y., Aldad, T., Taylor, H. *In utero* exposure to diethylstilbestrol (DES) or bisphenol-A (BPA) increases EZH2 expression in the mammary gland: An epigenetic mechanism linking endocrine disruptors to breast cancer. *Horm. Cancer* **2010**, *1*, 146–155.
38. Vire, E., Brenner, C., Deplus, R., Blanchon, L., Fraga, M., Didelot, C., Morey, L., van Eynde, A., Bernard, D., Vanderwinden, J. -M., *et al.* The polycomb group protein EZH2 directly controls DNA methylation. *Nature* **2006**, *439*, 871–874.
39. Avissar-Whiting, M., Veiga, K. R., Uhl, K. M., Maccani, M. A., Gagne, L. A., Moen, E. L., Marsit, C. J. Bisphenol A exposure leads to specific microRNA alterations in placental cells. *Reprod. Toxicol.* **2010**, *29*, 401–406.

40. Cho, H., Kim, S., Park, H. -W., Oh, M. -J., Yu, S., Lee, S., Park, C., Han, J., Oh, J. -H., Hwang, S., *et al.* A relationship between miRNA and gene expression in the mouse sertoli cell line after exposure to bisphenol A. *BioChip J.* **2010**, *4*, 75–81.
41. Chan Kang, S., Mu Lee, B. DNA methylation of estrogen receptor α gene by phthalates. *J. Toxicol. Environ. Health A* **2005**, *68*, 1995–2003.
42. Wu, S., Zhu, J., Li, Y., Lin, T., Gan, L., Yuan, X., Xu, M., Wei, G. Dynamic effect of di-2-(ethylhexyl) phthalate on testicular toxicity: Epigenetic changes and their impact on gene expression. *Int. J. Toxicol.* **2010**, *29*, 193–200.
43. Kundakovic, M., Champagne, F. A. Epigenetic perspective on the developmental effects of bisphenol A. *Brain Behav. Immun.* **2011**, *25*, 1084–1093.
44. Manikkam,M.,Guerrero-Bosagna,C.,Tracey,R.,Haque,M. M.,Skinner,M. K. Transgenerational actions of environmental compounds on reproductive disease and identification of epigenetic biomarkers of ancestral exposures. *PLoS One* **2012**, *7*, e31901.
45. Li, S. S. -L., Liu, Y. H., Tseng, C. N., Chung, T. L., Lee, T. Y., Singh, S. Characterization and gene expression profiling of five new human embryonic stem cell lines derived in Taiwan. *Stem Cells Dev.* **2006**, *15*, 532–555.
46. Chen, B. Z., Yu, S. L., Singh, S., Kao, L. P., Tsai, Z. Y., Yang, P. C., Chen, B. H., Li, S. S. -L. Identification of microRNAs expressed highly in pancreatic islet-like cell clusters differentiated from human embryonic stem cells. *Cell Biol. Int.* **2011**, *35*, 29–37.

CHAPTER 11

THE INFLUENCE OF DNA SEQUENCE ON EPIGENOME-INDUCED PATHOLOGIES

RICHARD B. MEAGHER and KRISTOFER J. MÜSSAR

REVIEW

CAUSE-AND-EFFECT AND EPIGENETIC RISK

The inheritance of numerous genetic risk factors for human and plant diseases as well as biotic and abiotic stress susceptibility phenotypes are well established [1-6]. Particular DNA mutations and their mechanistic effect on the timing, level, or quality of gene expression produce the risk of disease. Thus, a clear cause-and-effect relationship is established between the inherited aberrant genotype and the risk phenotype (that is, the increased chance or certainty of presenting a disease).

Epigenetics is cited as contributing to the risk of acquiring numerous diseases and aberrant phenotypes in human and plant populations based primarily on correlations between changes in chromatin structure and penetrance of the undesired phenotype [7-10]. There has been a growing suspicion, particularly since the 1980s, that - along with classical genetics - epigenetics is required to explain many complex phenotypes associated with disease [11,12]. The influences of age and environment (for example, chemicals, heat, nutrition, daylight) on various pathologies and the seemingly stochastic penetrance of developmental abnormalities are particularly difficult to interpret using purely molecular genetic models and are more easily explained by considering epigenetic control mechanisms [13-18]. However, few cause-and-effect relationships have been

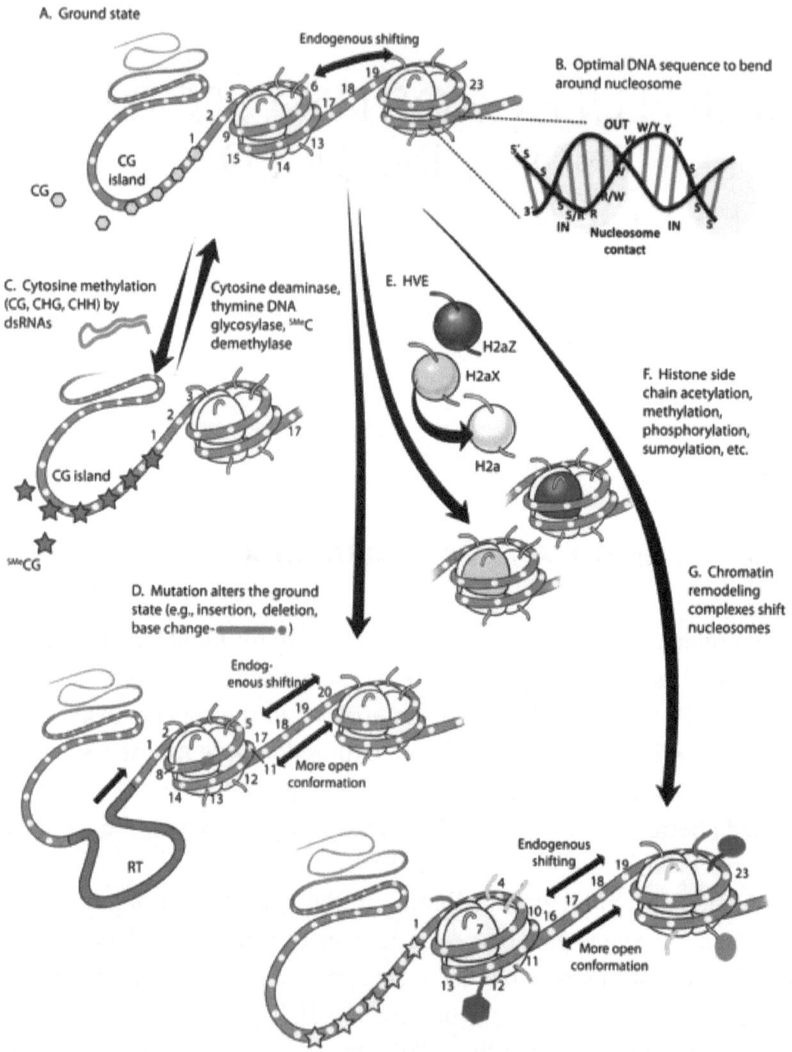

FIGURE 1. Summary of relationship between epitype and DNA sequence. A. Theoretical ground state for a chromatin structure comprised of naked DNA bound to two nucleosomes and an unbound upstream DNA region. Every 10 bp the approximately 2 bp of inward facing surface of the DNA helix has the potential to contact and bind nucleosomal histones (for example, yellow ovals numbered 1 to 23 for region surrounding one nucleosome, see **B**). Each nucleosome has the potential to bind 14 such 2 bp regions. **B**. One 10 bp region of the DNA helix with the consensus ((Y)RRRRRYYYYY(R) provides a bend for optimal nucleosome binding. Nucleotides that provide strong or weak

Figure 1 continued...
nucleosome binding are indicated (S=strong binding to G or C nucleotides, W=weak binding to A or T nucleotides, R=purine, Y=pyrimidine, IN identifies the surface facing the nucleosome, and OUT the surface facing away from the nucleosome). The strength of nucleosome binding and positioning to 147 bp stretches of DNA appears to be determined by the sum of affinities for 14 small sequences (yellow ovals, same as in **A**). **C**. Double stranded (ds) RNAs (for example, siRNA, piRNA, miRNA) program cytosine methylation for transgenerational inheritance and somatic inheritance in different tissues, while various enzymes remove 5MeC. **D**. Mutations such as single nucleotide polymorphisms (SNPs, red dot) and inserted retrotransposons (RT, red line) may alter nucleosome binding and the stochastic movement of nucleosomes. **E**. Histone variant exchange (HVE) by a subset of chromatin remodeling complexes (for example, SWR1) replaces common core histones (for example, H2A) with specialized protein sequence variants (for example, H2AZ, H2AX). **F**. A variety of histone post-translational modifications (PTMs) of primarily lysine and arginine residues at the N- and C-termini of core histones produce a diverse "histone code" for different nucleosomes. **G**. A large number of chromatin remodeling machines (for example, SWI/SNF, INO80) control nucleosome positioning, often moving nucleosomes in approximately 10 bp increments. Not shown is that the individual epitypes interact with each other to produce complex epitypes. For example, a subset of individuals may contain in their genome a retrotransposons that is targeted by small RNAs, which cause the hypermethylation or hypomethylation of adjacent sequences and alters gene expression (that is, the interaction of **C** and **D**).

established that prove that that particular inherited cis-linked chromatin structures (epitypes) are in fact useful in predicting the inherited risk of acquiring disease phenotypes. Exceptions are the epigenetic silencing of the *skeletal-muscle ryanodine-receptor gene (RYR1)* that causes congenital myopathies and the *MutL Homolog 1 gene (MLH1)* that causes increased risk of colorectal or endometrial tumors, which are discussed in the following section.

Inherited risk epitypes should evolve in populations in ways similar to the evolution of genotypes [19]. The problem is that the transgenerational inheritance of epigenetic controls is not well understood in any multicellular organism and often difficult to prove. This is particularly true in humans and agricultural crops, where the need for understanding epigenetic risk is the greatest [20-27]. Without knowledge about the molecular basis for the transgenerational inheritance or generational reprogramming of defined epigenetic risk factors that contribute to disease, it is difficult to design effective targeted therapeutics for humans or to knowledgably alter breeding programs for crops that will avoid the onset of a disease phenotype [28-32].

This study explores the cause-and-effect relationships among genotype, epitype and phenotype, where the epitype of a single gene or an entire genome is defined as its various cis-linked chromatin structures (Figure 1) [19]. Thus, epitype includes - but is not limited to - chromatin domain structures, such as large DNA loops, the position of all nucleosomes and of subclasses of nucleosomes with particular histone variant compositions (for example, H2A or H2AZ or H2AX), DNA cytosine methylation, and a myriad of histone post-translational modifications (PTMs) [33-35]. By focusing on epitype, we eliminate from consideration several other classes of epigenetic controls such as cell-to-cell communication by morphogens or the inheritance of cell surface patterning [36-40]. Addressing these other epigenetic controls would distract this discussion from a focus on the transgenerational inheritance of chromatin structures.

A working hypothesis that emerged from a preliminary examination of the inheritance and evolution of various epitypes [19] is that "genotype predisposes epitype" for most transgenerationally inherited chromatin structures. Only epitypes that are transgenerationally inherited at significant frequencies may contribute to the primary cause of inherited epigenetic risk. Within this hypothesis, epitype and the machinery that alters epitype are modifiers of the central dogma of molecular biology (DNA→RNA→Protein) influencing the activity of DNA and RNA, as shown in Figure 2A. In addition, we will discuss how particular DNA and RNA sequences strongly influence the penetrance of some epitypes and resulting phenotypes. By this view transgenerationally inherited epitypes are not acting at a higher level than or independent of DNA sequence in determining phenotype (for example, RNA and protein expression, disease phenotype).

It will be useful at this point to make the distinction between the transgenerational epigenetic inheritance among parents and offspring and the somatic inheritance between mother and daughter cells within developing tissues and organs [20,23,26,41-45]. The inheritance of epitypes between dividing somatic cells, such as the transmission of a histone PTM [46], is undoubtedly essential to tissue and organ development [47-49] and may be subject to various environmental influences that reveal a phenotype [50]; however, inheritance among somatic cells need not contribute causally to epigenetic inheritance across organismal generations. Again,

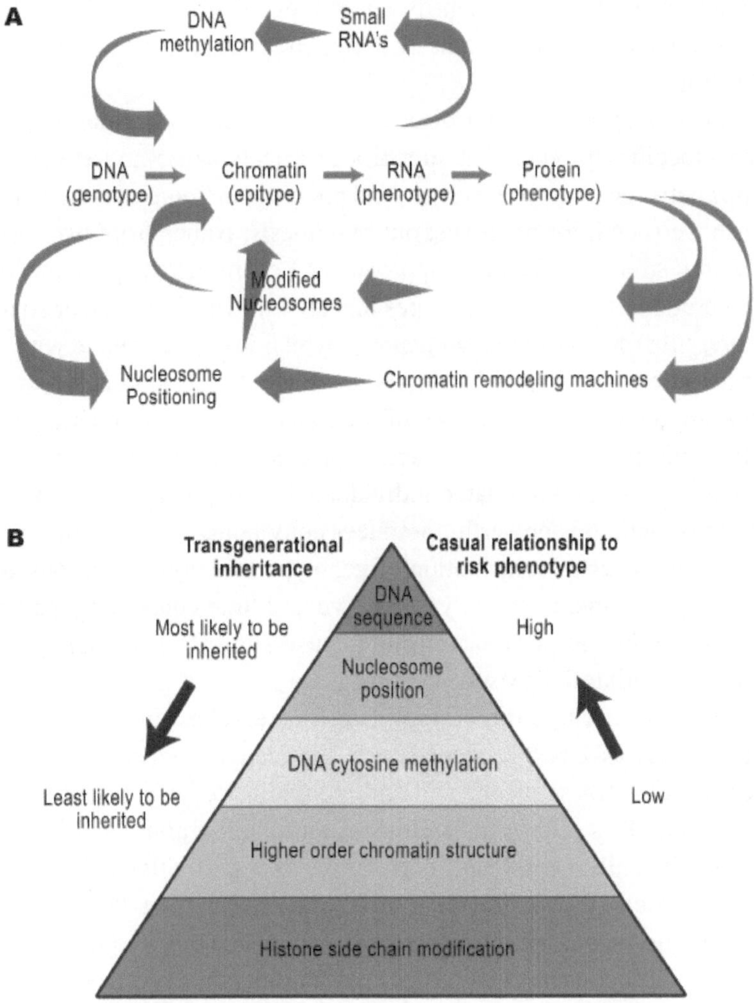

FIGURE 2. The relationships among genotype, epitype, and phenotype. A. The informational relationship and interaction of genotype, epitype and phenotype described in the context of the central dogma of molecular genetics. **B**. A pyramid illustrating the likelihood of different classes of epitypes being transgenerationally inherited and ranking the relative causal relationships of these epitypes to the risk of an aberrant phenotype.

we are interested herein in identifying epitypes that may be the primary cause of transgenerationally inherited epigenetic risk of acquiring a disease phenotype.

To test this hypothesis our discussion is focused on finding evidence for gene-specific epitypes that supports or rejects cause-and-effect relationships between genotype, epitype and phenotype. Some of the strongest evidence we found, for or against our hypothesis, comes from two different research strategies. The first approach (A) examines the penetrance of transgenerationally inherited epitypes that are known to activate or silence the expression of disease related gene(s), which in turn correlate with onset of the aberrant "disease" phenotypes. This direct approach requires the measurement of the frequency of the transgenerational inheritance of causative epitype(s), the relevant gene expression pattern(s), and the aberrant phenotype(s) among related individuals in a population known to be at risk. This method is powerful, produces convincing data, and in several cases reveals the clear contribution of genotype to epitype. But transgenerational measurements are very expensive and time consuming, particularly in the early stages of establishing cause-and-effect relationships to human or agricultural diseases.

The second and less direct approach (B) searches out epitypes that are duplicated, as DNA sequences are duplicated, and examines multiple copies of DNA sequence and epitype that have been evolutionarily co-conserved. This approach establishes an unambiguous, and in many cases a statistically significant, correlation of a particular epitype with highly reiterated DNA sequence motifs, and/or examines the conservation of an epitype among duplicated gene sequences. With this strategy, the evolutionary conservation of epitypes among conserved sequences is used as a filter to identify epitypes that were transgenerationally inherited [19,51]. In other words, those epitypes that are widely conserved in their sequence position across the genome or may be shown to evolve by gene duplication within a gene family have almost certainly been inherited through past generations. Again, only epitypes that are transgenerationally inherited have the potential to contribute causally to inherited risk. This second approach simplifies analyses, because the initial screening for likely transgenerationally inherited epitypes may be made within a single genome and in one generation. Conversely, an epitype that is not inherited after gene duplication is less likely to be closely and causally related to phenotype, even if its presence in an allele correlates well with the disease phenotype. Hence, epitypes not inherited via DNA sequence duplications are likely to be poor predictors of inhereted epigenetic risk. The disadvantage of this second genome-centered approach is precisely that it is not focused on finding associated risk phenotypes and during the early stages of analysis we are frequently left with very large datasets describing relationships among epitypes and genotypes without yet knowing correlated pathologies.

DIRECT MEASUREMENT OF TRANSGENERATIONAL EPIGENETIC INHERITANCE

Only a handful of studies have succeeded in fully demonstrating that the transgenerational transmission of an epitype produces changes in known target gene expression, which results in a disease or its risk of penetrance (that is, a causal relationship between genotype, epitype, and risk). Two of the best examples from humans concern chromatin structure at the *RYR1* and *MLH1*genes, resulting in muscle myopathies and cancer, respectively. However, the complexity of the data on these two systems highlights the problems that arise when trying to establish such cause-and-effect epigenetic relationships, particularly in humans.

(1) *RYR1:* Genetic mutations causing a loss of expression of *RYR1* function are associated with susceptibility to malignant hyperthermia and congenital myopathies (for example, central core disease, multiminicore disease) [52-54]. However, many individuals with core myopathy disease are known to be heterozygous for a mutant defective *ryr1* allele [54,55]. The epigenetic silencing of the otherwise functional *RYR1* allele appears to account for the loss of functional RYR1 protein expression. For example, among a sampling of 11 patients with the disease, six patients showed tissue-specific silencing of the maternally inherited functional *RYR1* allele, which apparently resulted from cytosine hypermethylation of that allele [56]. Treating skeletal-muscle myoblasts cultured from these patients with 5-aza-deoxycytidine, an inhibitor of cytosine methylation in newly replicated DNA, reactivates the

transcription of the epigenetically silenced, but otherwise functional allele. These data strongly support the view that hypermethylation is the primary cause of *RYR1* silencing and onset of an epigenetically determined form of the disease (Figure 2). However, the particular region(s) of DNA in which cytosine residues are methylated to cause gene silencing has not been identified in spite of intense efforts to identify it among three CG islands within the gene. This leaves open the possibility that an epigenetically controlled transacting factor is the causative agent [56]. Thus, for *RYR1* there is not yet a clear causal link between an aberrant genotype, epitype, and the silenced *RYR1* gene expression producing the disease.

(2) *MLH1:* The human *MLH1* gene encodes a homologue of the bacterial mismatch DNA repair protein MutL and, hence, *MLH1* is classified as a tumor suppressor. Hypermethylation of DNA cytosine residues and silencing of a particular functional *MLH1* alleles (for example, -93 single nucleotide polymorphism (SNP)) [57], when paired with a dysfunctional mutant allele of the same gene, correlates with relatively young individuals developing tumors of the colorectum or endometrium [27,58]. The tumors and tumor-derived cell lines from individuals with these hypermethylation epimutations fail to express MLH1 protein from this otherwise functional allele[59]. The hypermethylation of the potentially functional *MLH1* allele and its transcriptional silencing is found in most organs and tissues of individuals who also have hypermethylation of this *MLH1* allele in their tumors. Hence, one might expect that this *heritable epimutation* resulted from the transgenerational inheritance of this epitype. However, studies of the children of these individuals generally show loss of hypermethylation and loss of silencing of this *MLH1* allele in the first generation of transmission. Out of several individuals examined, only in one case was the epitype of hypermethylation and silencing inherited through the male parent to the individual with the disease. The *MHL1* silencing phenotype in females with colorectal cancer was associated with a particular CG island centered at −93 bp from the start of transcription in a particular *MHL1* allele containing a SNP, -93 SNP, in this region as illustrated for the more general case in Figure 1C,D [57]. While 5-aza-2'-deoxycytidine will reactivate the silenced allele in cultured cancer cell lines, demethylation is also correlated with a shift in nucleosome position and increased nucleosome

density in the promoter region Figure 1A,G [60]. In a very recent study, laser capture microdissection of the ovarian epithelium from ovarian tumors of cancer patients was used to analyze the cell type specific epitype and shows that the hypermethylation of MHL1 is an early somatic event in the malignant transformation of these cells [61]. Cogent to a theme of this article is the fact that the *MHL1* epitypes of aberrant nucleosome position and cytosine methylation appear to be dependent upon the genotype of the epigenetically silenced *MHL1* allele. Epimutations of other tumor suppressor genes including *MSH2, MSH6, PMS2*, and*BRCA1* have also been associated with colorectal cancers, but the cause-and-effect relationships with disease are less clear then they are for *MHL1* [62].

There are considerably more robust examples of the transgenerational epigenetic inheritance from model genetic organisms and wild plants, where it is easier to analyze aberrant epitypes and associated phenotypes through multiple generations. A few of the best cases with solid supporting evidence for a relationship between epitype and phenotype will be summarized.

(3) *AGOUTI:* In mice, the secreted AGOUTI peptide functions normally as a paracrine regulator of pigmentation. However, the dominant constitutive expression of the *AGOUTI* gene also targets changes in the hypothalamus and adipose tissues and this aberrant expression causes obesity. Hypomethylated, transcriptionally active dominant epialleles of the *agouti* gene may be maternally inherited through meiosis. Variation in the penetrance of different active epialleles generates a distribution of offspring from abnormal yellow (agouti) obese mice to darker mice with normal amounts of fat [63-65]. Several of the best characterized hypomethylated active and dominant alleles of *agouti* (*Agoutiiapy, Agoutiy, Agoutivy*) that are associated with a high penetrance of the yellow coat color and obesity phenotypes have promoter-containing retrotransposons positioned just upstream of the natural *Agouti* promoter [66,67]. For the best-studied alleles, these altered promoter structures are correlated with the hypomethylation of*agouti* and constitutive AGOUTI protein expression. However, a recent detailed examination of the DNA methylation profiles of active and silent alleles suggest that hypomethylation alone may not fully account for the complex ectopic expression of Agouti [18]. Nonetheless, the Agouti examples give reasonable support for the hypothesis

(Figure 1C,D) that *genotype predisposes epitype* and aberrant phenotype. It would not be surprising to find a shift in promoter nucleosome position resulting from the various retrotransposon insertions contributing to the causative epitype.

(4) *AXIN1-FUSED*: Axin1 is an inhibitor of Wnt (a hybrid of the names for *Wingless* and*Integration1*) signaling that regulates embryonic axis formation in deuterostome animals. In mice, Axin1 is the product of the mouse *Fused* locus. Some murine alleles of *Axin1-fused (Axin1Fu)* show variable and stochastic expression levels, where high expression of a hypomethylated allele correlates with an abnormal kinked tail. Highly penetrant *Axin1Fu* alleles contain an upstream retrotransposon or retrotransposon-mediated DNA rearrangement that alters chromatin structure and contributes to dominant transcript expression [68,69]. An active, highly penetrant mutant allele may be inherited maternally or paternally for multiple generations. Both cytosine hypomethylation and histone acetylation patterns are reported to correlate with increased *Axin1Fu* expression and risk of abnormal tail development [70-72]. The causal relationships between genotype, the DNA methylation epitype, gene expression, and the kinked tail phenotype are supported by the fact that methyl donor dietary supplementation of the mothers, a treatment known to increase DNA methylation, reduced *Axin1Fu* expression and halved the incidence of kinked tails. Conversely, treatment of mice with the histone hyperacetylation agent Trichostatin A increased *Axin1Fu* expression and the frequency of a kinked tail phenotype [72]. This same recent study examining the chromatin from blastocyst stage heterozygous *Axin1Fu/+* embryos shows that dimethylation of lysine-4 on histone H3 (H3K4Me2) as well as acetylation of lysine-9 on histone H3 (H3K9Ac) correlate with penetrant alleles [72]. By contrast, there was no correlation of blastocyst stage cytosine methylation with penetrant alleles. However, both the drug treatments and studies of development after the blastocyst stage only prove the importance of somatic epigenetic inheritance during tail development. Again, it is reasonable to propose that the presence of retrotransposon-mediated changes in DNA sequence, which are present in all the aberrantly expressed Axin1Fu alleles, is the primary cause of the transgenerational inheritance of epigenetic risk. A shift in nucleosome position in penetrant alleles could

affect downstream cytosine methylation and histone PTM, resulting in higher *Axin1Fu* gene expression and the kinked tail phenotype. By this view, genotype determines the nucleosomal epitype, which produces other aberrant hypomethylation and histone PTM epitypes, leading to increased gene expression and the novel kinked tail phenotype (Figure 1, Figure 2A).

(5) *CNR:* The tomato colorless non-ripening gene *CNR* encodes a homolog of the animal*SQUAMOSA* promoter binding protein (SPB box protein). CNR is essential to normal carotenoid biosynthesis and fruit ripening in the tomato and provides one of the best examples of a stable transgenerationally inherited epitype producing an abnormal phenotype. The natural epialleles of*CNR* in the tomato *Lycopersicon esculentum* contain 18 methylated cytosine residues (5MeCG or 5MeCHG, where H is C, A, or T) in a 286 bp contiguous region [73]. Hypermethylation of this region and silencing of the *CNR* gene leads to colorless tomatoes low in carotenoids (Figure 1C). Because the phenotype is relatively stable, these epialleles were originally mistaken as mutant alleles. The silenced *cnr* epiallele and active wild type *CNR* gene do not have any encoded DNA sequence differences for thousands of base pairs within or flanking this hypermethylated region. Thus, while there is no mutational basis for the change in epitype, the *CNR* gene is potentiated for a stochastic DNA methylation event, because it contains such a large number of strategically positioned cytosine residues in its sequence. While this example supports a link between the *CNR*gene sequence, epitype, and risk phenotype, there does not appear to be a particular genotype that predisposes the cytosine hypermethylation epitype. The significant question becomes, once the aberrant epitype is established, how is this hypermethylation epitype stably inherited through the germ line?

(6) *CYCLOIDEA*: The perennial plant in which *CYCLOIDEA* was first identified, *Linaria vulgaris*(Toadflax, Butter and Eggs), normally produces yellow and orange asymmetric flowers composed of three petals of different morphologies. "Mutant" plants are found in wild populations with aberrant abnormally symmetrical "peloric" flowers that are comprised of five evenly arrayed petals of similar morphology. Plants with these aberrant flowers were first characterized by Carl Linnaeus 260 years ago and collected as herbarium specimens [74]. The peloric floral phenotype is

produced by the hypermethylation and transcriptional silencing of the gene encoding a transcription factor CYCLOIDEA (CYC) [75]. Inheritance of the recessive peloric floral phenotype and silenced *cyc* epialalele is relatively stable, follows Mendelian segregation and, hence, appeared upon initial investigation to be a normal mutant allele. However, gene silencing always maps to a DNA polymorphic *cyc308G* allele with a single nucleotide polymorphism in an unmethylated region 308 nt downstream of the stop codon and never to the more common wild type *CYC308A* allele. Peloric individuals are homozygous recessive for the *cyc308G* allele with both copies being hypermethylated and completely silenced for RNA expression. Thus, it is reasonable to conclude that *genotype predisposes epitype*, gene silencing, and the peloric phenotype (Figure 1C,D).

(7) *Histone H3K4Me2 demethylase erases epigenetic memory in each generation:* A number of histone PTMs such as H3K4Me2 are acquired during transcription and are associated with active genes [76]. These epigenetic marks are removed at different stages in development by an H3K4Me2 demethylase, known as LSD1 in humans and SPR-5 in *Caenorhabditis elegans*(Figure 1F). Removal of the H3K4Me2 epitype prior to meiosis by SPR-5 in *Caenorhabditis elegans*is essential for maintaining an immortal germline [77,78]. Within two-dozen generations of worms acquiring the recessive null genotype these *spr-5* mutants have a brood size several-fold lower than wild type, with 70% of the worms being fully sterile. Homologs of LSD1 (SPR-5) are found throughout the four eukaryotic kingdoms and a number of these genes are known to be essential for normal organismal development [79-81]. The unmodified H3K4 epitype is essential and retention of the histone PTM causes aberrant development. However, there is as yet little evidence that this particular histone PTM epitype is normally preserved through meiosis or that genotype plays any role in determining the H3K4Me2 epitype at any particular locus.

(8) *Inheritance of quantitative epigenetic trait loci.* Two separate genome-wide epigenetic studies demonstrate that multi-generational inheritance of complex traits such as flowering-time, plant height, biomass, and bacterial pathogen resistance behave as quantitative epigenetic trait loci in*Arabidopsis thaliana* [22,82,83]. These studies used two independently derived sets of recombinant inbred lines (RILs), where one of the founding parents was a recessive null for one of two known genes necessary for

DNA cytosine methylation. For example, one study begins with a fourth generation plant homozygous defective *ddm1/ddm1* that is highly compromised in a number of phenotypic traits due to DNA hypomethylation. *DECREASED DNA METHYLATION1*(DDM1) is a Swi2/Snf2-like DNA-dependent ATPase chromatin remodeler required for most DNA cytosine methylation. The *ddm1/ddm1* line was backcrossed to wild type, and this heterozygous F1 *ddm1/DDM1* was backcrossed to wild type again and screened to obtain hundreds of separate DDM1/DDM1 lines. These lines were selfed to establish hundreds of epiallelic recombinant inbred plant lines (epiRILs) [22]. For several generations, approximately 30% of the *DDM1/DDM1* epiRILs displayed aberrant morphological phenotypes affecting flowering time and plant height, among other phenotypes. They assayed 22 epiRILs for the methylation of 11 candidate genes that are normally cytosine hypermethylated, but are hypomethylated in *ddm1*. Six alleles showed partial remethylation and five alleles were completely remethylated producing the identical complex epitype for this later gene set to wild type. Control genes that were previously unmethylated remained unmethylated.

In one particular example, Johannes and colleagues [22] followed the methylation sensitive *FWA*gene, for which the ectopic expression of the hypomethylated epiallele in *ddm1* parental plants produces strong late flowering phenotypes [84]. All of the 22 randomly selected epiRILs were now normally methylated at FWA and flowered at normal times. However, when they examined three extremely late flowering lines from among the population of hundreds of epiRILs (that is, plants that flowered after more than 48 days versus 33 days to flowering in wild type) these epiRILs were almost completely hypomethylated at *FWA* and expressed high levels of *FWA* transcripts, accounting for their phenotype. Hence, out of hundreds, only a few of the epiRILs escaped from the remethylation of *FWA,* when DDM1 was restored.

In summary, aberrant DNA methylation epitypes at many loci and the resulting changes in downstream molecular and developmental phenotypes appear to be transgenerationally inherited. Most genes regain wild type methylation patterns and phenotypes within a few generations and the restoration appears to be sequence specific. Hence, the genetic machinery necessary for the *de novo* remethylation of these completely unmethylated loci is encoded in the Arabidopsis genome and remethylation did not require hemi-methylated DNA templates to be newly inherited. These data suggest that genotype predisposes this global cytosine methylation epitype.

(9) *Reprogramming of DNA cytosine methylation by double stranded dsRNAs.* The 5′-methylation of DNA cytosine residues occurs in three sequence contexts: 5MeCG, 5MeCHG and 5MeCHH (Figure 1C). A number of DNA methyl-transferases (DMTs) are known to methylate DNA cytosine in the 5′ position. DMT1 efficiently propagates hemi-methylated symmetrical CG sequences and, hence, the somatic inheritance of islands of 5MeCG hypermethylation that may lead to gene silencing is not hard to explain. However, DNA methylation of all types is predominantly erased (that is, 80 to 90% loss of methylation) in germ line cells in the embryos of both plants and animals [85-87]. Hence, the reprogramming of CG, CHG, and CHH methylation and a mechanism for transgenerational inheritance of these epitypes has been of intense interest in recent years[88,89]. To simplify the discussion of the gene-specific DNA cytosine remethylation and subsequent inheritance of methylation, Richards [90] introduced three working categories: obligate, facilitative, and pure DNA methylation.

Epialleles in heterochromatic DNA that display obligate DNA cytosine methylation always remain methylated due to the presence of large numbers of transposable elements in various orientations producing dsRNA that promote a strong RNA interference response and adjacent target gene remethylation [91]. Genes within or closely adjacent to the centromer are good examples of obligate epialleles. Axin1Fu and AgoutiAy are typical examples of facilitative epialleles, because the presence of an upstream change in DNA sequence facilitates a seemingly stochastic epigenetic variation in methylation and phenotype. Because the wild type loci for these alleles lack an altered promoter element there is seldom any variation in the cytosine methylation epitype at the wild type loci. Pure epialleles are defined as those showing variation in cytosine methylation without a known genotypic cause and appear to be examples of *de novo* DNA cytosine methylation. If pure epialleles are truly independent of genotype, then they stand as strong evidence against our hypothesis. The well studied hypermethylation and silencing of wild-type *CNR* and *RYR1* alleles fit the definition of pure epialelles. Schmitz and colleagues [92] examined the complete methylome of 100 Arabidopsis lines propagated for 30 generations by single seed descent from a single parent. They observed that CG ↔ 5MeGC single methylation polymorphisms (SMPs) occurred at a 10,000-fold increased frequency per generation over the DNA base mutation rate, which they also measured (Figure 1D). While CG SMPs occurred primarily within gene bodies, large numbers of CHG and CHH SMPs occurred in flanking regions. Thus, novel inherited SMPs are generated at high frequencies and, if this remethylation is independent of DNA sequence, then pure epialleles are common.

One relevant question for this discussion is the following: are ostensibly pure epialleles truly independent of genotype, or are they simply facilitative epialleles for which we have not yet identified the associated cis- or trans-acting genes making dsRNAs that program inherited CG, CHG and CHH methylation epitypes? There is recent evidence supporting the latter interpretation that we now summarize.

Despite being generated through slightly different mechanisms, many classes of small RNAs (for example, siRNA, miRNA, piRNA) are known to template the remethylation of cytosine in different sequence-specific contexts (Figure 1C) for the transgenerational inheritance of gene silencing and or activation [89,93]. This general mechanism for reprogramming using different classes of small RNAs appears ancient in that it is found in all four eukaryotic kingdoms. These RNAs facilitate the remethylation of appropriate CG, GHG, and CHH sequences. But these data began to raise the question: does remethylation occur on a global genome-wide scale? To address the scope of remethylation, Teixeira and colleagues [94] examined the remethylation of numerous transposable element loci in DDM1/DDM1 epiRIL plants that had descended from an essentially unmethylated*ddm1/ddm1* plant backcrossed to wild type. Those loci that were remethylated after a few generations in the epiRILs contained cytosine rich gene sequences that were highly complementary to the sequence of siRNAs. Those loci with similar cytosine rich composition for which they could not identify complementary siRNAs remained hypomethylated. siRNAs attract RNA interference (RNAi) and DNA methylation machinery to complementary DNA sequences and thereby template sequence-related DNA methylation [95]. This shows that RNAi mechanisms are essential for the proper remethylation of much of the Arabidopsis genome. These and other data make it clear that, for a large number of repetitive elements in yeast, plants, and animals, the matching genotypes of structural genes and small RNAs predict a cytosine methylation phenotype. However, the study of Teixeira and colleagues [94] raises further questions about the biology, regulation, and timing of cytosine remethylation for both transgenerational and somatically inherited epitypes. Recent evidence suggests that in both plants and animals "nurse cells" may transfer hundreds of undefined small RNAs to adjacent egg or sperm germ cells to reprogram cytosine methylation [88,89,93]. For example, in mice in which 80% to 90% of the germline DNA methylation is erased for single copy genes at approximately day 11. 5 of embryo development (E11. 5). Remethylation of sperm DNA occurs in the embryo at approximately E16. 5 and is significantly directed by populations of 24 to 30 nt long piRNAs produced in adjacent cells in the pro-spermatogonia [96-98]. The identities of most of the plant and animal small RNAs transferred to developing germ cells are not yet known, but there is the real potential that large populations of RNAs may account for most transgenerational remethylation and perhaps even the apparent *de novo* methylation described by Schmitz and colleagues [92]. Appropriately positioned target sequences in these epialleles and thousands of expressed small RNAs would have to be inherited together for genotype to predispose the transgenerational inheritance of the global DNA methylation epitype.

(10) *Reprogramming epitype during somatic cell nuclear transfer.* In most of the above examples, genotype determines the likelihood, but not the certainty, of particular epitypes and phenotypes being displayed, because the same DNA sequence may be flexibly reprogrammed into many different chromatin conformations. It is fundamental to epigenetics that as cell types differentiate the same DNA sequence may display multiple epitypes and some epitypes may be more or less stable than others. An interesting example of a variety of epitypes descending from one genotype comes from research using somatic cell nuclear transfer (SCNT) to produce identical or genetically modified laboratory and farm animals. SCNT is achieved by transplanting a somatic cell nucleus into a functional embryonic cell capable of forming a viable organism. This technology has met with modest success, generating cloned mice, rabbits,

pigs, sheep, cows and more, but the efficiency of obtaining viable healthy offspring is low. Even if genetically modified embryos are established in surrogate mothers, developmental abnormalities and spontaneous abortions are common. A major limitation to obtaining relatively normal full-term development appears to be variations in epigenetic reprogramming of the transplanted nucleus [99-102]. The field of regenerative medicine faces similar problems with epigenetic reprogramming when trying to establish genetically altered lines of induced pluripotent stem cells by SCNT - for example, by transferring a somatic cell nucleus into an oocyte [103,104]. Without prior knowledge of the successes in producing cloned animals by SCNT, one would not necessarily expect that the new nuclear environment should correctly reprogram the donated nucleus. A known source of the reprogramming problem in the animal cloning field is that the transferred nucleus frequently loses a significant fraction of its DNA cytosine methylation and nucleosomal histone side chain methylation and acetylation relative to the more modified epitype of nuclei in native embryonic cells (Figure 1C,F) [105-109]. However, the surprising fact remains that some relatively healthy animals resembling the nuclear donor are obtained via SCNT and that genetic and epigenetic totipotency of the donor nucleus is re-established in the viable offspring. For appropriate reprogramming to take place on a genome-wide scale the donor DNA sequence must have the capacity to interact with the embryonic cellular environment and determine, albeit at low frequency, an epitype(s) compatible with full-term development. These results support the idea that during SCNT the donated DNA sequence predisposes much of its own epigenetic reprogramming.

EVOLUTIONARY CO-CONSERVATION OF DNA SEQUENCE AND CHROMATIN STRUCTURE FILTERS OUT TRANSGENERATIONALLY INHERITED EPITYPES

If genotype pedisposes epitype then a reasonable corollary is that some transgenerationally inherited chromatin structures should align with particular DNA sequence motifs and be passed on to duplicate gene copies. In this model, the range of possible epitypes for a sequence would evolve by gene duplication and mutation in parallel with genotype [19,51]. Rapidly evolving epitypes might only be conserved and identifiable among very recently duplicated genes examined among a limited number of related cell types or when examined statistically in comparisons of large numbers of aligned sequences, while slowly evolving highly conserved epitypes might be found among anciently duplicated genes and descended from a common ancestral protist sequence.

(1) *Short DNA sequence repeats such as RRRRRYYYYY determine the bending and positioning of DNA around the nucleosome.* More than 30 years ago, Trifonov and his colleagues [110,111] presented the case that gene sequence is fundamentally important to nucleosome positioning. He argued that the necessary high degree of bending of DNA as it wraps twice around and binds the nucleosome would be favored by particular 10. 5 bp repeat sequences of approximately 5 purines (R) followed by 5 pyrimidines (Y) (*RRRRRYYYYY*) (Figure 1B), or the inverse of this sequence, *YYYYYRRRRR*. He also found a good correlation for 10 bp repetitions of the dinucleotides GG, TA, TG, and TT in the modest compilation of 30,000 bp of DNA sequence from different eukaryotes available at that time. [a] Within the 10 bp motif these dinucleotides were proposed to help position nucleosomes. The statistical concept was a bit counterintuitive and slow to gain acceptance, because it was hard to reconcile the functional demands of sequences encoding proteins and regulatory regions with the proposed special sequence demands of nucleosome interaction. Recently, with access to nearly unlimited numbers of nucleosome-delimited 147 bp DNA sequence fragments and more advanced computational methods, it has become very clear that 14 repetitions of the 10. 5 bp repeat sequences *Y-RRRRRYYYYY-R* or *R-YYYYYRRRRR-Y* are statistically favored for nucleosome positioning. Regional differences in GC compositions in the genome favor particularly skewed repeats such as *T-AAAAATTTTT-A* or *C-GGGGGCCCCC-G* [112,113]. These consensus sequences are based on a statistical argument, and at the genome level any one dinucleotide such as *AA* or *GG* is seldom found in a particular position in the 147 bp repeat more than 30% of the time [114]. Because the inward facing helix of any one 147 bp of nucleosomal DNA fragment has 14 chances to contact the core of nucleosomal proteins, this mechanism requires only several correctly positioned dinucleotides contacting

the nucleosome to give sequence specificity to nucleosome positioning. Hence, there is in fact little conflict with conserved coding and regulatory sequences and the sequence constraints of nucleosome positioning. Furthermore, the most common classes of ATP-dependent chromatin remodeling machines, switch/sucrose nonfermentable (SWI/SNF) and imitation switch (ISW)2, move DNA in approximately 9 to 11 bp increments over the surface of a nucleosome, consistent with the importance of 10. 5 bp repeats in nucleosome binding [115]. These data strongly support a model where genotype predisposes possible nucleosome position epitypes. More particular support for this argument comes from examining the sequences for subsets of the nucleosomal DNA population binding nucleosomes containing histone variants H2AZ and CENH3.

(2a) *The geneome-wide positioning of H2AZ nucleosomes*. The histone variant H2AZ and likely other histone variants are inserted into assembled nucleosomes by histone variant exchange complexes (HVE) such as SWR1 (Figure 1E). Albert and colleagues [116] precisely aligned the sequences of thousands of 147 bp yeast nucleosomal DNA fragments enriched for histone variant H2AZ. Their data show conclusively that H2AZ nucleosome positioning on a genome-wide scale is strongly influenced by dinucleotide repeat patterns spaced 10 bp apart in the DNA sequence (Figure 1A,B). In particular, GC-rich dinucleotides are on the inside as the DNA helix wraps around the nucleosomal protein core, and AT-rich dinucleotides are on the outside. The preference for these nucleotide pairs at each of their 14 possible positions within any 147 bp nucleosomal fragment is only about 2 to 9%, and therefore any single nucleosomal fragment sequence is likely to vary significantly from the statistical consensus. However, it is clear that the H2AZ nucleosome position is determined by the overall pattern in the DNA sequence and, hence, H2AZ nucleosome position will be conserved following gene duplication. Similar results were obtained for genome-wide positioning of all nucleosomes from humans and Arabidopsis [117] and subsets of human nucleosomes specific to certain classes of genes [118]. In the total 147 bp nucleosomal fraction from Arabidopsis and humans, an AT-rich dinucleotide repeat is spaced every 10 bp and out of phase by 5 bp with a GC-rich dinucleotide repeat.

(2b) Further support for the concept that DNA sequence positions H2AZ nucleosomes comes from a comparison of duplicated genes in

Arabidopsis. A single peak of H2AZ enriched nucleosome(s) is found at the 5′ end of nearly half of all plant, animal, and fungal genes that have been examined [116,119-121]. In Arabidopsis, three related MADS box genes that regulate flowering time require normal H2AZ for full expression. In wild-type cells, all three MADS box genes show a striking bimodal distribution of H2AZ deposition, with peaks of H2AZ histone-containing nucleosomes at their 5′ and 3′ ends [122]. This pattern is quite distinct from the single 5′ spike of H2AZ observed for other MADS genes in humans, *Arabidopsis*, and yeast. These three genes are estimated to have diverged from a common gene ancestry in the eudicot plant lineage in the last 100 million years and stand alone in their own distinct clade, among more than 100 other MADS box genes in Arabidopsis that do not have a bimodal distribution of H2AZ nucleosomes. These data are consistent with the bimodal distribution of H2AZ being inherited following gene sequence duplication from an ancestral MADS gene [19].

(3) *The genome-wide positioning of CENH3 centromeric nucleosomes.* Recent experimental evidence demonstrates that CENH3 enriched centromeric nucleosome positions are determined by DNA sequence. Animal and plant centromeres are composed of a diverse variety of retroelements and repetitive satellites that generally appear unrelated in their DNA sequences. Numerous earlier studies of centromere and neocentromere sequences concluded that a distinct conserved DNA sequence was not essential to centromere activity. However, a very recent analysis of 100,000 centromeric histone CENH3 enriched nucleosomal DNA fragments from maize suggests that a 10 bp repeat of AA or TT dinucleotides contributes to determining the positioning of centromeric nucleosomes [123]. The CENH3 nucleosome specific sequence was not revealed until the 147 bp micrococcal nuclease protected DNA sequences were precisely aligned. The preference for AA or TT nucleotide pairs at each of the 14 positions within a typical 147 bp nucleosomal fragment was statistically significant. The likelihood of finding one of these dinucleotide pairs at any of the potential contact points ranges from 13% to 60% above the frequency at which other dinucleotides are found. Thus, CENH3 enriched nucleosomes are positioned by a variation on what is shown in Figure 1B, where the inward facing DNA base pairs that bind are generally AA or TT and would be classified as weak binding. This would indicate that any single

centromeric nucleosomal sequence may vary significantly from the statistical consensus for these nucleotide pairs. In this way, a subset of retroelements that are seemingly unrelated in sequence using standard sequence alignment methods may contain suitable sequence repeats that position centromeric nucleosomes. The human and Arabidopsis genomes each encode more than a dozen histone protein sequence variants for each of three classes of histones, H2A, H2B, and H3. Within each class a few subclass variants are easily identified as predating the divergence of plants, animals, and fungi from their more recent protist ancestors. Thus, it is reasonable to speculate that distinct DNA sequence patterns evolved in concert with each histone variant subclass to provide complex patterns of nucleosome positioning. If true, then DNA sequence would be responsible for the transgenerational positioning of most classes of nucleosomes.

(4) *Cytosine methylation in the human plasminogen gene family.* In an attempt to show that epitypes and associated phenotypes can evolve by gene duplication and divergence, Cortese and colleagues [51] compared promoter CG methylation patterns among the four duplicated gene members of the approximately 35-million-year-old human plasminogen (*PLG*) precursor gene family, encoding blood-clotting factors found only in hominids. Cytosine DNA methylation patterns are well conserved among seven CG sites located −171 to −378 nucleotides upstream from the start of transcription within all four *PLG* gene promoters (similar to Figure 1A,C). In liver, where transcripts for all four genes are expressed, one allelic copy of each gene pair is almost completely unmethylated at all seven sites. In heart muscle and in skeletal muscle, where the four *PLG* genes are turned off, nearly 100% of the seven sites are fully cytosine methylated on both alleles for all four genes. In other words, promoter cytosine methylation silences all gene copies in the two nonexpressing tissues examined, while hypomethylation of one copy of each*PLG* gene activates their expression in liver. The *PLG* data support the generational inheritance and conservation of the cytosine methylation epitype following gene duplication for recently duplicated genes that are co-expressed. Cortese and colleagues [51] also compared promoter CG methylation patterns among several members of the much older human T Box (TBX) gene family in which the most gene duplications date back 300 to 600 million years. No evidence was obtained for conserved CG methylation patterns among any pair-wise comparison

of *TBX* genes. Perhaps because the *TBX* genes are differentially expressed and the divergence events between genes are much more ancient, the lack of conserved CG methylation patterns is to be expected.

(5) *Histone side chain modifications in human segmental sequence duplications.* Barski and colleagues [124] published a ground-breaking genome-wide study on sequence specific location of 23 histone PTMs and a few other epitypes in purified human CD4+ T cells. From this dataset, Zheng [125] examined 14 distinct patterns of histone PTM in nucleosomes from 1,646 relatively recent (that is, less than approximately 25 million-year-old) segmental chromosome duplications (SDs). They found no significant evidence for the inheritance of these histone modifications between the original and derived loci. Specifically, the duplicated copy did not inherit the parental pattern of histone side chain methylation or acetylation (Figure 1F). Moreover, inheritance appears to be distinctly asymmetric for some of the modifications, such that there is a strong statistical bias toward histone methylation of one gene copy for each SD and not the other copy, beyond what might have occurred at random. Many of the asymmetrical histone modifications correlate with gene activation and repression, suggesting that active genes in the parent sequence are silenced in the duplicated loci, and *visa versa*. These data imply that histone PTM epitypes may not be the direct transgenerationally inherited "cause" of the phenotypes with which they are associated. Thus, these data on histone PTM epitypes at SDs do not support our working hypothesis. If these results are supported by more experimental studies, it will not mean that histone modifications are not useful epitypes for predicting risk, but that they may be further from the inherited cause of epigenome-induced pathologies than other epitypes such as nucleosome position and cytosine methylation. Histone PTMs are indeed important to somatic inheritance and development [46,126].

(6) *Nucleosome positioning and H3K4Me2 modifications in the HOXD cluster.* There are six genes at the *HOXD* gene cluster (that is, HOXD13, 11, 9, 8, 4, 3) covering approximately 100,000 bp on human Chromosome 2. In human sperm, there are one or two spikes of general nucleosome occupancy and H3K4Me2-enriched nucleosome occupancy within each of the promoters of these genes, whereas the approximate 100,000 bp of 5′ flanking region is relatively free of nucleosomes [127] (Figure 1A,F). Because

nucleosome positioning was performed using microarrays, the sequence specificity of H3K4Me2-enriched nucleosomes among these *HOXD*promoters cannot be determined from these data or compared to the results from Barski and colleagues [124] who did not find sequence specificity for histone H3K4Me2-enriched nucleosome binding. These results showing the conserved positioning of nucleosomes in *HOXD* promoters in human sperm are similar to those for H2AZ-enriched nucleosomes among the *FLC*-related *MADS*genes in Arabidopsis shoot tissue [122].

(7) *Higher-order chromatin structures.* Genes and regulatory sequences that are narrowly or widely spaced on a chromosome may interact productively through higher order chromatin structures such as solenoids, small and giant loops, and minibands [128-130]. For example, small concatenated DNA loops may be formed by re-association of the single strands of the poly (CA)-poly (TG) microsatellite at their base [131]. These small loops appear to impact the control of gene expression via binding to HMG-box proteins [131,132]. There is mounting evidence that interactions of distant intra- and inter-chromosomal domains provide epigenetic mechanisms to maintain specialized gene expression states [133-135]. Hence, the potential exists that higher order structures contribute to epigenetic control and are determined in part by DNA sequence.

SUMMARY FROM DIRECT AND INDIRECT ANALYSES OF EPIGENETIC INHERITANCE

An examination of several examples of the direct transgenerational inheritance of epitype and the epitypes of duplicated and/or conserved DNA sequences revealed the complexities of determining cause-and-effect relationships among genotype, epitype and phenotype. However, in balance, there are robust experimental data supporting the hypothesis that "genotype predisposes epitype," for some epitypes. In particular, it is becoming clear that a large fraction of, if not all, cytosine methylation is determined by gene sequence and the presence of paired sequence-specific complementary small RNAs that direct their transgenerational remethylation. Similarly, based on the sequences of H2AZ and CENH3 enriched nucleosomal fragments, nucleosome position appears strongly influenced by DNA sequence (Figure 1A,B,C). However, there is little evidence suggesting that DNA sequence determines the position of any of more than 20 different classes of histone PTM enriched nucleosomes (Figure 1F).

Based on this analysis, it is worth ranking the utility of various classes of epitype in estimating epigenetic risk. A risk pyramid linking the relationships of genotype and epitype with epigenetic risk phenotype is shown in Figure 2B. DNA sequence is placed at the apex, as the primary cause of inhereted epigenetic risk. This is followed by nucleosome position that appears to be directly dependent upon 10 bp repeats in DNA sequence and DNA cytosine methylation that is highly dependent upon *cis*-acting CG, CHG, and CHH sequences in the target gene and the sequence of trans-acting small RNAs. However, while histone PTM may be strongly correlated with epigenetically controlled phenotype, there is no evidence that any histone PTM is causal to transgenerationally inherited risk. Histone PTM epitypes may represent the effect of other epigenetic and genetic controls and may be principally important to somatic inheritance of epigenetic controls. The clear relationship between novel genotypes and many of the most robustly characterized inherited epitypes of nucleosome position and cytosine methylation is a recurrent theme in the literature of the most thoroughly studied genes under epigenetic control. This suggests that human and animal therapeutic treatments or plant and animal genetic breeding strategies that address harmful meiotically inherited epitypes should consider the possibility that there are genotypic causes predisposing these epitypes. If, for example, the environment of a developing somatic tissue (for example, obesity, stress, nutrients) is influencing RNA sequence directed cytosine remethylation and gene silencing, drugs targeting downstream histone PTM epitypes of that gene may be less effective than ones addressing remethylation. Strategies directed at controlling gene expression by altering histone PTM epitypes may be useful if they target the gene or genes producing the disease's phenotype. Finally, the undeniable influence of genotype on epigenetic controls leading to deleterious phenotypes has to be taken into account in a consideration of epigenetic risk, even if it confounds many current, working definitions of epigenetics.

DEFINING EPIGENETICS

We've summarized direct and indirect evidence that genotype predisposes epitype and that epigenetic controls are strongly influenced by DNA and RNA sequences (Figures 1 and 2). Our hypothesis and these supporting data may be viewed as contrary to some of the widely stated precepts of epigenetics. For example, Riggs and colleagues defined epigenetics as "*the study of mitotically and/or meiotically heritable changes in gene function that cannot be explained by changes in DNA sequence*" [34,136]. A rephrasing of this statement as "*the study of mitotically and/or meiotically inherited changes in gene function that cannot be explained by the classical central dogma of molecular genetics*" (Figure 2A) provides a working definition that is quite consistent with our deliberations. In David Nanney's seminal article describing epigenetic control systems, he states "*The term "epigenetic" is chosen to emphasize the reliance of these systems on the genetic systems*" and goes on to say "*epigenetic systems regulate the expression of the genetically determined potentialities*" [39]. Nanney's definitions of epigenetics are completely consistent with genotype predisposing inherited epitype, and with epitype modifying gene expression and risk phenotype.

THE INFLUENCE OF DNA SEQUENCE ON EPIGENOME-INDUCED PATHOLOGIES POINTS A WAY FORWARD

Understanding that genotype predetermines many inherited epitypes suggests a few useful strategies and concerns as we try to address epigenome-induced pathologies. First, we are in a better technical position than ever before to determine the influence of genotype on epitype. New rapid DNA sequencing and DNA bead array methods for identifying SNPs and 5MeC residues combined with a wide selection of treatments to chromatin (for example, ChIP, bisulfite, micrococcal nuclease) allow us to quantitatively determine the precise genome-wide sequence-specific positioning of every nucleosome, methylated cytosine residue, and dozens of distinct histone PTMs in a genome. These epitypes may be correlated with the risk of cancer, behavioral disorders, pathogen susceptibility, or the role of aging and environmental factors on risk, as examples. The lower costs of genome-wide approaches is enabling the epitypes of larger populations of humans, laboratory animals, and plants to be examined in order to identify the epigenetic causes of complex diseases such as obesity, lupus, or pathogen susceptibility [137-140]. Second, we are in a position to develop batteries of gene-specific epigenetic biomarkers for DNA methylation epitypes that are clearly associated with disease risk and may be predictive of the penetrance of pathology. For example, this is currently being done for systemic lupus erythematosus, myeloid leukemia, and breast cancer [138,141-143]. However, new technologies are needed if we are also to use nucleosome position and histone PTM epitypes as inexpensive epigenetic biomarkers for screening populations. Third, because the development of each plant and animal cell type in an organ system is under strong epigenetic control, it is essential that we examine epitypes in distinct cell types within organs. Most current epigenetic studies examine mixed cell types such as are present in whole organs and tissues (for example, blood, tumor, hypocampus, skeletal muscle, plant shoots or roots), wherein cell type-specific epitypes are blurred due to variation of epitypes among developmentally distinct cell types. For example, several orders of magnitude more statistically significant relationships were obtained between the cytosine methylation epitype of various genes with lupus when CD4+ T cells were examined as compared to the data obtained from mixed populations of white blood cells [138,144]. Technologies have been developed to access cell type-specific epitypes, including laser cell capture micro-dissection, fluorescent activated cell sorting (FACS) of dissociated fluorescently tagged cells, and the isolation of nuclei tagged in specific cell types (INTACT). These technologies enable the more precise determination of epitypes within individual cell types as has been shown for CD4+ T cells, primordial germ cells, ovarian epithelium, retinal cones, and plant root epithelial trichoblasts and atrichoblasts [61,124,145-148]. Fourth, therapeutic approaches to human epimutations that increase the risk of pathology, or plant breeding strategies to address epigenetic susceptibility to stress or disease, need to consider that molecular mechanisms may be obscurely hidden in DNA sequence motifs and/or the sequences of small RNAs that are imperfectly matched with their target genes (Figure 1). Current basic research is laying the course for using small RNAs to direct transcriptional gene silencing by promoter DNA methylation for therapeutics and crop improvement. For example, siRNA transgenes have been used for the methylation-based transcriptional silencing of the *Heparanase* gene in human cancer cells in culture [149] and to elucidate the mechanisms of small RNA-based transcriptional silencing in plants [150,151]. Unless we can develop therapeutic approaches, identifying genotypic influences on epigenetic risk may only add more diseases to the list of thousands for which we know the cause, but have no known cure. However, taking the numerous advances in epigenetics research altogether, it is reasonable to propose that during the next two decades effective therapeutic treatments will follow the dissection of the molecular mechanisms by which genotype and epitype interact to produce disease pathologies.

CONCLUSION

There is substantial evidence that altered epigenetic controls contribute to a variety of diseases ranging from cancer and developmental malformations to susceptibility to various forms of biotic and abiotic stress. We reviewed experimental genetic, epigenetic, cell biological, and biochemical data surrounding the transgenerational inheritance of several examples of well studied epigenome-induced pathologies and the contribution of conserved DNA sequence motifs to epitype. The preponderance of evidence suggests that genotypes predispose epitypes for most chromatin structures that are transgenerationally inherited and this relationship contributes to the penetrance of epigenetically controlled diseases. Genotypes influencing inherited epigenetic risk are often obscurely encoded in DNA sequence and small RNAs. Furthermore, the remethylation of DNA cytosine residues may only be reprogrammed at particular times in development and only in particular tissues such that a special effort may be required to identify and characterize these mechanisms. Some of the best characterized examples that were discussed herein suggest we are only just beginning to understand the molecular biology behind inherited epigenome-induced disorders. Finally, the paths to effective therapeutic development or to lowering epigenetic risk will be easier to trace out once we understand the mechanisms by which genotype predisposes epitype for a particular disease.

ENDNOTE

[a]Trifinov did not have nucleosome specific DNA sequence data available 30 years ago.

REFERENCES

1. Fumagalli M, Cagliani R, Riva S, Pozzoli U, Biasin M, Piacentini L, Comi GP, Bresolin N, Clerici M, Sironi M: Population *genetics* of IFIH1: ancient population structure, local selection, and implications for susceptibility to type 1 diabetes. *Mol Biol Evol* **2010**, 27:2555-2566.
2. Zhang B, Fackenthal JD, Niu Q, Huo D, Sveen WE, DeMarco T, Adebamowo CA, Ogundiran T, Olopade OI: Evidence for an ancient BRCA1 mutation in breast cancer patients of Yoruban ancestry. *Fam Cancer* **2009**, 8:15-22.
3. Hayano-Kanashiro C, Calderon-Vazquez C, Ibarra-Laclette E, Herrera-Estrella L, Simpson J:Analysis of gene expression and physiological responses in three Mexican maize landraces under drought stress and recovery irrigation. *PLoS One* **2009**, 4:e7531.
4. Zwonitzer JC, Bubeck DM, Bhattramakki D, Goodman MM, Arellano C, Balint-Kurti PJ: Use of selection with recurrent backcrossing and QTL mapping to identify loci contributing to southern leaf blight resistance in a highly resistant maize line. *Theor Appl Genet* **2009**, 118:911-925.
5. Pataky JK, Bohn MO, Lutz JD, Richter PM: Selection for quantitative trait loci associated with resistance to Stewart's wilt in sweet corn. *Phytopathology* **2008**, 98:469-474.
6. Wang G, Ellendorff U, Kemp B, Mansfield JW, Forsyth A, Mitchell K, Bastas K, Liu CM, Woods-Tör A, Zipfel C, de Wit PJ, Jones JD, Tör M, Thomma BP: A genome-wide functional investigation into the roles of receptor-like proteins in Arabidopsis. *Plant Physiol* **2008**, 147:503-517.
7. van Vliet J, Oates NA, Whitelaw E: Epigenetic mechanisms in the context of complex diseases. *Cell Mol Life Sci* **2007**, 64:1531-1538.
8. Stokes TL, Kunkel BN, Richards EJ: Epigenetic variation in Arabidopsis disease resistance. *Genes Dev* **2002**, 16:171-182.
9. Hitchins MP: Inheritance of epigenetic aberrations (constitutional epimutations) in cancer susceptibility. *Adv Genet* **2010**, 70:201-243.
10. Alvarez ME, Nota F, Cambiagno DA: Epigenetic control of plant immunity. *Mol Plant Pathol* **2010**, 11:563-576.
11. Holliday R: The inheritance of epigenetic defects. *Science* 1987, 238:163-170.
12. Nicolson GL: Cell surface molecules and tumor metastasis. Regulation of metastatic phenotypic diversity. *Exp Cell Res* 1984, 150:3-22.
13. Ferluga J: Possible organ and age-related epigenetic factors in Huntington's disease and colorectal carcinoma. *Med Hypotheses* 1989, 29:51-54.

14. Bjornsson HT, Fallin MD, Feinberg AP: An integrated epigenetic and genetic approach to common human disease. *Trends Genet* **2004**, 20:350-358.
15. Mehler MF: Epigenetic principles and mechanisms underlying nervous system functions in health and disease. *Prog Neurobiol* **2008**, 86:305-341.
16. Graff J, Mansuy IM: Epigenetic codes in cognition and behaviour. *Behav Brain Res* **2008**, 192:70-87.
17. Dolinoy DC, Huang D, Jirtle RL: Maternal nutrient supplementation counteracts bisphenol A-induced DNA hypomethylation in early development. *Proc Natl Acad Sci USA* **2007**, 104:13056-13061.
18. Cropley JE, Suter CM, Beckman KB, Martin DI: CpG methylation of a silent controlling element in the murine Avy allele is incomplete and unresponsive to methyl donor supplementation. *PLoS One* **2010**, 5:e9055.
19. Meagher RB: The evolution of epitype. *Plant Cell* **2010**, 22:1658-1666.
20. Slatkin M: Epigenetic inheritance and the missing heritability problem. *Genetics* **2009**, 182:845-850.
21. Martin C, Zhang Y: Mechanisms of epigenetic inheritance. *Curr Opin Cell Biol* **2007**, 19:266-272.
22. Johannes F, Porcher E, Teixeira FK, Saliba-Colombani V, Simon M, Agier N, Bulski A, Albuisson J, Heredia F, Audigier P, Bouchez D, Dillmann C, Guerche P, Hospital F, Colot V:Assessing the impact of transgenerational epigenetic variation on complex traits. *PLoS Genet* **2009**, 5:e1000530.
23. Jablonka E, Raz G: Transgenerational epigenetic inheritance: prevalence, mechanisms, and implications for the study of heredity and evolution. Q Rev Biol **2009**, 84:131-176.
24. Whitelaw NC, Whitelaw E: Transgenerational epigenetic inheritance in health and disease. *Curr Opin Genet Dev* **2008**, 18:273-279.
25. Xing Y, Shi S, Le L, Lee CA, Silver-Morse L, Li WX: Evidence for transgenerational transmission of epigenetic tumor susceptibility in Drosophila. *PLoS Genet* **2007**, 3:1598-1606.
26. Rakyan V, Whitelaw E: Transgenerational epigenetic inheritance. *Curr Biol* **2003**, 13:R6.
27. Morgan DK, Whitelaw E: The case for transgenerational epigenetic inheritance in humans. *Mamm Genome* **2008**, 19:394-397.
28. Febbo PG: Epigenetic events highlight the challenge of validating prognostic biomarkers during the clinical and biologic evolution of prostate cancer. *J Clin Oncol* **2009**, 27:3088-3090.
29. Haslberger A, Varga F, Karlic H: Recursive causality in evolution: a model for epigenetic mechanisms in cancer development. *Med Hypotheses* **2006**, 67:1448-1454.
30. Mills J, Hricik T, Siddiqi S, Matushansky I: Chromatin structure predicts epigenetic therapy responsiveness in sarcoma. *Mol Cancer Ther* **2011**, 10:313-324.
31. King GJ, Amoah S, Kurup S: Exploring and exploiting epigenetic variation in crops. Genome **2010**, 53:856-868.
32. Long Y, Xia W, Li R, Wang J, Shao M, Feng J, King GJ, Meng J: Epigenetic QTL mapping in Brassica napus. *Genetics* **2011**, 189:1093-1102.
33. Jullien PE, Berger F: Gamete-specific epigenetic mechanisms shape genomic imprinting. *Curr Opin Plant Biol* **2009**, 12:637-642.

34. Haig D: The (dual) origin of epi*genetics*. Cold Spring Harb Symp Quant Biol **2004**, 69:67-70.
35. Hoekenga OA, Muszynski MG, Cone KC: Developmental patterns of chromatin structure and DNA methylation responsible for epigenetic expression of a maize regulatory gene. *Genetics* **2000**, 155:1889-1902.
36. Waddington CH: Chapter 2. The Cybernetics of Development. In *The strategy of the genes: A Discussion of Some Aspects of Theoretical Biology*. Ruskin House, George Allen & Unwin LTD, London; 1957.
37. Beisson J, Sonneborn TM: Cytoplasmic inheritance of the organization of the cell cortex in Paramecium aurelia. *Proc Natl Acad Sci USA* 1965, 53:275-282.
38. Verstrepen KJ, Fink GR: Genetic and epigenetic mechanisms underlying cell-surface variability in protozoa and fungi. *Annu Rev Genet* **2009**, 43:1-24.
39. Nanney DL: Epigenetic control systems. *Proc Natl Acad Sci USA* 1958, 44:712-717.
40. Hardy KM, Kirschmann DA, Seftor EA, Margaryan NV, Postovit LM, Strizzi L, Hendrix MJ:Regulation of the embryonic morphogen Nodal by Notch4 facilitates manifestation of the aggressive melanoma phenotype. *Cancer Res* **2010**, 70:10340-10350.
41. Daxinger L, Whitelaw E: Transgenerational epigenetic inheritance: more questions than answers. *Genome Res* **2010**, 20:1623-1628.
42. Menon DU, Meller VH: Germ line imprinting in Drosophila: epigenetics in search of function. *Fly (Austin)* **2010**, 4:48-52.
43. Lange UC, Schneider R: What an epigenome remembers. *Bioessays* **2010**, 32:659-668.
44. Ho DH, Burggren WW: Epigenetics and transgenerational transfer: a physiological perspective. *J Exp Biol* **2010**, 213:3-16.
45. Youngson NA, Whitelaw E: Transgenerational epigenetic effects. *Annu Rev Genomics Hum Genet* **2008**, 9:233-257.
46. Margueron R, Reinberg D: Chromatin structure and the inheritance of epigenetic information. *Nat Rev Genet* **2010**, 11:285-296.
47. Wen B, Wu H, Shinkai Y, Irizarry RA, Feinberg AP: Large histone H3 lysine 9 dimethylated chromatin blocks distinguish differentiated from embryonic stem cells. *Nat Genet* **2009**, 41:246-250.
48. Sass GL, Pannuti A, Lucchesi JC: Male-specific lethal complex of Drosophila targets activated regions of the X chromosome for chromatin remodeling. *Proc Natl Acad Sci USA* **2003**, 100:8287-8291.
49. Kang JK, Park KW, Chung YG, You JS, Kim YK, Lee SH, Hong SP, Choi KM, Heo KN, Seol JG, Lee JH, Jin DI, Park CS, Seo JS, Lee HW, Han JW: Coordinated change of a ratio of methylated H3-lysine 4 or acetylated H3 to acetylated H4 and DNA methylation is associated with tissue-specific gene expression in cloned pig. *Exp Mol Med* **2007**, 39:84-96.
50. Morgan DK, Whitelaw E: The role of epigenetics in mediating environmental effects on phenotype. *Nestle Nutr Workshop Ser Pediatr Program* **2009**, 63:109-117. *Discussion* 117–109, 259–168
51. Cortese R, Krispin M, Weiss G, Berlin K, Eckhardt F: DNA methylation profiling of pseudogene-parental gene pairs and two gene families. *Genomics* **2008**, 91:492-502.
52. Clarke NF, Waddell LB, Cooper ST, Perry M, Smith RL, Kornberg AJ, Muntoni F, Lillis S, Straub V, Bushby K, Guglieri M, King MD, Farrell MA, Marty I, Lunardi J,

Monnier N, North KN: Recessive mutations in RYR1 are a common cause of congenital fiber type disproportion. *Hum Mutat* **2010**, 31:E1544-E1550.
53. Robinson RL, Carpenter D, Halsall PJ, Iles DE, Booms P, Steele D, Hopkins PM, Shaw MA:Epigenetic allele silencing and variable penetrance of malignant hyperthermia susceptibility. *Br J Anaesth* **2009**, 103:220-225.
54. Wilmshurst JM, Lillis S, Zhou H, Pillay K, Henderson H, Kress W, Müller CR, Ndondo A, Cloke V, Cullup T, Bertini E, Boennemann C, Straub V, Quinlivan R, Dowling JJ, Al-Sarraj S, Treves S, Abbs S, Manzur AY, Sewry CA, Muntoni F, Jungbluth H: RYR1 mutations are a common cause of congenital myopathies with central nuclei. *Ann Neurol* **2010**, 68:717-726.
55. Zhou H, Lillis S, Loy RE, Ghassemi F, Rose MR, Norwood F, Mills K, Al-Sarraj S, Lane RJ, Feng L, Matthews E, Sewry CA, Abbs S, Buk S, Hanna M, Treves S, Dirksen RT, Meissner G, Muntoni F, Jungbluth H: Multi-minicore disease and atypical periodic paralysis associated with novel mutations in the skeletal muscle ryanodine receptor (RYR1) gene. *Neuromuscul Disord* **2010**, 20:166-173.
56. Zhou H, Brockington M, Jungbluth H, Monk D, Stanier P, Sewry CA, Moore GE, Muntoni F:Epigenetic allele silencing unveils recessive RYR1 mutations in core myopathies. *Am J Hum Genet* **2006**, 79:859-868.
57. Chen H, Taylor NP, Sotamaa KM, Mutch DG, Powell MA, Schmidt AP, Feng S, Hampel HL, de la Chapelle A, Goodfellow PJ: Evidence for heritable predisposition to epigenetic silencing of MLH1. *Int J Cancer* **2007**, 120:1684-1688.
58. Hitchins MP, Wong JJ, Suthers G, Suter CM, Martin DI, Hawkins NJ, Ward RL: Inheritance of a cancer-associated MLH1 germ-line epimutation. *N Engl J Med* **2007**, 356:697-705.
59. Vilar E, Scaltriti M, Balmana J, Saura C, Guzman M, Arribas J, Baselga J, Tabernero J:Microsatellite instability due to hMLH1 deficiency is associated with increased cytotoxicity to irinotecan in human colorectal cancer cell lines. *Br J Cancer* **2008**, 99:1607-1612.
60. Lin JC, Jeong S, Liang G, Takai D, Fatemi M, Tsai YC, Egger G, Gal-Yam EN, Jones PA: Role of nucleosomal occupancy in the epigenetic silencing of the MLH1 CpG island. *Cancer Cell* **2007**, 12:432-444.
61. Ren F, Wang D, Jiang Y: Epigenetic inactivation of hMLH1 in the malignant transformation of ovarian endometriosis. *Arch Gynecol Obstet* **2012**, 285:215-221.
62. Wu PY, Fan YM, Wang YP: Germ-line epimutations and human cancer. *Ai Zheng* **2009**, 28:1236-1242.
63. Morgan HD, Sutherland HG, Martin DI, Whitelaw E: Epigenetic inheritance at the agouti locus in the mouse. *Nat Genet* **1999**, 23:314-318.
64. Martin DI, Cropley JE, Suter CM: Environmental influence on epigenetic inheritance at the Avy allele. *Nutr Rev* **2008**, 66(Suppl 1):S12-S14.
65. Morgan HD, Jin XL, Li A, Whitelaw E, O'Neill C: The culture of zygotes to the blastocyst stage changes the postnatal expression of an epigentically labile allele, agouti viable yellow, in mice. *Biol Reprod* **2008**, 79:618-623.
66. Perry WL, Copeland NG, Jenkins NA: The molecular basis for dominant yellow agouti coat color mutations. *Bioessays* **1994**, 16:705-707.
67. Michaud EJ, van Vugt MJ, Bultman SJ, Sweet HO, Davisson MT, Woychik RP: Differential expression of a new dominant agouti allele (Aiapy) is correlated with

methylation state and is influenced by parental lineage. *Genes Dev* **1994**, 8:1463-1472.

68. Ruvinsky A, Flood WD, Zhang T, Costantini F: Unusual inheritance of the AxinFu mutation in mice is associated with widespread rearrangements in the proximal region of chromosome 17. *Genet Res* **2000**, 76:135-147.
69. Vasicek TJ, Zeng L, Guan XJ, Zhang T, Costantini F, Tilghman SM: Two dominant mutations in the mouse fused gene are the result of transposon insertions. *Genetics* **1997**, 147:777-786.
70. Rakyan VK, Chong S, Champ ME, Cuthbert PC, Morgan HD, Luu KV, Whitelaw E:Transgenerational inheritance of epigenetic states at the murine Axin(Fu) allele occurs after maternal and paternal transmission. *Proc Natl Acad Sci USA* **2003**, 100:2538-2543.
71. Waterland RA, Dolinoy DC, Lin JR, Smith CA, Shi X, Tahiliani KG: Maternal methyl supplements increase offspring DNA methylation at Axin Fused. *Genesis* **2006**, 44:401-406.
72. Fernandez-Gonzalez R, Ramirez MA, Pericuesta E, Calle S, Gutierrez-Adan A: Histone modifications at the blastocyst Axin1FU locus mark the heritability of in vitro culture-induced epigenetic alterations in mice. *Biol Reprod* **2010**, 83:720-727.
73. Manning K, Tor M, Poole M, Hong Y, Thompson AJ, King GJ, Giovannoni JJ, Seymour GB: A naturally occurring epigenetic mutation in a gene encoding an SBP-box transcription factor inhibits tomato fruit ripening. *Nat Genet* **2006**, 38:948-952.
74. Gustafsson A: Linnaeus' peloria: the history of a monster. *TAG* 1979, 54:241-248.
75. Cubas P, Vincent C, Coen E: An epigenetic mutation responsible for natural variation in floral symmetry. *Nature* **1999**, 401:157-161.
76. Li B, Carey M, Workman JL: The role of chromatin during transcription. *Cell* **2007**, 128:707-719.
77. Katz DJ, Edwards TM, Reinke V, Kelly WG: A C. elegans LSD1 demethylase contributes to germline immortality by reprogramming epigenetic memory. *Cell* **2009**, 137:308-320.
78. Arico JK, Katz DJ, van der Vlag J, Kelly WG: Epigenetic patterns maintained in early Caenorhabditis elegans embryos can be established by gene activity in the parental germ cells. *PLoS Genet* **2011**, 7:e1001391.
79. Agger K, Christensen J, Cloos PA, Helin K: The emerging functions of histone demethylases. *Curr Opin Genet Dev* **2008**, 18:159-168.
80. Zhou X, Ma H: Evolutionary history of histone demethylase families: distinct evolutionary patterns suggest functional divergence. *BMC Evol Biol* **2008**, 8:294.
81. Jeong JH, Song HR, Ko JH, Jeong YM, Kwon YE, Seol JH, Amasino RM, Noh B, Noh YS:Repression of FLOWERING LOCUS T chromatin by functionally redundant histone H3 lysine 4 demethylases in Arabidopsis. *PLoS One* **2009**, 4:e8033.
82. Reinders J, Wulff BB, Mirouze M, Mari-Ordonez A, Dapp M, Rozhon W, Bucher E, Theiler G, Paszkowski J: Compromised stability of DNA methylation and transposon immobilization in mosaic Arabidopsis epigenomes. *Genes Dev* **2009**, 23:939-950.
83. Richards EJ: Quantitative epigenetics: DNA sequence variation need not apply. *Genes Dev* **2009**, 23:1601-1605.

84. Soppe WJ, Jacobsen SE, Alonso-Blanco C, Jackson JP, Kakutani T, Koornneef M, Peeters AJ:The late flowering phenotype of fwa mutants is caused by gain-of-function epigenetic alleles of a homeodomain gene. *Mol Cell* **2000**, 6:791-802.
85. Hajkova P, Erhardt S, Lane N, Haaf T, El-Maarri O, Reik W, Walter J, Surani MA: Epigenetic reprogramming in mouse primordial germ cells. *Mech Dev* **2002**, 117:15-23.
86. Popp C, Dean W, Feng S, Cokus SJ, Andrews S, Pellegrini M, Jacobsen SE, Reik W:Genome-wide erasure of DNA methylation in mouse primordial germ cells is affected by AID deficiency. *Nature* **2010**, 463:1101-1105.
87. Monk M, Boubelik M, Lehnert S: Temporal and regional changes in DNA methylation in the embryonic, extraembryonic and germ cell lineages during mouse embryo development. *Development* 1987, 99:371-382.
88. Feng S, Jacobsen SE, Reik W: Epigenetic reprogramming in plant and animal development. *Science* **2010**, 330:622-627.
89. Bourc'his D, Voinnet O: A small-RNA perspective on gametogenesis, fertilization, and early zygotic development. *Science* **2010**, 330:617-622.
90. Richards EJ: Inherited epigenetic variation–revisiting soft inheritance. *Nat Rev Genet* **2006**, 7:395-401.
91. Lippman Z, Gendrel AV, Black M, Vaughn MW, Dedhia N, McCombie WR, Lavine K, Mittal V, May B, Kasschau KD, Carrington JC, Doerge RW, Colot V, Martienssen R: Role of transposable elements in heterochromatin and epigenetic control. *Nature* **2004**, 430:471-476.
92. Schmitz RJ, Schultz MD, Lewsey MG, O'Malley RC, Urich MA, Libiger O, Schork NJ, Ecker JR:Transgenerational epigenetic instability is a source of novel methylation variants. *Science* **2011**, 334:369-373.
93. Lejeune E, Allshire RC: Common ground: small RNA programming and chromatin modifications. *Curr Opin Cell Biol* **2011**, 23:258-265.
94. Teixeira FK, Heredia F, Sarazin A, Roudier F, Boccara M, Ciaudo C, Cruaud C, Poulain J, Berdasco M, Fraga MF, Voinnet O, Wincker P, Esteller M, Colot V: A role for RNAi in the selective correction of DNA methylation defects. *Science* **2009**, 323:1600-1604.
95. Verdel A, Vavasseur A, Le Gorrec M, Touat-Todeschini L: Common themes in siRNA-mediated epigenetic silencing pathways. *Int J Dev Biol* **2009**, 53:245-257.
96. Hajkova P, Ancelin K, Waldmann T, Lacoste N, Lange UC, Cesari F, Lee C, Almouzni G, Schneider R, Surani MA: Chromatin dynamics during epigenetic reprogramming in the mouse germ line. *Nature* **2008**, 452:877-881.
97. van der Heijden GW, Castaneda J, Bortvin A: Bodies of evidence - compartmentalization of the piRNA pathway in mouse fetal prospermatogonia. *Curr Opin Cell Biol* **2010**, 22:752-757.
98. Hajkova P: Epigenetic reprogramming in the germline: towards the ground state of the epigenome. *Philos Trans R Soc Lond B Biol Sci* **2011**, 366:2266-2273.
99. Boiani M, Eckardt S, Leu NA, Scholer HR, McLaughlin KJ: Pluripotency deficit in clones overcome by clone-clone aggregation: epigenetic complementation? *EMBO J* **2003**, 22:5304-5312.

100. Whitworth KM, Prather RS: Somatic cell nuclear transfer efficiency: how can it be improved through nuclear remodeling and reprogramming? *Mol Reprod Dev* **2010**, 77:1001-1015.
101. Balbach ST, Esteves TC, Houghton FD, Siatkowski M, Pfeiffer MJ, Tsurumi C, Kanzler B, Fuellen G, Boiani M: Nuclear reprogramming: kinetics of cell cycle and metabolic progression as determinants of success. *PLoS One* **2012**, 7:e35322.
102. Hosseini SM, Hajian M, Forouzanfar M, Moulavi F, Abedi P, Asgari V, Tanhaei S, Abbasi H, Jafarpour F, Ostadhosseini S, Karamali F, Karbaliaie K, Baharvand H, Nasr-Esfahani MH:Enucleated ovine oocyte supports human somatic cells reprogramming back to the embryonic stage. *Cell Reprogram* **2012**, 14:155-163.
103. Noggle S, Fung HL, Gore A, Martinez H, Satriani KC, Prosser R, Oum K, Paull D, Druckenmiller S, Freeby M, Greenberg E, Zhang K, Goland R, Sauer MV, Leibel RL, Egli D:Human oocytes reprogram somatic cells to a pluripotent state. *Nature* **2011**, 478:70-75.
104. Pan G, Wang T, Yao H, Pei D: Somatic cell reprogramming for regenerative medicine: SCNT vs. iPS cells. *Bioessays* **2012**, 34:472-476.
105. Niemann H, Carnwath JW, Herrmann D, Wieczorek G, Lemme E, Lucas-Hahn A, Olek S:DNA methylation patterns reflect epigenetic reprogramming in bovine embryos. *Cell Reprogram* **2010**, 12:33-42.
106. Cui XS, Zhang DX, Ko YG, Kim NH: Aberrant epigenetic reprogramming of imprinted microRNA-127 and Rtl1 in cloned mouse embryos. *Biochem Biophys Res Commun* **2009**, 379:390-394.
107. Su J, Wang Y, Li R, Peng H, Hua S, Li Q, Quan F, Guo Z, Zhang Y: Oocytes selected using BCB staining enhance nuclear reprogramming and the in vivo development of SCNT embryos in cattle. *PLoS One* **2012**, 7:e36181.
108. Wang F, Kou Z, Zhang Y, Gao S: Dynamic reprogramming of histone acetylation and methylation in the first cell cycle of cloned mouse embryos. *Biol Reprod* **2007**, 77:1007-1016.
109. Jullien J, Pasque V, Halley-Stott RP, Miyamoto K, Gurdon JB: Mechanisms of nuclear reprogramming by eggs and oocytes: a deterministic process? *Nat Rev Mol Cell Biol* **2011**, 12:453-459.
110. Trifonov EN: Sequence-dependent deformational anisotropy of chromatin DNA. *Nucleic Acids Res* **1980**, 8:4041-4053.
111. Trifonov EN, Sussman JL: The pitch of chromatin DNA is reflected in its nucleotide sequence. *Proc Natl Acad Sci USA* **1980**, 77:3816-3820.
112. Frenkel ZM, Bettecken T, Trifonov EN: Nucleosome DNA sequence structure of isochores. *BMC Genomics* **2011**, 12:203.
113. Trifonov EN: Nucleosome positioning by sequence, state of the art and apparent finale. *J Biomol Struct Dyn* **2010**, 27:741-746.
114. Rapoport AE, Frenkel ZM, Trifonov EN: Nucleosome positioning pattern derived from oligonucleotide compositions of genomic sequences. *J Biomol Struct Dyn* **2011**, 28:567-574.
115. Zofall M, Persinger J, Kassabov SR, Bartholomew B: Chromatin remodeling by ISW2 and SWI/SNF requires DNA translocation inside the nucleosome. *Nat Struct Mol Biol* **2006**, 13:339-346.

116. Albert I, Mavrich TN, Tomsho LP, Qi J, Zanton SJ, Schuster SC, Pugh BF: Translational and rotational settings of H2A. Z nucleosomes across the Saccharomyces cerevisiae genome. *Nature* **2007**, 446:572-576.
117. Chodavarapu RK, Feng S, Bernatavichute YV, Chen PY, Stroud H, Yu Y, Hetzel JA, Kuo F, Kim J, Cokus SJ, Casero D, Bernal M, Huijser P, Clark AT, Krämer U, Merchant SS, Zhang X, Jacobsen SE, Pellegrini M: Relationship between nucleosome positioning and DNA methylation. *Nature* **2010**, 466:388-392.
118. Cui F, Zhurkin VB: Structure-based analysis of DNA sequence patterns guiding nucleosome positioning in vitro. *J Biomol Struct Dyn* **2010**, 27:821-841.
119. Schones DE, Cui K, Cuddapah S, Roh TY, Barski A, Wang Z, Wei G, Zhao K: Dynamic regulation of nucleosome positioning in the human genome. *Cell* **2008**, 132:887-898.
120. Meneghini MD, Wu M, Madhani HD: Conserved histone variant H2A. Z protects euchromatin from the ectopic spread of silent heterochromatin. *Cell* **2003**, 112:725-736.
121. Li B, Pattenden SG, Lee D, Gutierrez J, Chen J, Seidel C, Gerton J, Workman JL:Preferential occupancy of histone variant H2AZ at inactive promoters influences local histone modifications and chromatin remodeling. *Proc Natl Acad Sci USA* **2005**, 102:18385-18390.
122. Deal RB, Topp CN, McKinney EC, Meagher RB: Repression of flowering in Arabidopsis requires activation of FLOWERING LOCUS C expression by the histone variant H2A. Z. *Plant Cell* **2007**, 19:74-83.
123. Gent JI, Schneider KL, Topp CN, Rodriguez C, Presting GG, Dawe RK: Distinct influences of tandem repeats and retrotransposons on CENH3 nucleosome positioning. *Epigenetics Chromatin* **2011**, 4:3.
124. Barski A, Cuddapah S, Cui K, Roh TY, Schones DE, Wang Z, Wei G, Chepelev I, Zhao K:High-resolution profiling of histone methylations in the human genome. *Cell* **2007**, 129:823-837.
125. Zheng D: Asymmetric histone modifications between the original and derived loci of human segmental duplications. *Genome Biol* **2008**, 9:R105.
126. Margueron R, Trojer P, Reinberg D: The key to development: interpreting the histone code? *Curr Opin Genet Dev* **2005**, 15:163-176.
127. Hammoud SS, Nix DA, Zhang H, Purwar J, Carrell DT, Cairns BR: Distinctive chromatin in human sperm packages genes for embryo development. *Nature* **2009**, 460:473-478.
128. Dillon N: Gene regulation and large-scale chromatin organization in the nucleus. *Chromosome Res* **2006**, 14:117-126.
129. Murrell A, Heeson S, Reik W: Interaction between differentially methylated regions partitions the imprinted genes Igf2 and H19 into parent-specific chromatin loops. *Nat Genet* **2004**, 36:889-893.
130. Teller K, Solovei I, Buiting K, Horsthemke B, Cremer T: Maintenance of imprinting and nuclear architecture in cycling cells. *Proc Natl Acad Sci USA* **2007**, 104:14970-14975.
131. Gaillard C, Borde C, Gozlan J, Marechal V, Strauss F: A high-sensitivity method for detection and measurement of HMGB1 protein concentration by high-affinity binding to DNA hemicatenanes. *PLoS One* **2008**, 3:e2855.

132. Jaouen S, de Koning L, Gaillard C, Muselikova-Polanska E, Stros M, Strauss F:Determinants of specific binding of HMGB1 protein to hemicatenated DNA loops. *J Mol Biol* **2005**, 353:822-837.
133. Gondor A, Ohlsson R: Chromosome crosstalk in three dimensions. *Nature* **2009**, 461:212-217.
134. Deng W, Blobel GA: Do chromatin loops provide epigenetic gene expression states? *Curr Opin Genet Dev* **2010**, 20:548-554.
135. Sumida N, Ohlsson R: Chromosomal networks as mediators of epigenetic states: the maternal genome connection. *Epigenetics* **2011**, 5:297-300.
136. Riggs AD, Martienssen RA, Russo VEA: Introduction. In *Epigenetic mechanisms of gene regulation.* Edited by Russo VEA. Cold Spring Harbor Laboratory Press, Cold Spring Harbor; **1996**:1-4.
137. Wang X, Zhu H, Snieder H, Su S, Munn D, Harshfield G, Maria BL, Dong Y, Treiber F, Gutin B, Shi H: Obesity related methylation changes in DNA of peripheral blood leukocytes. *BMC Med* **2010**, 8:87.
138. Jeffries MA, Dozmorov M, Tang Y, Merrill JT, Wren JD, Sawalha AH: Genome-wide DNA methylation patterns in CD4+ T cells from patients with systemic lupus erythematosus. *Epigenetics* **2011**, 6:593-601.
139. Luna E, Bruce TJ, Roberts MR, Flors V, Ton J: Next-generation systemic acquired resistance. *Plant Physiol* **2012**, 158:844-853.
140. Yazbek SN, Spiezio SH, Nadeau JH, Buchner DA: Ancestral paternal genotype controls body weight and food intake for multiple generations. *Hum Mol Genet* **2010**, 19:4134-4144.
141. McDevitt MA: Clinical applications of epigenetic markers and epigenetic profiling in myeloid malignancies. *Semin Oncol* **2012**, 39:109-122.
142. Radpour R, Kohler C, Haghighi MM, Fan AX, Holzgreve W, Zhong XY: Methylation profiles of 22 candidate genes in breast cancer using high-throughput MALDI-TOF mass array. *Oncogene* **2009**, 28:2969-2978.
143. Deng D, Liu Z, Du Y: Epigenetic alterations as cancer diagnostic, prognostic, and predictive biomarkers. *Adv Genet* **2010**, 71:125-176.
144. Javierre BM, Fernandez AF, Richter J, Al-Shahrour F, Martin-Subero JI, Rodriguez-Ubreva J, Berdasco M, Fraga MF, O'Hanlon TP, Rider LG, Jacinto FV, Lopez-Longo FJ, Dopazo J, Forn M, Peinado MA, Carreño L, Sawalha AH, Harley JB, Siebert R, Esteller M, Miller FW, Ballestar E:Changes in the pattern of DNA methylation associate with twin discordance in systemic lupus erythematosus. *Genome Res* **2010**, 20:170-179.
145. Tilgner K, Atkinson SP, Yung S, Golebiewska A, Stojkovic M, Moreno R, Lako M, Armstrong L: Expression of GFP under the control of the RNA helicase VASA permits fluorescence-activated cell sorting isolation of human primordial germ cells. *Stem Cells* **2010**, 28:84-92.
146. Merbs SL, Khan MA, Hackler L, Oliver VF, Wan J, Qian J, Zack DJ: *Cell*-specific DNA methylation patterns of retina-specific genes. *PLoS One* **2012**, 7:e32602.
147. Deal RB, Henikoff S: A simple method for gene expression and chromatin profiling of individual cell types within a tissue. *Dev Cell* **2010**, 18:1030-1040.
148. Deal RB, Henikoff S: The INTACT method for cell type-specific gene expression and chromatin profiling in Arabidopsis thaliana. *Nat Protoc* **2011**, 6:56-68.

149. Jiang G, Zheng L, Pu J, Mei H, Zhao J, Huang K, Zeng F, Tong Q: Small RNAs targeting transcription start site induce heparanase silencing through interference with transcription initiation in human cancer cells. *PLoS One* **2012**, 7:e31379.
150. Jauvion V, Rivard M, Bouteiller N, Elmayan T, Vaucheret H: RDR2 partially antagonizes the production of RDR6-dependent siRNA in sense transgene-mediated PTGS. *PLoS One* **2012**, 7:e29785.
151. Matzke M, Kanno T, Daxinger L, Huettel B, Matzke AJ: RNA-mediated chromatin-based silencing in plants. *Curr Opin Cell Biol* **2009**, 21:367-376.

CHAPTER 12

NEXT GENERATION SEQUENCING BASED APPROACHES TO EPIGENOMICS

MARTIN HIRST and MARCO A. MARRA

INTRODUCTION

At its inception massively parallel sequencing was ill suited for the task of sequencing the human genome. Perhaps then it is not surprising that some of the first publications that utilized next-generation sequencing were directed at chromatin immuno-precipitation enriched fractions of the genome [1–3]. Since their introduction, short read massively parallel sequencing platforms have continued to improve at an exponential rate, generating longer sequences of better quality in ever increasing numbers [4]. The research community has leveraged these improvements to develop a diverse collection of sequence-based methodologies to probe the functional genome [4–6]. These methodologies can be broadly divided into protocols that profile transcribed regions of the genome and those that profile the processes regulating transcription. Transcriptional regulation is maintained through complex interactions between sequence specific transcription factors which act in short time frame, generally in response to specific cellular stimuli, and those which act on longer time scales in response to more generalized environmental and developmental signals. The study of the mechanistic features that control this latter category is called epigenetics and the study of how these marks are patterned across the genome is called epigenomics.

Epigenetic processes act on DNA and histones, the building blocks of nucleosomes [7]. In the mammalian genome DNA modification occurs exclusively on cytosine residues, at the 5′-position of the purine ring, in the form of either a methyl or hydroxymethyl group [8, 9]. Until recently, modification of mammalian genomic DNA was thought to be restricted to the context of CG dinucleotides known as 'CpGs'. However, recent epigenomic profiles have revealed that methylation is found in alternate contexts including CHG and CHH in pluripotent cell types [10].

The nucleosome is the fundamental unit of chromatin and is composed of two copies of each of the four core histones (H3, H4, H2A and H2B) around which 146 bp of DNA are wrapped [11, 12]. Histones are evolutionarily conserved proteins characterized by an accessible amino-terminal tail and a histone fold domain that mediates interactions between histones to form the nucleosome scaffold [13]. The N-termini of histone polypeptides are extensively modified by more than 60 different post-translational modifications including methylation, acetylation, phosphorylation and ubiquitination [14, 15]. Although the vast majority of these modifications remain poorly understood there has been significant progress in recent years understanding the roles that methylation and acetylation play in transcriptional regulation [16].

A prerequisite for understanding the role of epigenetics in development and disease is knowledgeable of the genome-wide distribution of epigenetic modifications in normal and diseased states. The availability of reference genome assemblies and massively parallel, next generation sequencing platforms has led to methods which provide high-resolution genome-wide epigenomic profiles. In this review, we will describe methodologies that have been developed to profile the epigenome using next generation sequencing platforms. We will discuss these in terms of library preparation techniques, sequence platforms and analysis. Current next generation sequencing approaches require that the collection of DNA fragments to be sequenced and flanked by standard nucleotide string to allow for clonal amplification or, in the case of the Helicos platform, direct sequencing. In this review we will refer to collections of such fragments as 'libraries' and the process to build such collections as library preparation.

LIBRARY PREPARATION

Library preparation for next generation sequencing can be broadly divided into two distinct processes. The first involves preparation of genomic DNA (gDNA) fragments, generally in the size range of a single nucleosome, followed by preparation of the fragments for sequencing. Preparation of genomic DNA fragments for next generation sequencing generally involves the addition of nucleotide sequences on the ends of the fragments that will hybridize to complementary sequences present on the matrix used to generated clonal copies of the library fragment for sequencing.

HISTONE MODIFICATION PROFILING

The N-terminal tails of histones are extensively modified in response to developmental and environmental signals [14, 15]. The predominant method for mapping these post-translational modifications genome-wide involves a technique known as chromatin immuno-precipitation (ChIP) [17]. In this method histones are either chemically coupled to DNA through the action of a cross-linking reagent such as formaldehyde [17] or released in their native form by the addition of a nuclease that, in the correct dilution, specifically digests gDNA at unprotected linker sequence [18]. Following gDNA fragmentation the protein/DNA mixture is subjected to immuno-precipitation using antibodies raised against the post-translational modification under study. In the process of immuno-precipitation, DNA fragments that are in association with histone peptides are co-purified and, following proteolytic digestion and DNA purification, subjected to library construction and direct sequencing (ChIP-seq) [2, 3, 19, 20].

Direct sequencing of ChIP enriched fractions has distinct advantages over competing hybridization based techniques. Among these is the ability to interrogate epigenomic marks in repetitive elements, which comprise ~45% of the human genome [21]. Early methods for ChIP followed by direct sequencing by either capillary sequencing [22, 23] or the 454 platform [24] included concatenation of short sequence tags derived from the immuno-precipitated fragments to effectively utilize the relatively longer reads provided by these sequencing methodologies. The development of massively parallel short read platforms such as the Genome Analyzer (Illumina Inc.) and SOLiD (Life Technologies) platforms negated the need for complicated library construction techniques and allowed for direct library construction from the immuno-precipitated products. The dominant platform utilized to date for ChIP-sequencing is the Illumina Genome Analyzer [1, 2, 3, 19, 20]. More recently the SOLiD platform [25, 26] has been applied in this area and a single reference is available outlining the application of the Heliscope Genetic Analysis platform [27] (Helicos BioSciences).

ChIP sequencing (ChIP-Seq) library construction for the Genome Analyzer or SOLiD next generation sequencing platforms is an implementation of standard methodologies for whole genome shotgun sequencing [28, 29]. In this method the ragged ends of the enriched fragmented DNA, typically in the low nanogram range, are repaired and platform specific adapters are ligated onto the resulting fragments on either blunt end (SOLiD) or A-tailed (Genome Analyzer) DNA fragments. Adapter ligated product is then PCR amplified using primers which hybridize to the adapter sequences and extend to include sequences which facilitate clonal amplification and sequencing. In addition to A-tailing the gDNA fragments, library preparation for the Genome Analyzer utilizes adapters that are only partially complementary introducing a 'fork' in the adapter that is subsequently resolved during PCR. Application of this structure allows for all adapted fragments to be PCR amplified. In contrast, the SOLiD platform involves the addition of two independent adapter sequences during ligation allowing 50% of adapted fragments to participate PCR amplification.

Recently, Goren *et al.* [27] reported the use of the Heliscope Genetic Analysis platform for a ChIP-seq study directed at limited cell populations. Library construction for the Heliscope platform differs significantly from competing next generation sequencing platforms in that PCR amplification is not required. Library construction involves a single step where immuno-precipitated gDNA fragments are A-tailed using a terminal transferase enzyme and dATP and, after capture to a proprietary substrate, directly sequenced.

DNA METHYLATION PROFILING

In contrast to histone modification profiling, a wide variety of approaches have been developed to profile DNA methylation utilizing next generation sequencing platforms. Approaches to profile DNA methylation genome-wide can be broadly divided into those that rely on methylation dependent enzymatic restriction, methyl-DNA enrichment and direct bisulfite conversion [21, 30]. Individual methods can also be combined to increase the resolution or efficiency of a single method. For example, a combination of MeDIP-seq and MRE-seq, to profile both the methylated and unmethylated fraction of the genome [31].

ENRICHMENT BASED METHODS

METHYLATED DNA IMMUNO-PRECIPITATION

Methylated DNA Immuno-precipitation sequencing (MeDIP-Seq) is an immuno-precipitation based technique where fragmented DNA is enriched based on its methylation content [32, 33]. Antibodies used in this technique are raised against a single stranded methyl-cytosine and thus the immuno-precipitation is performed in a denatured state [34]. To avoid over representation of repeat content in the subsequent library through preferential annealing of highly methylated genomic repeats, library construction is performed prior to the immuno-precipitation and amplified following enrichment by PCR.

At sufficient sequencing depths, on the order of two Gigabases aligned, MeDIP-seq is capable of identifying 70–80% of the 28 million CpGs in the human haploid genome at a resolution of 100–300 bases [21]. This is near to the expected frequency of methylation in the human genome [8, 9]. At saturating sequencing depths it may also be possible to annotate uncovered CpGs as non-methylated.

METHYLATED DNA BINDING DOMAIN SEQUENCING

Methylated DNA Binding Domain sequencing (MBD-seq) is similar in concept to MeDIP-seq where genomic fragments are enriched based on their methylation content [35]. In this technique bead immobilized recombinant methylated-CpG binding proteins MECP2 or MBD2 are used to enrich for methylated DNA fragments from a pool of genomic DNA fragmented by sonication to 100–300 bp in length. Following enrichment of methylated double stranded DNA fragments standard library construction techniques are utilized to generate a library representing the methylated fraction of the genome.

MBD-seq differs from MeDIP-seq in the application of multiple salt cuts during elution of the methyl-CpG containing DNA fragments bound to the immobilized methyl binding domain. In this technique, weakly methylated DNA fragments are eluted at lower salt concentrations compared with moderately or densely methylated DNA fragments (e. g. methylated CpG Islands). Thus it is possible to selectively enrich for weakly or densely methylated DNA fragments during immuno-precipitation, potentially reducing downstream sequencing costs. In the absence of selective enrichment, MBD-seq requires a similar degree of sequencing as MeDIP-seq and at this depth (2 Gigabases aligned) MBD-seq is capable of identifying 70–80% of the 28 million CpGs in the human genome at a resolution of 100–300 bases [21]. As with MeDip-seq, at saturating sequencing depths, it may also be possible to call any uncovered CpGs as non-methylated.

BISULFITE CONVERSION BASED METHODS

METHYLC-SEQ

The 'gold standard' for profiling methylated cytosine is bisulfite-mediated deamination of cytosine. This technique, discovered simultaneously by the Shapiro and Hayatsu groups in the early 1970s, relies on the selectivity of the bisulfite reaction to deaminate cytosine, but not 5-methylcytosine, to uracil which is subsequently read as thymidine during sequencing [36, 37]. Bisulfite-based methods detect hydroxylmethylation, but cannot distinguish it from methylation [38]. In the original methodology, bisulfite treated genomic regions were amplified by site specific PCR, cloned and subjected to Sanger sequencing [37]. Sequence reads were assessed individually and visualized as a matrix with the CpG content of each clone represented as a row. While this approach has been extremely valuable in the elucidation of the methylation status of discrete genomic regions it does not scale well and cannot be feasibly applied to whole genome studies. With the advent of next generation sequencing is it now possible to directly shotgun sequence bisulfite treated genomic DNA. In this method, library construction is performed prior to bisulfite treatment using adapters in which cytosines have been replaced by methyl-cytosines to protect them from deamination during the bisulfite treatment [39, 40]. Following the bisulfite treatment, a process that is performed under denaturing conditions, the library is PCR amplified using PCR primers that extend the adapter sequencing to allow for clonal amplification and sequencing. This technique, termed Methyl-C-seq or BS-seq, first performed genome-wide on the genome of the flowering plant Arabidopsis thaliana [39, 40], has recently been applied to the human genome [10]. To generate sufficient read coverage for the latter study over 200 lanes of Illumina Genome Analyzer sequence data, at a list cost of over $200 000 USD in reagents, was required. However, recent advances in throughput and efficiency of next generation sequencing platforms have reduced the costs associated with such an experiment dramatically. It is expected that in the fall of 2010 the cost for such an experiment will have dropped 20-fold to the range of the $10 000 USD (reagents only).

TARGETED BISULFITE SEQUENCING

The high cost of sequencing a sodium bisulfite converted genome has spurred the development of strategies for enriching genomic regions of interest followed by bisulfite sequencing [3, 41–44]. Two general strategies have emerged. In the first, coined reduced representation bisulfite sequencing (RRBS), the genome is digested by the methylation insensitive restriction enzyme Msp1 and size selected to generate a fragment library within the range of next generation sequencing platforms (typically in the 100–300 bp) [41]. The size selected material becomes the input for the library construction using methylated adapters and subjected to the bisulfite conversion analogous to the procedure used for the full genome bisulfite shotgun sequencing [39]. Due to the selective nature of the method, RRBS covers only 12% of CpGs genome-wide however these CpGs are highly enriched within CpG islands [21].

Alternatively, genome enrichment can be performed using molecular inversion probes and PCR following bisulphite conversion of the genome [43, 44]. Molecular inversion probes can be designed to include all possible combinations of the cytosines and uracils or to avoid CpGs to mitigate specificity loss associated with the cytosine to uracil conversion. Once amplified the targeted regions can be directly sequenced on a next generation sequencing platform following standard techniques. Publications utilizing these methods typically target in the range of 1000s of CpGs, or 0. 2% of genome-wide CpGs [43, 44].

METHYL-SENSITIVE RESTRICTION BASED METHODS

Various strategies have been developed to profile the unmethylated fraction of the genome using restriction enzymes that are sensitive the CpG methylation state. Protocols involving a single methyl-sensitive restriction enzyme (HpaII) enzymatic digestion (HpaII; HELP-seq, Methyl-seq and MSCC) as well as multiple digestions (HpaII, AciI, Hinc6I; MRE-seq) have been developed [31, 43, 45, 46]. The protocol involves the digestion of the genomic DNA by one or more methyl-sensitive enzymes followed by size selection, pooling where appropriate and library construction. Minor modifications of the standard library construction procedures are used to account for the nature of the overhangs generated by the restriction digests. The use of additional restriction enzymes during the digestions increases the diversity of fragments in the library and thus allows for an increase in the total number of CpGs that can be queried. In these methods, the methylation status of 1–2 million CpGs are assessed.

Restriction based methodologies pose unique challenges during sequencing on the Illumina Genome Analyzer and SOLiD sequencing platforms. Enzymatic restriction skews the nucleotide representation on the terminal ends of the fragments that are subjected to library construction. During the initial stages of sequencing, this nucleotide bias can lead to the generation of poor quality focal maps, a key step in massively parallel next generation sequencing. This can be avoided by including a balanced nucleotide adapter onto the fragment ends or, in the case of the Genome Analyzer, starting base calling after the restriction site.

METHOD INTEGRATION

Individual methods may also be combined to increase coverage and/or efficiency. For example, MeDIP-seq and MRE-seq may be combined to profile both the methylated and un-methylated fractions of the genome simultaneously [21, 31]. Bisulfite conversion can be combined with an enrichment strategy (for example MBD-seq or MeDIP-seq) to provide increased resolution of methyl-cytosines in the immuno-precipitated fraction.

DIRECT DETECTION

Recent advances in sequencing technology have raised the possibility of the direct detection of DNA modifications. In the forefront of these efforts is Pacific Biosciences who have recently demonstrated an ability to directly detect DNA methylation during single-molecule, real-time (SMRT) DNA sequencing, a technique for studying nucleic acid sequence and structure [47–49]. Similarly, Oxford Nanopore Technologies has published proof of concept data for the direct detection of the 5-methylcytosine [50]. At the appropriate scale, these techniques offer the exciting possibility of the direct, de novo detection of the DNA methlyation genome-wide.

NEXT GENERATION SEQUENCING PLATFORMS

The majority of published epigenomic studies utilizing next generation sequencing have been generated on an Illumina Genome Analyzer. This is in part due its early adoption by the field as well as the flexibility of library preparation and base space massively parallel outp ut. While there were some early examples of epigenomic data sets generated on the comparably longer read 454 platform [24], these have largely been replaced by methods on the comparatively shorter read platforms. Conceptually, the SOLiD platform from Life Technologies is equally well suited to sequencing epigenomic libraries and more recently research groups have begun to publish ChIP-seq data sets using this platform [25, 26]. There is a single report of the application of the Heliscope Genetic Analysis platform to ChIP-seq studies [27] and proof of concept methylation data sets have been published by Pacific Biosciences and Oxford Nanopore Technologies [47, 50].

SECOND GENERATION SEQUENCING PLATFORMS

ILLUMINA GENOME ANALYZER

The Genome Analyzer is a synchronous sequence-by-synthesis platform that leverages reversible dye terminators [28]. Libraries of DNA fragments are clonally amplified on the surface of a flow cell (closed microscope slide) on to which modified oligos complementary to the sequence of the PCR primers utilized in library construction have been grafted. Sequencing is performed by the stepwise application of reagents, single nucleotide incorporation, flushing of excess reagents and imaging. The images are subsequently analyzed to generate a focal map for each clonally derived cluster and then used to call bases on each cycle. A typical Illumina Genome Analyzer run can currently generate 30 million reads per lane, 210 million per flowcell at read lengths up to 100 bases. In the spring of 2010, a higher throughput version of the Illumina Genome Analyzer was released. The specifications for this instrument, called the HiSeq2000, indicate over 60 million reads per lane, 500 million per flow cell can be achieved.

To facilitate unambiguous alignment of sequence reads within genomic repeat regions paired-end sequencing can be performed. In this implementation, a second read is generated on the clonally amplified cluster using a sequencing primer that anneals to the opposing adapter. To achieve this, the clonally derived read cluster, rendered single stranded during the first round of sequencing, is regenerated by PCR on the flowcell surface. Sequencing is performed as above utilizing the focal map generated from the first read to associate the two sequence reads together. A similar strategy can also be employed to read a sequence barcode added to the adapter during library construction. This so-called "third read" enables pooling of multiple libraries in a single flowcell lane.

SOLiD

The SOLiD platform is a synchronous sequencer utilizing a sequence by ligation approach [29]. In this platform, libraries of fragments are clonally amplified on the surface of a 1 micron bead on which an oligo complementary to one of the two adapters used in the library construction is covalently bound. Clonal amplification is achieved by limiting dilution of the fragment library during PCR, (emPCR) which is performed as an emulsion generated by mechanical whipping of an aqueous solution containing PCR reagents, amplification beads, the library and oil. Following emPCR 'loaded' beads are enriched by hybridization of the alternate adapter to complementary oligos covalently attached to a polystyrene bead. Enriched beads are subsequently attached to the surface of a glass slide and the sequencing is performed by the stepwise application of reagents, ligation of labeled probes, flushing of excess probes and imaging. The images are subsequently analyzed to generate a focal map for each bead and call the transition of bases generated during the ligations. A SOLiD4 slide can currently generate ~600 million clonal reads at reads lengths up to 50 bases. As with the Illumina Genome Analyzer, paired-end sequencing and barcoding methodologies have also been developed for the SOLiD platform.

454 GENOME SEQUENCER FLX

The 454 Genome Sequencer FLX is a pyro-sequencing platform [51]. Similar to the SOLiD platform, the 454 FLX leverages emPCR to clonally amplify library fragments onto the surface of a bead. Following enrichment sequencing is performed by depositing beads onto the surface of a micro-fabricated slide that contains 1. 6 million small reaction chambers. Single beads are sequenced in each micro-chamber by the stepwise addition of the nucleotides in a fixed order followed by imaging. Nucleotide incorporation is monitored for each micro-well by a chemi-illuminescent signal generated as a by-product of nucleotide incorporation. A 454 Genome Sequencer FLX can currently generate 1 million reads at read lengths up to 400 bp. Due to the limited number of reads and high cost/read compared with other next generation sequencing platforms, the 454 FLX platform is generally not used for epigenomic studies.

THIRD GENERATION SEQUENCING PLATFORMS

Third generation sequencing platforms are distinct from their forebearers, in that they are designed to sequence DNA at the level of a single molecule. The advantages of such an approach include a much simplified library generation process, massively parallel sequencing at long read lengths and importantly the lack of the repeated PCR amplifications prior to sequencing. A testament to human ingenuity is the diverse number of such platforms under development. Examples include; Helicos, the first company to provide a single molecule sequencer using an sequencing-by-synthesis and imaging approach [52]; Pacific Biosciences, which sequences DNA in real time by imaging fixed DNA polymerases [49] and Oxford Nanopore Technology, developing a sequencing platform based on the current changes induced by nucleotides as they pass through an alpha hemolysin nanopore [50]. Early versions of some of these platforms have already been used in proof of concept epigenomic studies [47, 50]. However, full realization of their potential is perhaps 2–5 years away from common use.

ANALYSIS

Analysis of epigenomic data sets generated on a next generation sequencing platform remains a significant challenge. This is due, in part, to the relatively short period of time for which these data sets have been available compounded by the rapid rate of change of next generation sequencing platforms.

Analysis of epigenomic data sets generated by next generation sequencing platforms can be broken into four steps, the results of which can considered analysis levels (Figure 1). Data generated from a next generation sequence platform consist of strings of bases (Illumina Genome Analyzer, 454 FLX) or color space base transitions (SOLiD) along with associated quality scores. The first step in analysis is to align this primary, level 0, data to a reference genome assembly to generate a level 1 data set consisting of the genomic coordinates of the alignments and strand on the reference genome. A number of specialized aligners have been developed to map the tens of millions reads generated in a single experiment to a mammalian sized reference genome (for review see ref. [53]). The majority of widely adopted aligners use a 'seed and extend' based algorithm where a sub-string contained within the read is rapidly aligned to either a hash table (MAQ [54], SOAP [55], SHRiMP [56], ZOOM [57] and BFAST [58]) or more recently a suffix array generated from Burrows–Wheeler transform of the reference genome (BOWTIE [59], BWA [60] and SOAP2 [61]). Once a match is found the read is 'extended' up to the maximum read length on the genome to attempt to uniquely place the read within the genome. Reads that cannot be placed uniquely are either randomly placed on the genome or ignored for downstream analyses. Within the last year the output of such alignments has largely been standardized on the SAM/BAM file format [62]. Bisulfite treated DNA requires specialized alignment to account for the C to T conversion. Several short read alignment algorithms are available that can be configured for bisulfite converted DNA alignment including, BSMAP [63], Pash [64], RMAP [65], ZOOM [57] and BS Seeker [66]. A recent comparison of these aligners concluded that, despite minor differences in speed and accuracy, aligner choice is unlikely to have a significant impact on overall analysis [21]. Following read alignment, level 1 data may be viewed directly by converting the read alignments into read density maps and displaying the result on a genome browser or further processed through segmentation.

Segmentation methods attempt to transform raw sequence alignments into regions of signal and background (level 3, Figure 1). In general, segmentation tools attempt to model the expected behavior of the epigenomic mark (for a recent review of segmentation methods see [67]). For immuno-precipitation based methodologies two main strategies have emerged. The first, used primarily for epigenomic marks that tend to be

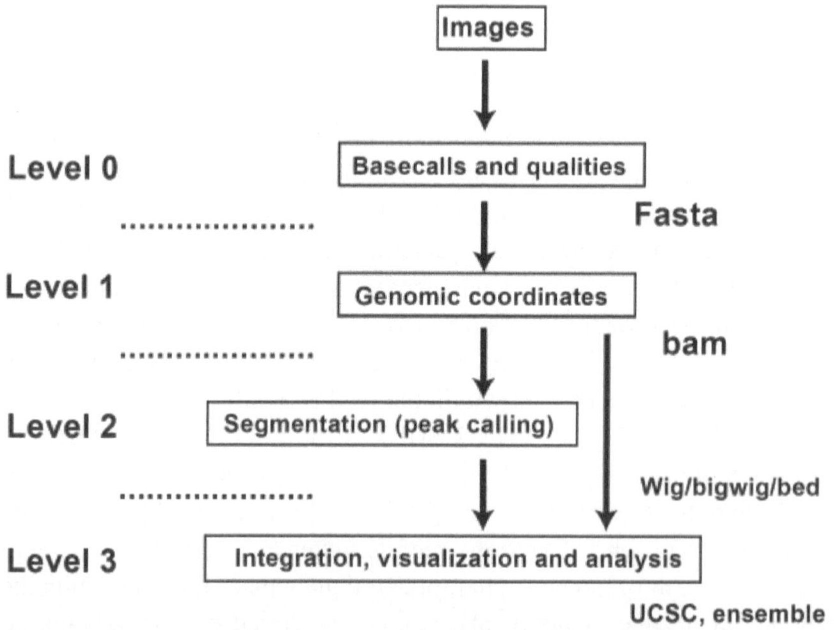

FIGURE 1. Analysis process flow. Images generated during the sequencing process are converted to base (Illumina Genome Analyzer) or color (SOLiD) space strings and associated qualities. This process is performed on instrument and the output (level 0) consists of fastq or csfasta and QV_qual files for the Genome Analyzer or SOLiD respectively. Level 0 data is aligned to the reference genome assembly using an optimized short read aligner (see text). The output of the alignment process (Level 1) is a file containing the sequences, qualities and alignment coordinates relative to the reference genome. The bam file format is currently the standard for level 1 data (53). The alignment can be directly converted to a standardized file format for visualization (for example, wig, bigwig or bigbed [81]). Alternatively, level 1 data can be transformed by a segmentation algorithm that attempts to model the behavior of the epigenomic mark under study and correct for background signal (level 2). The output of the level 2 can be subsequently viewed using a web-based or stand alone web browser and integrated with additional data types (level 3).

punctuate in their genomic distribution such as H3K4me3 or H3K9Ac, attempts to build 'peaks' of enrichment by modeling individual fragments within the library. Regions of enrichment are defined by oriented read sets that are computational extended by the insert size of the fragment library. Examples of such tools are Findpeaks [68], ERANGE [20], GLITR [69] and PeakSeq [70]. The second attempts to model more broadly distributed (spreading) chromatin modifications such as H3K9me3 or H3K36me3, by dividing the genome into windows of defined size and enumerating either the raw or normalized number of reads which align within the windows. Examples of binning tools are CisGenome [71] and ChromaSig [72]. In addition attempts have been made to combine the attributes of a binning and peak calling methodologies into a single algorithm [73].

An important consideration for segmenting ChIP-seq data sets is the use of a control signal for normalization and background estimation. A control signal is typically derived from sequencing either the sheared input DNA that was used for the immuno-precipitation or a non-specific immune-precipitate (IgG). Here the idea is to control for incorrect mappings (e. g. Read Stacks) driven by genome miss-assembly and/or polymorphisms and background signal generated from the shearing process itself— open chromatin would be expected to be more readily sheared by sonication than closed chromatin, for example. One of the main differences between segmentation tools is in how this is approached, but the general idea is to subtract the signal obtained in the control from experimental track, thus normalizing the signal to the background.

The diversity of segmentation tools currently available is a natural consequence of the rapid advances being made in the field. It is outside the scope of this review to provide a detailed breakdown of the various segmentation tools available (for an excellent current review on this please see [67]). However, researchers undertaking epigenomic studies utilizing next generation platforms need to be cognizant of the differences to make an informed decision on which tool would be most suitable for their data set. Overtime, as was the case with microarray analysis, it is expected that standardized tools, accepted by the majority of the community, will be employed in epigenomic research.

An additional consideration for any next generation sequencing based epigenomic method is how deeply to sample each library. As the sequencing depth increases, the number of unique reads covering a particular region should approach the total possible reads present in the library for each enriched region. Such a point, referred to as 'saturation', occurs when further sequencing fails to discover additional regions above background. Sequencing beyond saturation improves confidence in the observations and increases the coverage of events, though at greater cost per event covered. Thus, sequencing below or up to saturation may be sufficient, for example when maximizing the number of samples analyzed, while sequencing beyond saturation increases coverage and improves confidence.

There a number of stand-alone and web-based options available for visualization of aligned or segmented epigenomic data sets (for review see [74]). The most mature and widely used are the Genome Browsers maintained by the University of California Santa Cruz [75] and Ensemble [76]. These 'first generation' genome browsers enable visualization of genome-wide data sets as linear tracks provided in the context of genome annotations. While extremely powerful for manual genome 'browsing' and focused visualizing on a gene-by-gene basis linear browser become unwieldy when large numbers of individual tracks are visualized at once. In addition these tools do not provide a capacity for larger scale integrative analysis of epigenomic data sets. While a number of informatic platforms designed for global, genome-wide analysis are currently in development, and few early versions have been published [77], the majority of genome-wide analysis of next generation sequencing based epigenomic data sets require custom scripting capabilities.

FUTURE PERSPECTIVES

Next generation sequencing has brought epigenomic studies to the forefront of current research. The past 5 years has seen dramatic increases in the stability, throughput and quality of next generation sequencing. This exponential rate of change is expected to continue as third generation sequencing platforms become available. However the underlying molecular biology supporting epigenomic experiments is likely to remain largely unchanged. Thus the effective interpretation of data sets generated from diverse laboratories using common epigenomic techniques requires the development and adoption of standards. These standards reach through from the molecular biology to sequencing, analysis and metadata included in public data submissions.

Perhaps in no other area would the epigenomic community benefit more than from the standardization of the affinity reagents used for ChIP-seq experiments, on which the bulk of current epigenomic studies rely. Currently, a diverse collection of vendors provide affinity reagents of various sensitivities and specificities. Moreover a large fraction of these resources are non-renewable polyclonals and as such cannot be used as ongoing standards in the field. Large scale epigenomic projects such as the NIH Epigenomics Roadmap [78] and ENCODE [79] have recognized this limitation and have programs targeted at the generation of renewable standardized affinity reagents. However until these are fully developed and become widely available it is critical that individual researchers undertaking epigenomic studies fully characterize the affinity reagents used in their laboratories. In this regard, arrays of modified peptides representing commonly targeted histone post-translational modifications have recently become available and should be used to assess both the false positive (cross reactions) and false negative profiles (antibody recognition blocked by adjacent modification) [80].

Equally important is the development and standardization of computational methods to process and display epigenomic data sets. Key to this effort is the development of computational derived quality metrics, similar to base quality calls used in genomic studies, for enrichment based epigenomic profiles. Ideally such a metric would provide a researcher with an understanding of the overall level of enrichment of an experiment. If widely adopted, such a common metric would allow for the meaningful comparisons between experiments. Finally as the scale of epigenomic data sets continues to increase information associated with data submissions needs to be standardized. Information related to the antibody, including vendor and lot, as well as experimental conditions are critical to enable meta-analyses of these rich data sets in the future.

REFERENCES

1. Robertson G, Hirst M, Bainbridge M, et al. Genome-wide profiles of STAT1 DNA association using chromatin immunoprecipitation and massively parallel sequencing. *Nat Methods.* **2007**; 4:651–7.
2. Barski A, Cuddapah S, Cui K, et al. High-resolution profiling of histone methylations in the human genome. *Cell* **2007**; 129:823–37.

3. Mikkelsen TS, Ku M, Jaffe DB, et al. Genome-wide maps of chromatin state in pluripotent and lineage-committed cells. *Nature* **2007**; 448:553–60.
4. Ansorge WJ. Next-generation DNA sequencing techniques. *N Biotechnol.* **2009**; 25:195–203. [PubMed]
5. Morozova O, Hirst M, Marra MA. Applications of new sequencing technologies for transcriptome analysis. *Annu Rev Genomics Hum Genet.* **2009**; 10:135–51.
6. Wold B, Myers RM. Sequence census methods for functional genomics. *Nat Methods.* **2008**; 5:19–21.
7. Bernstein BE, Meissner A, Lander ES. The mammalian epigenome. *Cell* **2007**; 128:669–81.
8. Bird A. DNA methylation patterns and epigenetic memory. Genes Dev. **2002**; 16:6–21.
9. Kriaucionis S, Heintz N. The nuclear DNA base 5-hydroxymethylcytosine is present in Purkinje neurons and the brain. *Science* **2009**; 324:929–30.
10. Lister R, Pelizzola M, Dowen RH, et al. Human DNA methylomes at base resolution show widespread epigenomic differences. *Nature* **2009**; 462:315–22.
11. Kornberg RD, Lorch Y. Twenty-five years of the nucleosome, fundamental particle of the eukaryote chromosome. *Cell* **1999**; 98:285–94.
12. Kornberg RD. Chromatin structure: a repeating unit of histones and DNA. *Science* **1974**; 184:868–71.
13. Luger K, Mader AW, Richmond RK, et al. Crystal structure of the nucleosome core particle at 2. 8 A resolution. *Nature* **1997**; 389:251–60.
14. Bernstein BE, Meissner A, Lander ES. The mammalian epigenome. *Cell* **2007**; 128:669–81.
15. Kouzarides T. Chromatin modifications and their function. *Cell* **2007**; 128:693–705.
16. Hon GC, Hawkins RD, Ren B. Predictive chromatin signatures in the mammalian genome. *Hum Mol Genet.* **2009**; 18:R195–201.
17. Solomon MJ, Larsen PL, Varshavsky A. Mapping protein-DNA interactions in vivo with formaldehyde: evidence that histone H4 is retained on a highly transcribed gene. *Cell* **1988**; 53:937–47.
18. O'Neill LP, Turner BM. Histone H4 acetylation distinguishes coding regions of the human genome from heterochromatin in a differentiation-dependent but transcription-independent manner. *EMBO J.* **1995**; 14:3946–57.
19. Robertson AG, Bilenky M, Tam A, et al. Genome-wide relationship between histone H3 lysine 4 mono- and tri-methylation and transcription factor binding. *Genome Res.* **2008**; 18:1906–17.
20. Johnson DS, Mortazavi A, Myers RM, Wold B. Genome-wide mapping of in vivo protein-DNA interactions. *Science* **2007**; 316:1497–502.
21. Harris RA, Ting Wang, Cristian Coarfa, et al. Sequence-based profiling of DNA methylation: comparisons of methods and catalogue of allelic epigenetic modifications. *NBiotechnol.* **2010**; 28:1097–105.
22. Chen J, Sadowski I. Identification of the mismatch repair genes PMS2 and MLH1 as p53 target genes by using serial analysis of binding elements. *Proc Natl Acad Sci USA* **2005**; 102:4813–8.
23. Wei CL, Wu Q, Vega VB, et al. A global map of p53 transcription-factor binding sites in the human genome. *Cell* **2006**; 124:207–19.

24. Ng P, Tan JJ, Ooi HS, et al. Multiplex sequencing of paired-end ditags (MS-PET): a strategy for the ultra-high-throughput analysis of transcriptomes and genomes. *Nucleic Acids Res.* **2006**; 34:e84.
25. Tallack MR, Whitington T, Yuen WS, et al. A global role for KLF1 in erythropoiesis revealed by ChIP-seq in primary erythroid cells. *Genome Res.* **2010**; 20:1052–63.
26. Motallebipour M, Ameur A, Reddy BMS, et al. Differential binding and co-binding pattern of FOXA1 and FOXA3 and their relation to H3K4me3 in HepG2 cells revealed by ChIP-seq. *Genome Biol.* **2009**; 10:R129.
27. Goren A, Ozsolak F, Shoresh N, et al. Chromatin profiling by directly sequencing small quantities of immunoprecipitated DNA. *Nat Methods.* **2010**; 7:47–9.
28. Bentley DR, Balasubramanian S, Swerdlow HP, et al. Accurate whole human genome sequencing using reversible terminator chemistry. *Nature* **2008**; 456:53–9.
29. McKernan KJ, Peckham HE, Costa GL, et al. Sequence and structural variation in a human genome uncovered by short-read, massively parallel ligation sequencing using two-base encoding. *Genome Res.* **2009**; 19:1527–41.
30. Fouse SD, Nagarajan RP, Costello JF. Genome-scale DNA methylation analysis. *Epigenomics* **2010**; 2:105–17.
31. Maunakea AK, Nagarajan RP, Bilenky M, et al. Conserved role of intragenic DNA methylation in regulating alternative promoters. *Nature* **2010**; 466:253–7.
32. Jacinto FV, Ballestar E, Esteller M. Methyl-DNA immunoprecipitation (MeDIP): hunting down the DNA methylome. *Biotechniques* **2008**; 44:35,37,39 passim.
33. Wilson IM, Davies JJ, Weber M, et al. Epigenomics: mapping the methylome. *Cell Cycle* **2006**; 5:155–8.
34. Sano H, Royer HD, Sager R. Identification of 5-methylcytosine in DNA fragments immobilized on nitrocellulose paper. *Proc Natl Acad Sci USA* **1980**; 77:3581–5.
35. Serre D, Lee BH, Ting AH. MBD-isolated Genome Sequencing provides a high-throughput and comprehensive survey of DNA methylation in the human genome. *Nucleic Acids Res.* **2010**; 38:391–9.
36. Hayatsu H. Discovery of bisulfite-mediated cytosine conversion to uracil, the key reaction for DNA methylation analysis–a personal account. *Proc Jpn Acad Ser B Phys Biol Sci.* **2008**; 84:321–30.
37. Frommer M, McDonald LE, Millar DS, et al. A genomic sequencing protocol that yields a positive display of 5-methylcytosine residues in individual DNA strands. *Proc Natl Acad Sci USA* **1992**; 89:1827–31.
38. Huang Y, Pastor WA, Shen Y, et al. The behaviour of 5-hydroxymethylcytosine in bisulfite sequencing. *PLoS One* **2010**; 5:e8888.
39. Lister R, O'Malley RC, Tonti-Filippini J, et al. Highly integrated single-base resolution maps of the epigenome in Arabidopsis. *Cell* **2008**; 133:523–36.
40. Cokus SJ, Feng S, Zhang X, et al. Shotgun bisulphite sequencing of the Arabidopsis genome reveals DNA methylation patterning. *Nature* **2008**; 452:215–9.
41. Meissner A, Gnirke A, Bell GW, et al. Reduced representation bisulfite sequencing for comparative high-resolution DNA methylation analysis. *Nucleic Acids Res.* **2005**; 33:5868–7.
42. Smith ZD, Gu H, Bock C, et al. High-throughput bisulfite sequencing in mammalian genomes. *Methods* **2009**; 48:226–32.

43. Ball MP, Li JB, Gao Y, et al. Targeted and genome-scale strategies reveal gene-body methylation signatures in human cells. *Nat Biotechnol.* **2009**; 27:361–8.
44. Deng J, Shoemaker R, Xie B, et al. Targeted bisulfite sequencing reveals changes in DNA methylation associated with nuclear reprogramming. *Nat Biotechnol* **2009**; 27:353–60.
45. Brunner AL, Johnson DS, Kim SW, et al. Distinct DNA methylation patterns characterize differentiated human embryonic stem cells and developing human fetal liver. *Genome Res.* **2009**; 19:1044–56.
46. Oda M, Glass JL, Thompson RF, et al. High-resolution genome-wide cytosine methylation profiling with simultaneous copy number analysis and optimization for limited cell numbers. *Nucleic Acids Res.* **2009**; 37:3829–39.
47. Flusberg BA, Webster DR, Lee JH, et al. Direct detection of DNA methylation during single-molecule, real-time sequencing. *Nat Methods.* **2010**; 7:461–5.
48. Korlach J, Bjornson KP, Chaudhuri BP, et al. Real-time DNA sequencing from single polymerase molecules. *Methods Enzymol.* **2010**; 472:431–55.
49. Eid J, Fehr A, Gray J, et al. Real-time DNA sequencing from single polymerase molecules. *Science* **2009**; 323:133–8.
50. Clarke J, Wu HC, Jayasinghe L, et al. Continuous base identification for single-molecule nanopore DNA sequencing. *Nat Nanotechnol.* **2009**; 4:265–70.
51. Leamon JH, Lee WL, Tartaro KR, et al. A massively parallel PicoTiterPlate based platform for discrete picoliter-scale polymerase chain reactions. *Electrophoresis.* **2003**; 24:3769–77.
52. Harris TD, Buzby PR, Babcock H, et al. Single-molecule DNA sequencing of a viral genome. *Science* **2008**; 320:106–9.
53. Flicek P, Birney E. Sense from sequence reads: methods for alignment and assembly. *Nat Methods.* **2009**; 6:S6–12
54. Li H, Ruan J, Durbin R. Mapping short DNA sequencing reads and calling variants using mapping quality scores. *Genome Res.* **2008**; 18:1851–8.
55. Li R, Li Y, Kristiansen K, Wang J. SOAP: short oligonucleotide alignment program. *Bioinformatics.* **2008**; 24:713–4.
56. Rumble SM, Lacroute P, Dalca AV, et al. SHRiMP: accurate mapping of short colorspace reads. *PLoS Comput Biol.* **2009**; 5:e1000386.
57. Lin H, Zhang Z, Zhang MQ, et al. ZOOM! Zillions of oligos mapped. *Bioinformatics.* **2008**; 24:2431–7.
58. Homer N, Merriman B, Nelson SF. BFAST: an alignment tool for large scale genome resequencing. *PLoS One.* **2009**; 4:e7767.
59. Langmead B, Trapnell C, Pop M, Salzberg SL. Ultrafast and memory-efficient alignment of short DNA sequences to the human genome. *Genome Biol.* **2009**; 10:R25.
60. Li H, Durbin R. Fast and accurate long-read alignment with Burrows-Wheeler transform. *Bioinformatics* **2010**; 26:589–95.
61. Li R, Yu C, Li Y, et al. SOAP2: an improved ultrafast tool for short read alignment. *Bioinformatics* **2009**; 25:1966–7.
62. Li H, Handsaker B, Wysoker A, et al. The Sequence Alignment/Map format and SAMtools. *Bioinformatics* **2009**; 25:2078–9.
63. Xi Y, Li W. BSMAP: whole genome bisulfite sequence MAPping program. *BMC Bioinformatics* **2009**; 10:232.

64. Coarfa C, Milosavljevic A. Pash 2. 0: scaleable sequence anchoring for next-generation sequencing technologies. *Pac Symp Biocomput.* **2008**:102–13.
65. Smith AD, Chung WY, Hodges E, et al. Updates to the RMAP short-read mapping software. *Bioinformatics* **2009**; 25:2841–2.
66. Chen PY, Cokus SJ, Pellegrini M. BS Seeker: precise mapping for bisulfite sequencing. *BMC Bioinformatics* **2010**; 11:203.
67. Pepke S, Wold B, Mortazavi A. Computation for ChIP-seq and RNA-seq studies. *Nat Methods* **2009**; 6:S22–32.
68. Fejes AP, Robertson G, Bilenky M, et al. FindPeaks 3. 1: a tool for identifying areas of enrichment from massively parallel short-read sequencing technology. *Bioinformatics* **2008**; 24:1729–30.
69. Tuteja G, White P, Schug J, Kaestner KH. Extracting transcription factor targets from ChIP-Seq data. Nucleic Acids Res. **2009**; 37:e113.
70. Rozowsky J, Euskirchen G, Auerbach RK, et al. PeakSeq enables systematic scoring of ChIP-seq experiments relative to controls. *Nat Biotechnol.* **2009**; 27:66–75.
71. Ji H, Jiang H, Ma W, et al. An integrated software system for analyzing ChIP-chip and ChIP-seq data. *Nat Biotechnol.* **2008**; 26:1293–300.
72. Hon G, Ren B, Wang W. ChromaSig: a probabilistic approach to finding common chromatin signatures in the human genome. *PLoS Comput. Biol.* **2008**; 4:e1000201.
73. Hawkins RD, Hon GC, Lee LK, et al. Distinct epigenomic landscapes of pluripotent and lineage-committed human cells. *Cell Stem Cell* **2010**; 6:479–91.
74. Nielsen CB, Cantor M, Dubchak I, et al. Visualizing genomes: techniques and challenges. *Nat Methods* **2010**; 7:S5–15.
75. Rhead B, Karolchik D, Kuhn RM, et al. The UCSC Genome Browser database: update **2010**. *Nucleic Acids Res.* **2010**; 38:D613–9.
76. Flicek P, Aken BL, Ballester B, et al. Ensembl's 10th year. *Nucleic Acids Res.* **2010**; 38:D557–62.
77. Bock C, Von Kuster G, Halachev K, et al. Web-based analysis of (Epi-) genome data using EpiGRAPH and Galaxy. *Methods Mol Biol.* **2010**; 628:275–96.
78. Bernstein BE, Stamatoyannopoulos JA, Costello JF, et al. The NIH Roadmap Epigenomics Mapping Consortium. *Nat Biotechnol.* **2010**; 28:1045–8.
79. Birney E, Stamatoyannopoulos JA, Dutta A, et al. Identification and analysis of functional elements in 1% of the human genome by the ENCODE pilot project. *Nature* **2007**; 447:799–816.
80. Zhang Y, Jurkowska R, Soeroes S, et al. Chromatin methylation activity of Dnmt3a and Dnmt3a/3L is guided by interaction of the ADD domain with the histone H3 tail. *Nucleic Acids Res.* **2010**; 38:4246–53.
81. Kent WJ, Zweig AS, Barber G, et al. BigWig and BigBed: enabling browsing of large distributed datasets. *Bioinformatics* **2010**; 26:2204–7.

CHAPTER 13

DESIGN AND ANALYSIS ISSUES IN GENE AND ENVIRONMENT STUDIES

CHEN-YU LIU, ARNAB MAITY, XIHONG LIN, ROBERT O. WRIGHT, and DAVID C. CHRISTIANI

INTRODUCTION

Although some diseases are predominantly environmental or genetic, both environmental and genetic factors play an important role in most common or complex human diseases. One of the major challenges of exploring mechanisms and treatment of complex diseases is that neither purely environmental factors, nor purely genetic factors can fully explain the observed estimates of disease incidence and progression. To correctly model risk estimates, we must measure genetics and environment together in the same studies. Recent advances in human genomics have made it possible to study tens of thousands of genes simultaneously and incorporate their interactions with the environment. In this review, we discuss design and analysis issues for gene-environmental interactions studies.

Traditional study designs have been used to study gene-environment interaction, including cohort and case–control studies. However some designs tend to favor the measurement of genetic over environmental factors. For example, because genotypes do not vary over time, case–control studies have been more common than cohort studies for studying genetic associations. Genotypes can always be presumed to precede phenotype and the efficiency of a case–control design over a cohort design in determining genetic main effects is well known. Several other methods, such as family-based and case-only studies have also been used, but like case–control studies, sampling is still predicated on the presence of the disease phenotype. Some of the earlier discussions of these study designs in studying genes and environment can be found in Caparaso et al. [1], Langholz et al. [2] and Garcia-Closas et al. [3]. We focus below on design and analysis issues in studying gene-environment interactions in environmental epidemiological studies including recent developments.

HOW GENETIC AND ENVIRONMENTAL FACTORS WORK TOGETHER TO AFFECT PHENOTYPES

The detection of a gene-environment interaction likely depends on more than the measurement of a genotype and an exposure. Even a cumulative index of exposure to the environmental factor may not be sufficient. It is well known that environmental exposures vary over time, but what is frequently not considered is that gene expression also varies over time. Human development consists in large part on the timed expression and silencing of specific genes in specific cells at specific life stages. From a purely biological perspective it is difficult to conceive of a gene-environment

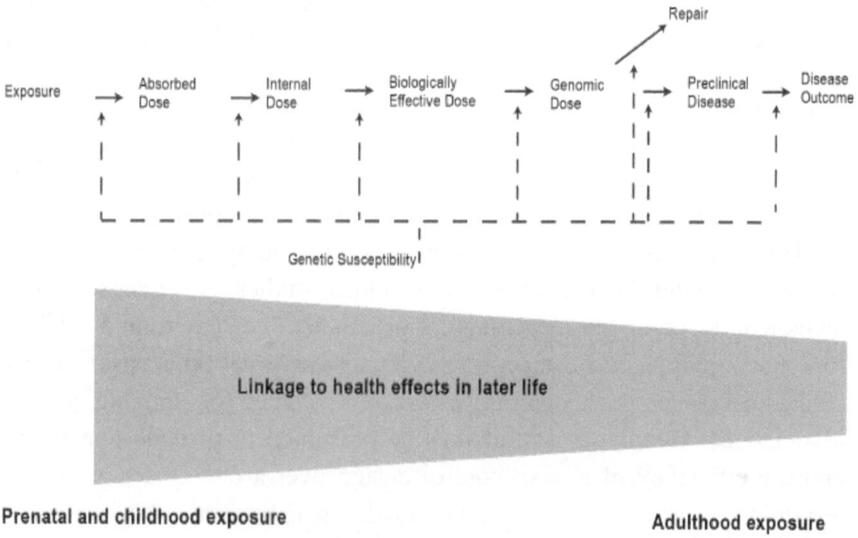

FIGURE 1. The integrated paradigm of genetic susceptibility in environmental disease development in different life stage. The exposure effects during critical developmental period (prenatal and childhood exposure) are highlighted

interaction occurring when the environmental exposure occurs during a life stage when the gene is not expressed. An overly simplistic example might be a chemical which inhibits growth by interacting with a variant in a growth factor gene. Chemical exposure at age 25 years cannot affect final height, while exposure in childhood can. In the field of toxicology, the concept of critical developmental windows of exposure has developed over the last 30 years. Rather than considering a chemical as having a single dose response curve for toxicity, chemicals appear to have different dose response curves depending on the life stage at which exposure occurs. For example, i*n utero* diethylstilbesterol exposure is associated with vaginal cancer in offspring, while mothers who took the drug do not appear to be at risk. In effect, gene-environment interaction may be conceived as a 3-way interaction, in which the time of the exposure is the 3rd factor. Alternatively one can consider environmental exposure as a time-varying covariate and study gene and time-varying-environment interactions by considering lag effects. As shown in Figure 1, we have integrated the time of the exposure in the paradigm by highlighting different exposure effects during each life stage. Direct measures of personal exposure, in particular biomarkers of exposure, provide insights into chemical, social or physical factors to specific individuals. The use of biomarkers of effect in epidemiologic studies allows researchers to study intermediate phenotypes (Figure 1) [4-6]. For example, glycosylated hemoglobin, a measure of chronic serum glucose, can be used to study diabetic risk factors with more power than a study focused on clinical diabetes. In spite of these potential advantages, the results of biomarker measurements sometimes can confuse the investigators a lot. Different conclusions may arise due to the differences of specimen kinds, collection and processing methods, laboratory error, and individual variation in the biomarker levels over time [7]. The usefulness of a biomarker is strongly depending on the specificity, sensitivity, assay reliability, and cost [8].

Another approach, instead of studying unknown effects, is by taking advantage of the established associations between genetic variations and exposure intermediate phenotypes. These genetic variations can mimic the modifiable exposure effects and serve as a surrogate to test the association between exposure and disease. This method has been referred to as 'Mendelian randomization', which provides an approach for making causal

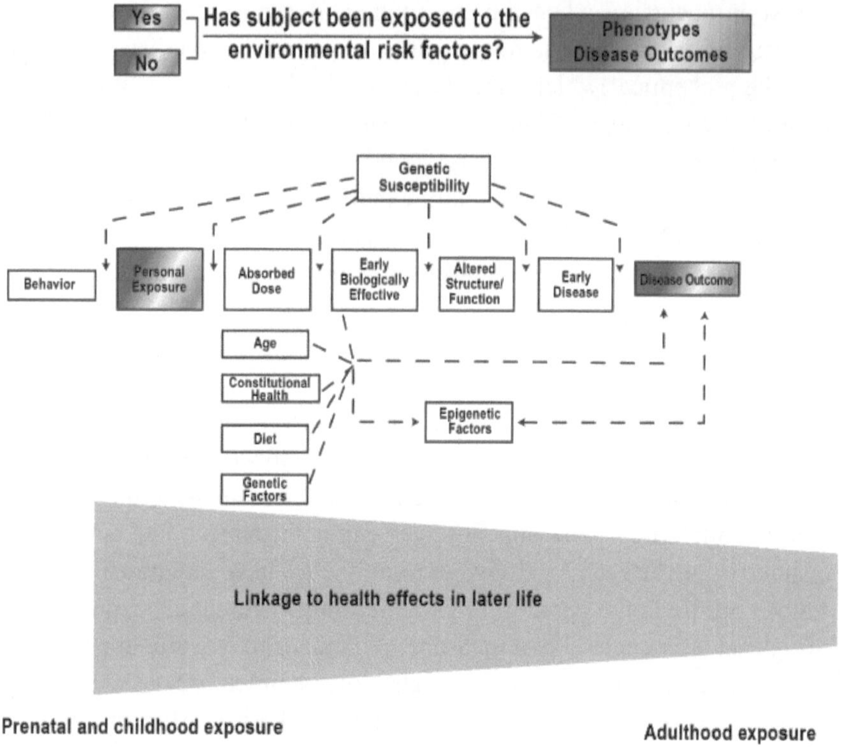

FIGURE 2. The expanded environmental genomic paradigm

inferences about the exposure by using the nature of randomly assigned genotypes from parents to offspring before conception [9,10]. However, as well with all genetic association studies, potential confounding effects by population stratifications and other limitations can still occur [10,11]. Careful study conduction and thorough verification remains essential before considering the causality.

EPIGENETICS

The role of epigenetics has been increasingly recognized as a mechanism of gene-environment interaction. Epigenetics refers to changes in gene function without altering DNA sequence. These changes may last for several generations [12]. Epigenetic mechanisms include alterations in DNA methylation, histone modification, and microRNA [13,14]. The toxic effects of exposure for several environmental chemicals, such as metals, particulate air pollution, benzene, endocrine-disrupting chemicals and reproductive toxicants, have been found to be mediated by epigenetic mechanisms[15]. Epigenetic alterations may be induced by environmental exposure, particularly in early development [16]. This field remains particularly compelling because a number of epigenetic events have been recognized as tissue-specific and reversible, which may help explain why exposures affect specific organs and the complexity of individual susceptibility among the exposed population. Epigenetic data, such as DNA methylation, can also be collected for each of the study designs described above. Epigenetic modifications provide a plausible link between the environment and alterations in gene expression that might lead to change of disease phenotypes. An increasing number of animal studies provide evidence of the role of environmental epigenetics both in disease susceptibility and in heritable environmentally induced transgenerational alterations in phenotype [17]. Thus, incorporating and analyzing epigenetic data in G-E statistical analysis has become immensely important. Epigenetic mechanisms in somatic cells also provide a potential explanation of how early life environmental exposures can program long-term effects in chronic disease susceptibility [18,19]. This expanded environmental genomic paradigm is shown in Figure2.

STUDY DESIGN ISSUES

CONFOUNDING AND SELECTION BIAS

When designing epidemiologic studies, issues of feasibility, efficiency, expense, and potential sources of bias must be considered. Perhaps the most feasible and efficient design is the case–control design, especially when studying rare diseases. A *case–control* study is conducted to collect data on environmental exposures retrospectively, and collects biomarkers after disease diagnosis of the cases. While genotypes are static and not prone to differential bias, the assessment of environment retrospectively is fraught with potential recall bias. Unfortunately, while biomarkers of exposure can reduce such bias, these measures rarely can reconstruct past exposure and may be affected by the *current disease status*, which may be one of the great challenges of retrospective studies. A fundamental requirement of a case–control study design is that cases and controls should be selected from the same population [20]. Population-based incident cases allow investigators to maximize the generalizability of the findings. Selection bias is generally a concern in case–control studies [21]. While the assessment of gene-environment interactions will not be subject to selection bias if participation does not differ by genotype conditional on exposure and disease status [22]. This assumption may seem reasonable for most genes and exposures, with the possible exception of (1) alleles that influence behavior, such as *aldehyde dehydrogenase*polymorphisms and alcohol exposure [23]; or (2) population stratification; or (3) alleles and exposure risk factor that influence disease detection. For example, in populations where prostate-specific antigen (PSA) screening is commonly performed, higher PSA levels often trigger for prostate biopsy and may increase early diagnosis of prostate cancer [24,25]. Differential prostate cancer screening and detection with respect to obesity [26,27] and PSA associated genes [28,29] may cause selection or detection bias. For fatal diseases, since only some of the incident cases may be available for interviewing, survivor bias can occur if genotypes or exposure status differ by survival time.

Observational epidemiological studies often suffer from confounding bias due to measured and unmeasured confounders. An example of genetic confounding bias is population stratification. Population stratification can occur in ethnically mixed populations and can lead to spurious (i. e. non-causal) associations if both the baseline disease incidence and the allele frequency vary by ethnicity [30].

Although most bias due to population stratification can be eliminated by following the rules of well-designed, well-conducted study and matching or adjusting on ethnicity, this may not apply to populations whose ancestors recently mixed, such as African or Hispanic Americans [31,32]. Several genomic control approaches have been used to attempt distinguishing the ethnicity by genotyping markers that are unrelated to disease and known to have different allele frequency in ancestral populations [33,34].

Fully distinguishing the observed association from population stratification bias, can be achieved by replication of consistent findings from multiple well-designed studies in different populations or family-based study design which preclude stratification [32]. Unlike the traditional case–control studies based on unrelated individuals, family-based studies are immune to population stratification bias [35,36]. Family-based studies of gene–environment interaction sometimes may be more powerful than population-based studies [37]. However, the application could be limited by shared environment among family members and the difficulties to collect DNA samples from family members than from unrelated cases and controls, especially for long latency or late-onset diseases. Family-based studies generally have less power for genetic main effects than do case–control studies. Besides, family-based studies usually collect environmental exposure information retrospectively and may have similar problems in exposure assessment as retrospective case–control studies. The over sampling of intact families would also not be expected to represent social environments in the general population. Another approach is to use the *case-only* method to study gene-environment interaction. This approach does not allow evaluation of the main effects of the genotype alone or the exposure alone, but only their interaction [38,39]. The case-only design requires an assumption of gene-environment independence in the general population [40,41].

The *prospective cohort* study requires study subjects to be recruited before the onset of disease. This approach has the advantage of prospective collection of environmental information and biomarkers, which both precede the disease and will be unaffected by recall bias [42]. Effective follow-up should minimize selection bias secondary to attrition, one can estimate the disease incidence rate, and the inference for an underlying cohort is often well defined. Analysis of data from cohort studies is subject to bias due to loss of follow-up. As incidence rates of most diseases are low, even with many years of follow-up a cohort study often requires collection of an *extremely* large number of individuals before the onset of disease and a sufficient follow-up time, which simultaneously lead to extraordinary cost increase (i. e. by completing follow-up and data collection, including the data of baseline characteristics, exposure, and genotyping data). Hence, prospective studies are considerable challenges for diseases with low incidence rate. Risk-based sampling is being used to increase the power of prospective studies by enrolling first-degree relatives of probands, such as the Sister Study for breast cancer risk [43,44] or the on-going Early Autism Risk Longitudinal Investigation (EARLI) study for autism risk. For common pediatric diseases such as asthma, obesity, and some adverse birth outcomes, a prospective cohort study will be extremely valuable to identify environmental risk factors as well as evaluate gene-environment interaction mechanisms [45,46]. Prospective cohort studies on a national scale [47] or by pooling data from existing prospective cohorts [48] should be conducted to ensure sufficient power in gene-environmental studies. The U. S. Congress, through the Children's Health Act of 2000, authorized the National Institute of Child Health and Human Development (NICHD) "to conduct a national longitudinal study of environmental influences (including physical, chemical, biological, and psychosocial) on children's health and development" [49]. The National Children's Study is a 21-year prospective cohort study of 100,000 US-born children. Environmental exposures, including chemical, physical, biological, and psychosocial exposure, will be assessed repeatedly during pregnancy and childhood in children's homes, schools, and communities. The National Children's Study will provide great opportunities to gene-environment interactions for common pediatric diseases.

MEASUREMENTS OF EXPOSURE AND EFFECTS BY LIFE STAGE

EXPOSURE BIOLOGY

Measurement errors, such as misclassification of genotypes or exposure status, can exist regardless of study design. Measurement of environmental exposures have been a great challenge in epidemiologic studies due to the complex pattern of long-term exposures and the need to collect accurate and repeated individual exposure data in large populations [50]. Misclassification of exposure generally leads to attenuation of the main effects when the error is non-differential [51]. Non-differential misclassification can also bias away from the null in some circumstances, including (1) if the exposure is multilevel (>2 levels), the intermediate levels of exposure could be biased away from null [52,53]; (2) if the misclassifications are correlated with other errors [54,55]; (3) if the measured exposure do not change monotonically with the true exposure [53,56]. However, in the estimation of multiplicative gene-environment interaction effect, Garcia-Closas *et al.* [57] showed that under a set of conditions typically satisfied in studies of gene-environment interactions, both differential and non-differential misclassification of a binary environmental factor biases a multiplicative interaction effect toward the null value. These conditions are that: (1) the environmental exposure is independent of the genotype among the controls, and (2) exposure misclassification is non-differential to the genotype. This result is also true for misclassification of genetic factors.

The use of questionnaires for exposure assessment relies on personal memory and has the potential for recall bias. Several technologies have been developed to improve measurements of environmental exposures. To incorporate qualitative and quantitative changes of environmental exposures, such as atmospheric conditions and topography, over time and space, as well as individuals' diverse demographic characteristics, lifestyles, activity patterns, geographic information systems (GIS)/global positioning system (GPS), personal monitoring, and biomonitoring are now being used in environmental epidemiology. Combined geospatial tools with statistical models allow investigators to model the transport of the pollutants from source to residence, e.g., using wind speed, temperature, and traffic density in addition to measurements from the central site, to estimate an individual-level exposure as well. Direct exposure monitoring includes personal monitoring by measuring toxics on or near the body, such as measuring air pollutants exposure levels at the breathing zone, or by sampling biological properties, such as the measurement of urinary 1-hydroxypyrene (1-OHP) as a biomarker of short-term polycyclic aromatic hydrocarbon (PAH) exposure [58]. Biomarkers of exposure are biological indicators of exogenous agents within the biological system, or other event in the biological system related to the exposure. With stringent quality control, these monitoring data hold great promise for improving exposure assessment by providing objective individual-level measurements. Biomarkers can be used to reflect the effects of earlier exposures and the association between exposure and disease at the molecular level [4-6]. Examples of intermediate biomarkers include chromosomal alterations, DNA, RNA and protein expression. In response to exposure, patterns of gene expressions, proteins, or metabolic profiles in cells and tissues change can serve as biomarkers for exposure or effect. These dynamic features however, make their interpretation in human studies challenging. Single measurement may not be reliable especially in those investigating long-term chronic effects. Incorporating long-term monitoring data with different exposure assessment techniques is needed to provide an integrated view of exposure in complex exposure–disease relationships [59,60].

DEVELOPMENTAL LIFE STAGE AND GENE-ENVIRONMENT INTERACTIONS

Measuring environment has added complexity beyond issues of measurement error or selection bias. Even measuring cumulative exposure prospectively may be insufficient to capture gene-environment interaction. This is because human development occurs in life stages during which gene expression undergoes radical yet temporary changes. Environmental exposures might alter the timing of normal developmental regulation of gene expression or the gene product expressed solely at a specific life stage may interact with the environmental exposure. In particular, during prenatal life and childhood, critical biological events occur that establish the number, connections and proper function of cells within given tissues. As an example, changes in gene expression could be modulated through DNA promoter methylation or chromatin remodeling, which may be induced by environmental exposure, particularly in early development [16]. Toxicological studies show that the central nervous system is especially vulnerable to toxic injury [61] and epidemiological studies clearly show an association between adverse neurodevelopment and *in utero* exposure to chemicals such as methyl mercury [62,63], PCBs [64], while exposure later in life demonstrates less toxicity. Epidemiological studies of chemicals typically show a large variance around the effect estimate for the dose–response relationship. While many factors contribute to this variance, including measurement error in exposure and/or phenotype, it is likely that the timing of the exposure and variant genetic factors that modify the response to toxicants contribute significantly to the observed variance. Genetic variants that produce gene-environment interactions may only do so when the exposure corresponds to a critical developmental window during which that gene is highly expressed. This is a fundamental concept in developmental biology that is often overlooked in epidemiologic studies. Indeed the concept of fetal origins of adult diseases demonstrates the critical nature of exposure *timing* in producing later health effects (e.g., the association of maternal smoking during pregnancy and reduced fetal growth [65], obesity [55], decreased lung function [66] and diabetes [67] in the offspring). Although a prospective study can address timing of exposure in a clearly unbiased manner, it is still challenging to assess the details of exposure timing and risk as the critical window likely differs for different phenotypes and for different exposures. It is also not possible to know with certainty what the critical exposure window is *a priori* (i. e. *in utero* vs. childhood vs. puberty). The difficulties in assessing the effects of exposure by timing present in carefully designed observational studies and even trial results. An example is the initial report from Women's Health Initiative (WHI) randomized trial and epidemiologic data on the risk of coronary heart disease (CHD) and the menopausal hormone therapy. Large observational studies include Nurses' Health Study (NHS) suggested a reduced risk of CHD among postmenopausal hormone therapy [8,68] while WHI randomized trial found increased risk of CHD among women assigned to the menopausal hormone therapy compared to the placebo group [69]. Hernán et al. re-analysis of the Nurses' Health Study and concluded that most of the difference could be attributed to the age distribution at the time of initiation of hormone therapy and length of follow-up [70].

Unfortunately, for most adult diseases, an unbiased reconstruction of childhood exposure is difficult, if not impossible. Thus, a major limitation of adult epidemiologic research will continue to be the inability to reconstruct childhood factors that predict disease. At least some of the difficulty in finding gene-environment interactions for adult disease is likely that the relevant exposure may have occurred in childhood, and a measure of cumulative exposure, while preferable to cross-sectional measures, cannot capture exposure during the critical developmental life stage predisposing to disease.

STATISTICAL ANALYSIS ISSUES FOR GENE-ENVIRONMENT STUDIES

LONGITUDINAL STUDIES

In order to incorporate exposure effects by life stage, gene-environment interaction may be conceived as a 3-way interaction, in which the time of the exposure is the 3rd factor. In general the gene-environment interaction as a function of time can be modeled by considering a general nonparametric model

$$Yij=f(Gij,Eij,tij)+eij,$$

where Y_{ij} is the response of interest of the i-th subject at the j-th time point t_{ij}; G_{ij} and E_{ij} are the genetic and environmental covariates measured at t_{ij}, and e_{ij} are random errors. Here the function f(.) models the combined effect of gene, environment and any possible interactions as function of time. Note that the formulation above can incorporate multiple genetic and environmental variables and thus has potential to model gene-gene interactions as well as gene-environment interactions involving several genes as well. For such general model of longitudinal data, Zhang [71,72] presented multivariate adaptive spline smoothing based estimation methods. For high-dimensional data, such as GWAS studies directly applying such methods for a large number of SNPs is undesirable. Zhu et al. [73] adapted the multivariate spline methodology for GWAS: Specifically, the procedure starts by starting with a model containing only intercept (the simplest model) and then gradually growing the model by adding terms (e.g., individual SNPs, SNP-SNP interaction) that minimizes a weighted least squares criteria. Finally the end model is selected via a backward step by deleting one least significant term at a time from the model.

Another popular and useful approach for modeling factors that change over time is the varying coefficient modeling strategy. Specifically for G-E interaction, one can consider the time-varying coefficient model

$$Yi(tij)=\beta 0(tij)+Gij\beta G(tij)+Eij\beta E(tij)+Gij*Eij\beta GE(tij)+eij,$$

where t_{ij} denotes the time point for the j-th measurement of the i-th subject; G_{ij} and E_{ij} are the genetic and environmental covariates measured at t_{ij}; $\beta_G(.)$, $\beta_E(.)$ and $\beta_{GE}(.)$ are unknown gene, environment and G-E interaction effect, respectively, depending on time. Note that this is a generalization of the conventional two-way G-E interaction model $Y_i(t_{ij}) = \beta_0 + G_{ij}\beta_G + E_{ij}\beta_E + G_{ij}E_{ij}\beta_{GE} + e_{ij}$ with non-time-varying effects. Depending on the data at hand, one could also consider different version of this model in various ways, e.g., $\beta_G(t_{ij}) = \beta_G$ corresponds to the model where one assumes that only the intercept, the environment effect and G-E interaction effect vary over time but the gene effect does not. There is a rich literature on varying coefficient models discussing estimation and testing procedures, e.g., Hoover, Rice, Wu and Yang [74] and Wu and Chiang [75] among many others. The coefficient $\beta_{GE}(.)$ reflects the G-E interaction effect as it changes over time. Thus, if the G-E interaction is prominent at a specific window of time but dormant in others, plotting this coefficient function over time could potentially reveal such patterns.

CASE–CONTROL STUDIES

Case–control studies are commonly used in studying for genes and environment. Case–control studies sample disease subjects (cases, D=1) and healthy subjects (controls, D=0), and retrospectively collect information about genes (G) and environment (E). The data from a case–control study can be used to compute three odds ratios (ORs), using subjects who are unexposed and have typical genotypes as they occur in nature (also known as wild type) (E=G=0) as the reference group: OR_{11} for subjects with both the gene and the exposure (E=G=1), OR_{10} for subjects with only the exposure

(E=1, G=0), and OR_{01} for subject with the only gene (E=0, G=1). Then under the multiplicative interaction model, the null hypothesis of no interaction can be written as $OR_{11} = OR_{01} \times OR_{10}$. Thus, to test for GxE interaction, one defines the interaction odds-ratio as $OR_I = OR_{01} \times OR_{10}/OR_{11}$ and tests for H0: $OR_I = 1$.

Logistic regression is commonly used for analysis of case–control studies, especially in the presence of covariates. A typical logistic model for assessing gene-environment interaction is

$$\text{logit}(p)=\beta0+\beta1G+\beta2E+\beta3G{*}E+\beta4\mathbf{X} \quad (1)$$

where p is the population disease probability and X is a vector of covariates. As subjects are sampled based on the case–control status and cases are over-sampled, the likelihood depends on distribution of the independent variable (G, E and X) in the population and the case–control sampling probability. Hence the intercept β_0 cannot be estimated from the case–control sample. However, Cornfield [77] and later Prentice and Pyke [78] showed that one can estimate all the regression coefficients β except for the intercept using the ordinary logistic regression likelihood as if the data were obtained in a prospective study.

Under model (1), the OR of (G, E) versus (G_0, E_0) is then given by $\exp\{\ \beta_1(G - G_0) + \beta_2(E - E_0) + \beta_3(GE- G_0E_0)\}$. In the presence of gene-environment interaction, the OR of disease and gene depends on exposure. For example, consider the case when both G and E are binary. The covariate X adjusted OR of D and G in the unexposed group (E=0) is $\exp(\beta_1)$ and the OR of D and G in the exposed group (E=1) is $\exp(\beta_1 + \beta_3)$. The interaction $OR_I = \exp(\beta_3)$. The null hypothesis H_0: $\beta_3 = 0$ constitutes a no gene-environment interaction. Note that no assumption about the distribution of gene (G), environment (E) and covariates X, e.g., independence of gene and environment, is made in logistic regression.

Several advanced models have been developed to incorporate gene-environment interactions. Selinger-Leneman et al. [79] explored the conditions under which accounting for gene-environment interaction enhances the ability to detect the genetic effects in complex diseases. Chatterjee, et al. [80] developed a maximum score based testing procedure for main gene effects in the presence of possible gene and environment interaction using parametric models. Kraft et al. [81] applied a two degree-of-freedom likelihood ratio test for the association between a disease and a genetic locus, allowing for the possibility that the genetic effect may be modified by an environmental factor. Maity et al. [82] developed more flexible statistical tests for genetic main effects in presence of possible gene-gene and gene-environment interactions using a semiparametric method.

Nevertheless, one should be aware that the case–control method may not be applicable for association studies in some situations, such as in the presence of population stratification that can not be estimated from the data. It is useful to complement case–control studies with family studies using genetic analytic techniques such as segregation and linkage methods [83].

CASE ONLY STUDIES

An important matter in case–control studies is the choice of control group. An inappropriate choice of controls, e.g., hospital based controls or shared controls for different studies, may result in erroneous findings, e.g., due to population stratification. To address this problem, several approaches have been developed, see e.g., [40]. One of these approaches to assess G-E interaction is the case-only design where one uses only cases (D=1).

A key assumption to study G-E interaction on D in the case-only design is that the distributions of gene and environment are independent. Examples of such situations are the cases when an environmental factor is not directly controlled by individual behaviors, e.g., air pollution. Specifically, in the absence of covariates, under model (1), assuming rare disease, Pr(D=0 | G, E) is approximately 1. Assuming that G and E are binary and independent in the population, it can be shown that the OR relating exposure and genotype in cases only is

$$pr(G=1,E=1||D=1)pr(G=0,E=0||D=1)/[pr(G=0,E=1||D=1)\,pr(G=1,E=0||D=1)]=\exp(\beta 3).$$

Thus, one can estimate the effect of the G-E interaction term approximately correctly without performing a logistic regression of D. This approach can also be applied in logistic models in the presence of covariates [39]. Under the assumption of the independence of gene and environment, the case-only analysis yields a smaller standard error when estimating the interaction term β_3, thus increasing power to detect GxE interaction [39]. Umbach and Weinberg [84] conjectured that imposing the gene and environment independence assumption in studies where controls are available could also improve precision for estimating main effects. They also investigate the power gain in detecting GxE interaction via simulation studies and find that in several parameter configurations considerable precision advantages can accrue by estimating the interaction term using G-E independence assumption. They find that sometimes the variance of the interaction term can be reduced by more than two-fold, even near the null value $\beta_3=0$. Thus, in situations where the key independence assumption is met, a study analyzed with G-E independence assumption may need considerably fewer subjects than one analyzed with the full model without G-E independence assumption to achieve the same power for detecting gene-environment interaction. Several researchers exploit the assumption of gene-environmental independence in the population to develop more powerful statistical tests for gene and environment interactions in more complex settings, see e.g., [84-86].

However, one should exercise caution when applying case-only analysis, as it makes a strong assumption that G and E are independent in the population, possibly conditioning on covariates. If the distribution of G and E depend on each other, the case-only design will yield a biased estimate of the interaction term β_3 In addition, it only estimates the interaction term β_3 and cannot estimate the main effects β_1 and β_2. In practice, the assumption of G-E independence in the population may not hold. For example, the genetic variants in a smoking pathway may affect the degree of addiction. In such scenario, a case-only study for studying the effects of genes and lung cancer risk would not be applicable. Further, the validity of a case-only study also hinges on the assumption that there is no hidden population stratification in the study population. Wang and Lee [87] showed that if a population stratification exists, then case-only studies may be biased, and the bias involves the coefficient of variation of the exposure prevalence odds, the coefficient of variation of the genotype frequency odds, and the correlation coefficient between the exposure prevalence odds and the genotype frequency odds. In other words, a case-only study may be biased if a systematic difference is present in either genotype frequencies or exposure prevalence between subpopulations.

CASE-PARENT AND CASE-SIB DESIGN

In a 'case-sib' design, each case is matched to one or more unaffected siblings [88-90]. Compared to the case–control design, this design has the advantage that cases and controls are perfectly matched on the ethnic background, thus this design reduces the bias due to population stratification.

In the 'case-parent' design, the parents of cases are used as a sort of control group to study genetic markers that could be associated with disease risk or be in linkage disequilibrium with alleles at a neighborhood locus. Genotypes are obtained from each case and his/her two parents, while environmental data are required only from cases [41]. Similar to the case-sib design, this design provides a perfect control for ethnic confounding. The main effect of environmental factors cannot be assessed in the case-parent design, but analysis of genetic main effects and G- E interactions can be conducted. Umbach and Weinberg [91] proposed an association test, which examines the joint effects of gene and environment using case-parent trios. The case-parental control method requires the availability of genotypic information on both parents of cases, although the EM algorithm can be used to maximize the likelihood if some genotypes are missing and the method has been extended to situations where only one parent is available [92]. Witte et al. [90] and Gauderman et al. [89] compared the relative efficiency of the case-sib and case-parent designs to the matched case–control design for estimation of genetic main effects. They also provided some comparisons of efficiency for estimation of the G × E interaction effect. They found that because of overmatching on genotype, the use of sibling controls leads to estimates of genetic relative risk that are approximately half as efficient as those obtained with the use of population controls, while relative efficiency for cousin controls is approximately

90%. However, they also find that for a rare gene, the sibling-control design can lead to improved efficiency for estimating a $G \times E$ interaction effect.

GENOME-WIDE ASSOCIATION STUDIES

A genome-wide association study involves scanning tens of thousands of genetic markers (SNPs) across the genome to identify the genetic variations that are associated with a disease or a trait[93,94]. Such studies are particularly useful in finding common genetic variations that contribute to common and complex diseases, such as heart disease, cancer, and diabetes. Compared to linkage analysis, GWAS can be more powerful in detecting genes associated with modest increases in disease risk [95]. In the past few years, GWAS have been successful in identifying over a hundred common genetic variants that are associated with complex diseases (http:// www. genome. gov/gwastudies *webcite*).

In a traditional case–control GWAS, one observes a disease outcome D, environmental exposure E, and the genotypes of M SNPs spanning the genome, with g_1, g_2, ..., g_M denoting the genotypes at the M loci. Illumina and Affymetrix provide common genotyping platforms for GWAS, where the genotypes of a million or more SNPs can now be simultaneously measured. Several models can be used for the pattern of inheritance of the genetic susceptibility. Under the dominant model, subjects with genotype g = AA or Aa are genetically susceptible, that is, they are at either increased or decreased risk compared to the baseline group (g = aa). This structure can be captured by defining the genetic covariate G such that G = 0 for g = aa, and G = 1 for g = AA or g = Aa. Under the recessive model, we have G = 1 for g = AA and G = 0 otherwise. Under the co-dominant model, one can use two dummy variables or an additive model (G=0,1,2) to model the genetic effect. Let p and q = 1- p be the probabilities of observing A and a respectively. Assuming Hardy-Weinberg equilibrium, the distribution of genotypes g in the population is given by pr(g = AA) = p^2, pr (g = Aa) = 2pq, and pr(g = aa) = q^2. Hardy-Weinberg equilibrium should be checked when the genotype data are cleaned.

GWAS studies have primarily focused on detecting the main gene effect by fitting the traditional logistic main effect model for each SNP G_j separately as

$$\text{logit}(p)=\beta 0+\beta 1Gj+\beta 2E+\beta 4X, \quad (2)$$

where X is a vector of covariates and often also includes a few principal components to control for population stratification [96]. A correction for multiple comparisons, such as the Bonferroni correction, or modified Bonferroni correction [97], is often used to control for the genome-wide type I error. Several multi-locus tests have been proposed to improve the power in GWAS studies[98]. Top SNPs from GWAS are then selected for validation in independent samples.

To study gene-environment interaction in GWAS, one can fit model (1) for each SNP separately and test for H_0: $\beta_3 = 0$ and use the Bonferroni correction to adjust for multiple comparisons. A main challenge in using GWAS to test for G-E interaction is that most GWAS have limited power to detect gene-environment interaction on the genome wide scale after accounting for multiple comparisons. One might consider using the case-only analysis to increase the analysis power. However, the case-only analysis relies on the strong assumption of the independence of gene and environment in the population, which might be not reasonable across all SNPs that are scanned in a GWAS.

Several approaches have been recently proposed to improve power for detecting gene-environment interactions in GWAS. Kraft et al. [81] proposed to screen for top genes in the presence of possible gene-environment interactions using a 2-df test for testing for the main genetic effect and G×E interactions jointly. They showed that under a variety of parameter settings, the 2-df test was often more powerful than a test of the main effect or the traditional test for G×E interactions.

Assuming binary E, Murcray et al. [99] proposed a two-step approach where they first use a likelihood ratio test of the association between G and E based on the logistic model $\text{pr}(E = 1 \mid G) = a_0 + a_1G$ and test for H_0: $a_1=0$. This corresponds to the standard case-only test for the G×E interaction.

One then screens for the significant genes with p–values below a threshold. In the second step, the screened SNPs are then tested using the standard G x E interaction test under model [1] with correction for multiple comparisons. They showed this two-stage test is more powerful than the standard one-stage test for the gene-environment interaction (H_0: β_3=0) using model [1]. The added power of this two-stage procedure derives from the fact that the multiple comparison in the second step is performed only based on the genes chosen in the first step, not the entire set of genes. As shown by Murcray et al., this two-step method can be almost twice powerful than the traditional one-step procedure if the G-E independence assumption is valid for a large fraction of G-E combinations under study. However, the power gain of the procedure diminishes as the total number of genetic markers increases [100].

Mukherjee and Chatterjee [101] proposed a 1-stage inferential procedure on G-E interactions using an empirical Bayes-type shrinkage estimation approach. They estimate the interaction using a weighted average of the case-only and case–control estimators where the weights are based on the difference between the two estimators and the variance of the robust case–control estimate. This estimator is shown to be robust to the departure from the G-E independence assumption. The associated test can gain efficiency and power when the assumption of G-E independence in satisfied in the underlying population but also preserves Type I error when the independence assumption is violated.

COHORT STUDIES

In prospective cohort studies, a sample of healthy subjects in a pre-specified cohort of subjects are recruited, environmental and lifestyle data and Biological samples, such as blood, are obtained at baseline (the start of the study), and the subjects are then followed prospectively over time for disease onset or quantitative traits. Questionnaire data and biological sample may be also updated over time. As Clayton and McKeigue [102] state, "The rationale for setting up cohort studies of genetic effects on disease risk is based on the argument that, because cohort studies can measure environmental exposures before disease onset, they are better than the case–control design for study of gene-environment interactions. Study of such interactions is thought to make detection of genes that influence disease risk easier, to allow individuals at high risk to be identified for targeted intervention, and to advance understanding of biological pathways leading to disease. "

For binary D, E, and G, one can now estimate disease risks and subsequently estimate relative risks (RRs). Parallel to the odds ratio calculations in case–control studies, one can define four RRs. For example, using the non-exposed and non-high-risk genotype (G=E=0) as the reference group: RR_{11} is the RR comparing the exposed and high-risk genotype group (G=E=1) and is estimated as h(a+c)/(f+h)a, similarly one can define RR_{10}, RR_{01}, and RR_{00}. Under the multiplicative interaction model, the null hypothesis of no G-E interaction can be written as H_0: $RR_{11} = RR_{01} \times RR_{10}$. This is equivalent to testing H_0: $OR_{11} = OR_{01} \times OR_{10}$ and can proceed with logistic regression.

One major limitation of cohort studies is that rare events will not occur at sufficient frequency so that most cohort studies may not record sufficient numbers of cases for rare diseases and might have only marginal power for common diseases [102]. Cohort studies can be used to study gene effects and gene-environment interactions for disease progression and censored time-to-event data using survival analysis techniques, e.g., the Cox model [103], and for longitudinal phenotypes using mixed models and GEEs [104].

NESTED CASE–CONTROL DESIGN AND CASE-COHORT STUDIES

Epidemiologic cohort studies and disease prevention trials typically require the follow-up of several thousand subjects for a number of years before yielding useful results, and hence can be prohibitively expensive. To address this issue, a pseudo case–control design can be used to reduce the number of subjects for whom covariate data are required (see for example, [105-108]), where each subject developing disease is matched to one or more subjects without disease at the same point in 'time' using incidence-density sample. Henceforth, relative risks are estimated using a matched case–control analysis. In this setup, one only requires the covariate measurements for only cases and their matched controls. This is the so-called 'case-control nested within a 'cohort' design.

However, intuitively the alignment of each selected control subject to its matched case could be inefficient, since that subject may also properly serve as a member of the comparison group for cases occurring at a range of other times. In the context of a disease prevention trial, it is often desirable to have a subset of the trial cohort for whom covariate data are analyzed on an ongoing basis in order to monitor intervention effectiveness and compliance. The case–control approach is not well suited to this purpose since covariate histories are only assembled following case occurrence. As an alternative, Prentice [78] proposed a 'case-cohort' design which involves the selection of a random sample (or a stratified random sample) of the entire cohort, and the assembly of covariate histories only for this random subcohort and for all cases. The subcohort in a given stratum constitutes the comparison set of cases occurring at a range of failure times. The subcohort also provides a basis for covariate monitoring during the course of cohort follow-up. Very similar designs have also been proposed by Kupper, McMichael & Spirtas [109] and Miettinen [110]. These more efficient designs have started being used to study gene-environment interactions in cohort studies [111].

The statistical efficiency for a case-cohort study over a nested case–control study is small. Wacholder [112] pointed out that nested case–control designs have a small to moderate advantage for studies with substantial late entry or censoring. Case-cohort studies gain small advantage in studies with little late entry or censoring. However, a major practical advantage of the case-cohort studies is the ability to use the same subcohort for several outcomes such as different subtypes of disease [112]. If one intends to compare the risk factors of different outcomes then adjustments of significance levels and confidence intervals are required due to multiple comparisons and to account for possible correlations between outcomes [113]. However, if the main focus is on the evaluation of risk factors for each disease separately then no such adjustment is required.

TWO-STAGE DESIGNS AND BIASED SAMPLING

In many situations, the exposure of interest and the disease endpoint can both be rare and studies of their relationship between them require a very large number of samples, and hence can be very expensive. In such cases, a two-stage design, originally proposed by White [114], can be employed. A major assumption in this scenario that the exposure information is already available for a large sample of controls and cases in the screening stage. Complete covariate and genotype information is then collected only on a subsample, where the sampling fraction can depend jointly on disease and exposure status. For example, in case of a rare exposure, one can oversample those who are more likely to have exposure and perform the genotyping on a more informative subset of subjects. A similar approach can be taken where a specific rare genotype is of interest and exposure is expensive to record. White [114], assuming the exposure and disease status are both binary, presented a procedure to derive valid estimates of odds ratio by incorporating the information from the first both stages sample and the sampling proportions for the second stage. Cain and Breslow [115] extended this approach by allowing for a multilevel exposure variable, and any number and type of any covariates. There is a large recent literature on analysis of two-stage case–control designs using more efficient inverse probability weighted estimation procedures and semiparametric efficiency procedures [116]. Weinberg and Wacholder [117] developed designs of case–control studies with biased sampling for more general cases. They developed and presented analytic techniques and estimation procedures. They show that unbiased estimation procedure of the main and the interaction effects are possible assuming given the sampling fractions are known for the second stage sampling. From the simulation studies of Weinberg and Wacholder, it is seen that the effect of the screening/matching factor in the stage 1 sampling can often be estimated with better precision compared to completely sampling. In addition, the main and interaction effects can be also estimated more efficiently compared to random sampling. The advantage of this design appears to improve the efficiency of estimation of the interaction coefficient; the efficiency gain could be as large as 250%. The efficiency gain is however dependent of the odds ratio relating the exposure and genotype to the disease.

POWER AND SAMPLE SIZE CONSIDERATIONS

Gene-environmental studies often require large sample sizes to detect interactions compared to studies for detecting main gene and environmental associations. Thus, power and sample size considerations are critical. There have been several publications about sample size and power calculations in G-E studies (Table 1). The software QUANTO developed by Gauderman [37] is convenient for power and sample size calculations for a range of gene-environmental designs.

TABLE 1.

Summarized publications regarding sample size and power calculations in gene-environment studies

Source	Design
Yang et. al. [118]	Case-only
Cai and Zheng [119]	Case-cohort
Schaid [41]	Matched case–control
Gauderman [37]	Case-sibling
	Case-parent
Lubin and Gail [120]	Unmatched case–control, Multivariate regression models for odds ratio
Hwang et al. [121]	Unmatched case–control, binary genetic and environmental factors
Foppa and Spiegelman [122]	Unmatched case–control, binary genetic factor and an environmental exposure with multiple categories

DISCUSSION

With advancement in human genetics and risk assessment, current research has shown that the interplay between genes and environment is critical to disease risk and progression. Consequently, more research efforts need to be directed towards investigating the genetic basis of individual susceptibility and the role of the genome and epigenome, to various environmental agents. The methodological issues raised above are focused on the "how to" approaches to assessing gene-environment interactions.

All individuals are exposed to a variety of hazardous agents and chemicals in the environment. However, genetic pathways are thought to have evolved for minimizing the adverse effects from these environmental insults. Genes expressed in these pathways, referred to broadly as environmentally responsive genes, exhibit heritable variability that may be associated with altered efficiency of the pathway. Hence, gene-environment investigation needs to go beyond individual genes to investigate the roles of genetic pathways and networks.

Several research programs were launched to promote and facilitate research in environmentally responsive genes. In the 1990's, the National Institute of Environmental Health Sciences (NIEHS), of the U. S. National Institutes of Health, initiated a multiyear project entitled the NIEHS Environmental Genome Project (EGP). The focus of the NIEHS EGP is on common sequence variations, referred to as genetic polymorphisms, in environmentally responsive genes. The NIH-wide Genes, Environment and Health Initiative (GEI) was launched in February 2006 to support research that will lead to the understanding of genetic contributions and gene-environment interactions in common disease. Numerous scientific advances have been made through these initiatives.

More advanced statistical and computationally efficient methods need to be developed to investigate the interplay of genes and environment in human diseases. Data integration is becoming more and more important. More interdisciplinary research by integrating molecular biological knowledge, environmental sciences, bioinformatics and computational biology, and statistical and computational methods is likely to advance research in genes and environment. More research is needed in several emerging research areas in genes and environment, such as exposure biology for identifying new biomarkers for better measuring exposures, mediation (causal inference) analysis, e.g., for effects of environment of disease phenotypes through epigenetic markers, statistical methods for high-dimensional data analysis for genes and environment, and risk prediction using genes and environment. It should be noted that the process of translating genetic and 'omic research into practice in environmental and occupational health is considered to be in an early phase. Thus, most research findings from genetic susceptibility studies should be communicated with caution to the general public at this time. Policy research on genes and environment deserves more attention.

CONCLUSION

In conclusion, we are entering an exciting period of research and knowledge generation about gene-environment interactions. The potential for combining basic bench work with human population studies opens up many opportunities to examine the health effects of complex environmental exposures. The challenges for the next decade for human population work in this field include maintaining rigorous epidemiologic study design, improving environmental exposure analysis, advancing genomic technology and knowledge, and expanding the necessary analytic and computational tools for high-throughout "-omic" and environmental data, and the concomitant policy and ethical implications.

REFERENCES

1. Caporaso N, Rothman N, Wacholder S: Case-control studies of common alleles and environmental factors. *J Natl Cancer Inst Monogr* **1999**, 26:25–30.
2. Langholz B, et al: Cohort studies for characterizing measured genes. *J Natl Cancer Inst Monogr* **1999**, 26:39–42.
3. Garcia-Closas M, et al: Inference issues in cohort and case–control studies of genetic effects and gene–environment interactions. In *Human genome epidemiology: a scientific foundation for using genetic information to improve health and prevent disease.* Edited by Khoury JLWBM. Oxford: Oxford University Press; **2004**.
4. NRC, N. R. C: Biological markers in environmental health research. *Environ Health Perspect* **1987**, 74:3–9.
5. Perera FP, Weinstein IB: Molecular epidemiology and carcinogen-DNA adduct detection: new approaches to studies of human cancer causation. *J Chronic Dis* **1982**, 35(7):581–600.
6. Rothman N, Stewart WF, Schulte PA: Incorporating biomarkers into cancer epidemiology: a matrix of biomarker and study design categories. *Cancer Epidemiol Biomarkers Prev* **1995**, 4(4):301–311.
7. Little RR, Sacks DB: HbA1c: how do we measure it and what does it mean? *Curr Opin Endocrinol Diabetes Obes* **2009**, 16(2):113–118.
8. Hemstreet GP 3rd, et al: Biomarker risk assessment and bladder cancer detection in a cohort exposed to benzidine. *J Natl Cancer Inst* **2001**, 93(6):427–436.

9. Gray R, Wheatley K: How to avoid bias when comparing bone marrow transplantation with chemotherapy. *Bone Marrow Transplant* **1991**, 7(Suppl 3):9–12.
10. Davey Smith G, Ebrahim S: 'Mendelian randomization': can genetic epidemiology contribute to understanding environmental determinants of disease? *Int J Epidemiol* **2003**, 32(1):1–22.
11. Cui Y, et al: Nanowire nanosensors for highly sensitive and selective detection of biological and chemical species. *Science* **2001**, 293 (5533):1289–1292.
12. Russo VEA MR, Riggs AD: *Epigenetic mechanisms of gene regulation.* Cold Spring Harbor, NY: Cold Spring Harbor Laboratory Press; **1996.**
13. Allis CD, Jenuwein T, Reinberg D: *Epigenetics.* Cold Spring Harbor: Cold Spring Harbor Laboratory Press; **2007.**
14. Chuang JC, Jones PA: Epigenetics and microRNAs. *Pediatr Res* **2007**, 61(5 Pt 2):24R–29R.
15. Baccarelli A, Bollati V: Epigenetics and environmental chemicals. *Curr Opin Pediatr* **2009**, 21(2):243–251.
16. Fleming JL, Huang TH, Toland AE: The role of parental and grandparental epigenetic alterations in familial cancer risk. *Cancer Res* **2008**, 68(22):9116–9121.
17. Jirtle RL, Skinner MK: Environmental epigenomics and disease susceptibility. *Nat Rev Genet* **2007**, 8(4):253–262.
18. Gluckman PD, Hanson MA: Developmental origins of disease paradigm: a mechanistic and evolutionary perspective. *Pediatr Res* ***2004***, 56(3):311–317.
19. Waterland RA, Michels KB: Epigenetic epidemiology of the developmental origins hypothesis. *Annu Rev Nutr* **2007**, 27:363–388.
20. Rothman KJ, Greenland S: Case-control studies. In: Rothman KJ, Greenland S (eds). *Modern Epidemiology*. Ed. 2. Philadelphia. PA: Lippincott-Raven **1998**, 93–114.
21. Last J, Spasoff R, Harris S: *A dictionary of epidemiology.* New York, NY: Oxford University Press; **2001.**
22. Morimoto LM, White E, Newcomb PA: Selection bias in the assessment of gene-environment interaction in case–control studies. *Am J Epidemiol* **2003**, 158(3):259–263.
23. Yokoyama A, et al: Reliability of a flushing questionnaire and the ethanol patch test in screening for inactive aldehyde dehydrogenase-2 and alcohol-related cancer risk. *Cancer Epidemiol Biomarkers Prev* **1997**, 6(12):1105–1107.
24. Stamey TA, et al: Prostate-specific antigen as a serum marker for adenocarcinoma of the prostate. *N Engl J Med* **1987**, 317(15):909–916.
25. Thompson IM, et al: Operating characteristics of prostate-specific antigen in men with an initial PSA level of 3. 0 ng/ml or lower. *JAMA* **2005**, 294(1):66–70.
26. Freedland SJ, Platz EA: Obesity and prostate cancer: making sense out of apparently conflicting data. *Epidemiol Rev* **2007**, 29:88–97.
27. Skolarus TA, Wolin KY, Grubb RL 3rd: The effect of body mass index on PSA levels and the development, screening and treatment of prostate cancer. *Nat Clin Pract Urol* **2007**, 4(11):605–614.
28. Ahn J, et al: Variation in KLK genes, prostate-specific antigen and risk of prostate cancer. *Nat Genet* **2008**, 40(9):1032–1034, author reply 1035–6.
29. Wiklund F, et al: Association of reported prostate cancer risk alleles with PSA levels among men without a diagnosis of prostate cancer. *Prostate* **2009**, 69(4):419–427.
30. Reich DE, Goldstein DB: Detecting association in a case–control study while correcting for population stratification. *Genet Epidemiol* **2001**, 20(1):4–16.

31. Kittles RA, et al: CYP3A4-V and prostate cancer in African Americans: causal or confounding association because of population stratification? *Hum Genet* **2002**, 110(6):553–560.
32. Thomas DC, Witte JS: Point: population stratification: a problem for case–control studies of candidate-gene associations? *Cancer Epidemiol Biomarkers Prev* **2002**, 11(6):505–512.
33. Khoury MJ, Yang Q: The future of genetic studies of complex human diseases: an epidemiologic perspective. *Epidemiology* 1998, 9(3):350–354.
34. Shriver MD, et al: Ethnic-affiliation estimation by use of population- specific DNA markers. *Am J Hum Genet* **1997**, 60(4):957–964.
35. Laird NM, Horvath S, Xu X: Implementing a unified approach to family- based tests of association. *Genet Epidemiol* **2000**, 19(Suppl 1):S36–S42.
36. Weinberg CR, Umbach DM: Choosing a retrospective design to assess joint genetic and environmental contributions to risk. *Am J Epidemiol* **2000**, 152(3):197–203.
37. Gauderman WJ: Sample size requirements for matched case–control studies of gene-environment interaction. *Stat Med* **2002**, 21(1):35–50.
38. Begg CB, Zhang ZF: Statistical analysis of molecular epidemiology studies employing case-series. *Cancer Epidemiol Biomarkers Prev* **1994**, 3(2):173–175.
39. Piegorsch WW, Weinberg CR, Taylor JA: Non-hierarchical logistic models and case-only designs for assessing susceptibility in population-based case–control studies. *Stat Med* **1994**, 13(2):153–162.
40. Khoury MJ, Flanders WD: Nontraditional epidemiologic approaches in the analysis of gene-environment interaction: case–control studies with no controls! *Am J Epidemiol* **1996,** 144(3):207–213.
41. Schaid DJ: Case-parents design for gene-environment interaction. *Genet Epidemiol* **1999,** 16(3):261–273.
42. Albert PS, et al: Limitations of the case-only design for identifying gene- environment interactions. *Am J Epidemiol* **2001**, 154(8):687–693.
43. Weinberg CR, et al: Using risk-based sampling to enrich cohorts for endpoints, genes, and exposures. *Am J Epidemiol* **2007**, 166(4):447–455.
44. Medlin J: Sister study hopes to answer breast cancer questions. *Environ Health Perspect* **2001**, 109(8):A368–A369.
45. Manolio TA: Cohort studies and the genetics of complex disease. *Nat Genet* **2009**, 41(1):5–6.
46. Manolio TA, Bailey-Wilson JE, Collins FS: Genes, environment and the value of prospective cohort studies. *Nat Rev Genet* **2006**, 7(10):812–820.
47. Collins FS, Manolio TA: Merging and emerging cohorts: necessary but not sufficient. *Nature* **2007**, 445(7125):259.
48. Willett WC, et al: Merging and emerging cohorts: not worth the wait. *Nature* **2007**, 445(7125):257–258.
49. CHA, Children's Health Act **2000**: Public Law 106–310 (codified at 42 USC 201), **2000.**
50. Morgenstern H, Thomas D: Principles of study design in environmental epidemiology. *Environ Health Perspect* **1993**, 101(Suppl 4):23–38.
51. Carroll RJ, Crainiceanu C, Ruppert D, Stefanski LA: *Measurement Error in Nonlinear Models: A Modern Perspective.* Chapman and Hall; **2006**.
52. Dosemeci M, Wacholder S, Lubin JH: Does nondifferential misclassification of exposure always bias a true effect toward the null value? *Am J Epidemiol* **1990**, 132(4):746–748.

53. Weinberg CR, Umbach DM, Greenland S: When will nondifferential misclassification of an exposure preserve the direction of a trend? *Am J Epidemiol* **1994**, 140(6):565–571.
54. Chavance M, Dellatolas G, Lellouch J: Correlated nondifferential misclassifications of disease and exposure: application to a cross-sectional study of the relation between handedness and immune disorders. *Int J Epidemiol* **1992**, 21(3):537–546.
55. Kristensen P: Bias from nondifferential but dependent misclassification of exposure and outcome. *Epidemiology* **1992**, 3(3):210–215.
56. Es G: Re: "Does nondifferential misclassification of exposure always bias a true effect toward the null value?" (Letter). *Am J Epidemiol* **1991**, 134(4):440–441.
57. Garcia-Closas M, Thompson WD, Robins JM: Differential misclassification and the assessment of gene-environment interactions in case–control studies. *Am J Epidemiol* **1998**, 147(5):426–433.
58. Jacob J, Seidel A: Biomonitoring of polycyclic aromatic hydrocarbons in human urine. *J Chromatogr B Analyt Technol Biomed Life Sci* **2002**, 778(1–2):31–47.
59. Lioy PJ: Measurement methods for human exposure analysis. *Environ Health Perspect* **1995**, 103(Suppl 3):35–43.
60. Weis BK, et al: Personalized exposure assessment: promising approaches for human environmental health research. *Environ Health Perspect* **2005**, 113(7):840–848.
61. Rodier PM: Environmental causes of central nervous system maldevelopment. *Pediatrics* **2004**, 113(4 Suppl):1076–1083.
62. Amin-Zaki L, et al: Intra-uterine methylmercury poisoning in Iraq. *Pediatrics* **1974**, 54(5):587–595.
63. Marsh DO, et al: Fetal methylmercury poisoning: clinical and toxicological data on 29 cases. *Ann Neurol* **1980**, 7(4):348–353.
64. Tilson HA, Jacobson JL, Rogan WJ: Polychlorinated biphenyls and the developing nervous system: cross-species comparisons. *Neurotoxicol Teratol* **1990**, 12(3):239–248.
65. Agrawal A, et al: The effects of maternal smoking during pregnancy on offspring outcomes. *Prev Med* **2010**, 50(1–2):13–18.
66. Wiencke JK, et al: Individual susceptibility to induced chromosome damage and its implications for detecting genotoxic exposures in human populations. *Cancer Res* **1991**, 51(19):5266–5269.
67. Montgomery SM, Ekbom A: Smoking during pregnancy and diabetes mellitus in a British longitudinal birth cohort. *BMJ* **2002**, 324(7328):26–27.
68. Angerer J, Ewers U, Wilhelm M: Human biomonitoring: state of the art. *Int J Hyg Environ Health* **2007**, 210(3–4):201–228.
69. Chatterjee BB, et al: The location of personal sampler filter filter heads. *Am Ind Hyg Assoc J* **1969**, 30(6):643–645.
70. Hernan MA, et al: Observational studies analyzed like randomized experiments: an application to postmenopausal hormone therapy and coronary heart disease. *Epidemiology* **2008**, 19(6):766–779.
71. Zhang HP: Multivariate adaptive splines for analysis of longitudinal data. *J Comput Graph Stat* **1997**, 6:74–91.
72. Zhang HP: Analysis of infant growth curves using multivariate adaptive splines. *Biometrics* **1999**, 55:452–459.
73. Zhu W, et al: A genome-wide association analysis of Framingham Heart. Study longitudinal data using multivariate adaptive splines. *BMC Proc* **2009**, 3(Suppl 7):S119.

74. Hoover DR, et al: Nonparametric smoothing estimates of time-varying coefficient models with longitudinal data. *Biometrika* **1998**, 85:809–822.
75. Wu CO, Chiang C-T: Kernel smoothing on varying coefficient models with longitudinal dependent variable. *Statistica Sinica* **2010**, 10(**2000**):433–456.
76. Mukherjee B, et al: Tests for gene-environment interaction from case–control data: a novel study of type I error, power and designs. *Genet Epidemiol* **2008**, 32(7):615–626.
77. Cornfield J: A statistical problem arising from retrospective studies. In Neyman, J. (ed.). *Proceedings of the Third Berkeley Symposium on Mathematical Statistics and Probability* **1956**, 4:135–148.
78. Prentice RL, Pyke R: Logistic disease incidence models and case–control studies. *Biometrika* **1979**, 66:403–411.
79. Selinger-Leneman H, et al: Does accounting for gene-environment (GxE) interaction increase the power to detect the effect of a gene in a multifactorial disease? *Genet Epidemiol* **2003**, 24(3):200–207.
80. Chatterjee N, et al: Powerful multilocus tests of genetic association in the presence of gene-gene and gene-environment interactions. *Am J Hum Genet* **2006**, 79(6):1002–1016.
81. Kraft P, et al: Exploiting gene-environment interaction to detect genetic associations. *Hum Hered* **2007**, 63(2):111–119.
82. Maity, A. Lin, X: Powerful tests for detecting a gene effect in the presence of possible gene-gene interactions using garrotte kernel machines. *Biometrics* **2011**, 67(4):1271–1284.
83. Khoury MJ, Beaty TH: Applications of the case–control method in genetic epidemiology. *Epidemiol Rev* **1994**, 16(1):134–150.
84. Umbach DM, Weinberg CR: Designing and analysing case–control studies to exploit independence of genotype and exposure. *Stat Med* **1997**, 16 (15):1731–1743.
85. Chatterjee N, Kalaylioglu Z, Carroll RJ: Exploiting gene-environment independence in family-based case–control studies: increased power for detecting associations, interactions and joint effects. *Genet Epidemiol* **2005**, 28(2):138–156.
86. Chatterjee, N, Carroll, RJ: Semiparametric maximum likelihood estimation exploiting gene-environment independence in case–control studies. *Biometrika* **2005**, 92:399–418.
87. Wang LY, Lee WC: Population stratification bias in the case-only study for gene-environment interactions. *Am J Epidemiol* **2008**, 168(2):197–201.
88. Curtis D: Use of siblings as controls in case–control association studies. *Ann Hum Genet* **1997**, 61:319–333.
89. Gauderman WJ, Witte JS, Thomas DC: Family-based association studies. *J Natl Cancer Inst Monogr* **1999**, 26:31–37.
90. Witte JS, Gauderman WJ, Thomas DC: Asymptotic bias and efficiency in case–control studies of candidate genes and gene-environment interactions: basic family designs. *Am J Epidemiol* **1999**, 149(8):693–705.
91. Umbach DM, Weinberg CR: The use of case-parent triads to study joint effects of genotype and exposure. *Am J Hum Genet* **2000**, 66(1):251–261.
92. Flanders WD, Khoury MJ: Analysis of case-parental control studies: method for the study of associations between disease and genetic markers. *Am J Epidemiol* **1996**, 144(7):696–703.
93. Kraft P, Cox DG: Study designs for genome-wide association studies. *Adv Genet* **2008**, 60:465–504.

94. McCarthy MI, et al: Genome-wide association studies for complex traits: consensus, uncertainty and challenges. *Nat Rev Genet* **2008**, 9(5):356–369.
95. Risch N, Merikangas K: The future of genetic studies of complex human diseases. *Science* **1996**, 273(5281):1516–1517.
96. Price AL, et al: Principal components analysis corrects for stratification in genome-wide association studies. *Nat Genet* **2006**, 38(8):904–909.
97. Moskvina V, Schmidt KM: On multiple-testing correction in genome-wide association studies. *Genet Epidemiol* **2008**, 32(6):567–573.
98. Browning BL, Browning SR: Efficient multilocus association testing for whole genome association studies using localized haplotype clustering. *Genet Epidemiol* **2007**, 31(5):365–375.
99. Murcray CE, Lewinger JP, Gauderman WJ: Gene-environment interaction in genome-wide association studies. *Am J Epidemiol* **2009**, 169(2):219–226.
100. Chatterjee N, Wacholder S: Invited commentary: efficient testing of gene-environment interaction. *Am J Epidemiol* **2009**, 169(2):231–233. discussion 234–5.
101. Mukherjee B, Chatterjee N: Exploiting gene-environment independence for analysis of case–control studies: an empirical Bayes-type shrinkage estimator to trade-off between bias and efficiency. *Biometrics* **2008**, 64(3):685–694.
102. Clayton D, McKeigue PM: Epidemiological methods for studying genes and environmental factors in complex diseases. *Lancet* **2001**, 358(9290):1356–1360.
103. Kalfleisch JD PR: *The Statistical Analysis of Failure Time Data.* New York, NY: Wiley; **2002.**
104. Diggle P, Heagerty P, Liang KY, Zeger S: *Analysis of Longitudinal Data.* New York: Oxford University Press; **2002.**
105. Breslow NE, Lubin JH, Marek P, Langholtz B: Multiplicative models and cohort analysis. *J. Am. Statist. Assoc.* **1983**, 78:1–12.
106. Lubin JH, Gail MH: Biased selection of controls for case–control analyses of cohort studies. *Biometrics* **1984**, 40(1):63–75.
107. Whittemore AS: The Efficiency of Synthetic Retrospective Studies. *Biom. J.* **1981**, 23:73–78.
108. Whittemore AS, McMillan A: Analyzing occupational cohort data: application to U. S. uranium miners. In *Environmental Epidemiology: Risk Assessment.* Edited by Prentice RL, Whittemore AS. Philadelphia: SIAM; **1982**:65–81.
109. Kupper LL, McMichael AJ, Spirtas R: A hybrid epidemiologic study design useful in estimating relative risk. *J Am Stat Assoc* **1975,** 70:524–528.
110. Liu et al. *Environmental Health* **2012**, 11:93 ; http://www. ehjournal. net/content/11/1/93.
111. Miettinen O: Design options in epidemiologic research. An update. *Scand J Work Environ Health* **1982**, 8(Suppl 1):7–14.
112. Bureau A, et al: Estimating interaction between genetic and environmental risk factors: efficiency of sampling designs within a cohort. *Epidemiology* **2008**, 19(1):83–93.
113. Wacholder S: Practical considerations in choosing between the case-cohort and nested case–control designs. *Epidemiology* **1991**, 2(2):155–158.
114. Langholz B, Thomas DC: Nested case–control and case-cohort methods of sampling from a cohort: a critical comparison. *Am J Epidemiol* 1990, 131(1):169–176.
115. White JE: A two stage design for the study of the relationship between a rare exposure and a rare disease. *Am J Epidemiol* **1982**, 115(1):119–128.

116. Cain KC, Breslow NE: Logistic regression analysis and efficient design for two-stage studies. *Am J Epidemiol* **1988**, 128(6):1198–1206.
117. Breslow NE, et al: Improved Horvitz-Thompson Estimation of Model Parameters from Two-phase Stratified Samples: Applications in Epidemiology. *Stat Biosci* **2009**, 1(1):32.
118. Weinberg CR, Wacholder S: The design and analysis of case–control studies with biased sampling. *Biometrics* **1990**, 46(4):963–975.
119. Yang Q, Khoury MJ, Flanders WD: Sample size requirements in case-only designs to detect gene-environment interaction. *Am J Epidemiol* **1997**, 146(9):713–720.
120. Cai J, Zeng D: Sample size/power calculation for case-cohort studies. *Biometrics* **2004**, 60(4):1015–1024.
121. Lubin JH, Gail MH: On power and sample size for studying features of the relative odds of disease. *Am J Epidemiol* **1990**, 131(3):552–566.
122. Hwang SJ, et al: Minimum sample size estimation to detect gene- environment interaction in case–control designs. *Am J Epidemiol* **1994**, 140(11):1029–1037.
123. Foppa I, Spiegelman D: Power and sample size calculations for case–control studies of gene-environment interactions with a polytomous exposure variable. *Am J Epidemiol* **1997**, 146(7):596–604.

APPENDIX 1

USING AUSTIN BRADFORD HILL'S GUIDELINES IN GENETICS AND GENOMICS

SARA GENELETTI, VALENTINA GALLO, MIQUEL PORTA, MUIN J. KHOURY, and PAOLO VINEIS

There are some general aspects to consider when tackling cause-effect relationships in genetics. First, most associations for individual genetic variants and common chronic diseases have weak to modest effects. Empirical findings show that even for fairly well established associations, the effect sizes are weak to modest; i.e., relative risks are usually under 2, and often between 1.2 and 1.6) [1]. Generally speaking, the stronger the association between a risk factor and a disease, the more likely it is that the association is causal, because confounding and other biases are unlikely to explain it away. However, in genetics the penetrance of an individual genetic variant associated with a disease depends on the interactions of the variant with external exposures, the internal environment, or other genetic variants. In spite of the etiologic complexity of common diseases and the resulting weak effects of individual genetic variants, theoretical work suggests that the combination of as few as 20 common variants with weak to moderate effect sizes, when put together as a system of variants (or genomic profiles), can account for a substantial attributable fraction of the disease in the population [2]. On the other hand, a large number of rare variants each contributing (or causing) a strong disease risk may also be a plausible explanation. The potential rarity of highly-penetrant variants, the weakness of common associations, and the frequency of complex gene-environment interactions pose severe challenges to the statistical power to find marginal effects of single gene variants on risks for common diseases. In fact, the strength of the association with the gene (main effect) may be low while the gene-exposure interaction is strong. This may be more convincing evidence of the truly causal nature of the association, given the available biological knowledge on environmental influences on gene expression.

Consistency in genetic studies was traditionally poor in the "candidate gene" era, with few associations confirmed in more than one study [3], but this has changed rapidly with genome-wide association studies (GWAs). More than 600 stable replicated hits have been reported in 2007 and 2008 from GWAs, due to an in-built, strong process of replication of findings. One advantage of GWAs is that they are published only if the results are replicated in 3-4 or more independent studies. As a result in genetic epidemiology there is now a widely accepted requirement for "internal" consistency. A similar approach would be invaluable in non-genetic epidemiology but is currently not practiced. Poor replication for candidate genes is related to multiple factors, including type 1 errors ("false positives") and publication bias, as well as to methodological issues as biases in the selection of cases and controls, exposure assessment errors, and confounding.

In addition, the expression of genes is so dependent on the surrounding circumstances (other genes, internal environment, e.g., immunological and nutritional status [4], external physical environment, gene expression), that the same main clinical effect of a gene variant is difficult to capture in different studies conducted under different conditions. In fact, such main effects may not be identical in different studies that are conducted in actual -sometimes, very different- human contexts; a genuine heterogeneity of human genetic effects across population groups -and individuals- is to be expected on the basis of knowledge on how biological, clinical and environmental processes jointly cause disease in humans. An example of the influence of study design is the investigation of gene-disease associations in founder populations, in which the effect of a genetic variant is likely to be higher than the average across all populations [5]. Another example is familial aggregation studies, where familial disease risks are influenced not only by the genetic mutations or variants of interest, but also by other genetic and epigenetic processes; if the latter are overlooked, the penetrance of the former may be overestimated [6–8].

To some extent it is reasonable to hope that genetic associations are specific, thus facilitating causal inference. For example, 5-HTT variants have been associated specifically with bipolar disorder, probably because of the role of the gene in serotonin metabolism [9]. But expectations of specificity may disregard biological knowledge (e.g., on cofactors, multiple causes and effects) that makes unspecificity more plausible. A potential problem in the use of specificity as a criterion for causality is that many genetic variants belong to metabolic, inflammatory, homeostatic and other pathways that could influence multiple disease processes. This is an extension of the concept of pleiotropy that we see in single gene disorders. For example, MTHFR variation involves folic acid and methylation pathways that may have potential relevance to the genesis of many disease outcomes, as birth defects, cardiovascular disease and cancer [10]. The same is likely to be true for DNA repair genes [11]. This issue has long been observed in non genetic epidemiology in relation to some common risk factors, such as socio-economic status or cigarette smoking, which are associated with many disease outcomes. The value of specificity increases with increasing knowledge about the constituents of the exposure (e.g., PAHs and other carcinogens for cigarette smoking), and of its biological or environmental effects. For example, on the basis of functional knowledge, only bladder cancer, and perhaps colon cancer, may be expected to be associated with NAT2 variants [12–14]. Such postulated associations are biologically plausible because there is evidence that aromatic amines or heterocyclic aromatic amines, which are metabolised by NAT2, are involved in bladder or colon carcinogenesis. Nevertheless, NAT2 associations are also observed with breast and lung cancer and mesothelioma [15, 16], without evidence of biological plausibility. This unexpected nonspecificity may be true and due, for instance, to a pleiotropic effect of the exposure; or the apparent association with the outcome (in this case, other than bladder cancer) may be confounded by yet unknown factors. Similar situations are encountered in clinical medicine and non genetic epidemiology; for example, the early observation of an inverse association between hormone replacement therapy (HRT) and mortality due to accidents and violence, which was of the same magnitude as that originally found for cardiovascular mortality [17]. This prompted a debate on the causal nature of the association between HRT and cardiovascular mortality, as no plausible biological reason for the protective effect of HRT on violent death could be argued.

Temporality is also relevant to the study of the genotypes; since gene variants are inherited and do not change after conception, they precede the onset of disease indeed. In addition the temporal pattern with which a particular variant/mutation manifests itself can be relevant. In Huntington's disease, for example, there is the phenomenon of "anticipation" (younger age of disease onset in one generation than in the previous) depending on the number of the repeated triplets in the gene (which tend to increase in the offspring). For acquired genetic alterations (e.g., somatic mutations) temporality is also important; in persons living in normal conditions the timing of occurrence of the mutation often cannot be observed directly. A collection of archived specimens may help, as can knowledge on the usual course of events gained from molecular pathology studies. For epigenetic mechanisms temporality is even more crucial, but it is beyond the purpose of this article [4, 6–8].

In genomics, the possibility of observing a dose-response gradient depends on the model of genotype-phenotype relationships. Even for a diallelic system at one locus, there could be recessive, dominant or codominant models. The biologic model for the action of numerous alleles at different loci is more complex and is essentially unknown for most common diseases. Only if the genetic model is codominant can a dose-response be observed. However, a different kind of dose-response is observable if we consider the cumulative effect of multiple genes or SNPs. Both the risk of lung cancer and the levels of DNA damage can increase approximately linearly with an increasing number of "at risk" gene variants [11, 18]. Gene copy number variation can lead to more complex dose-response relationships. Quantitative continuous markers used in epigenetics (promoter methylation) and transcriptomics (gene expression) may be analyzed in search of dose-response effects (linear or non-linear).

In genetics experimental evidence comes mainly from animal studies in which knock-out organisms are used in order to have a pure genetic disease model. This directly tests the effect of the absence or presence of specific genetic factors on the organism. Extrapolation of the results of these experiments to humans is challenging due to differences between humans and the knock-out organisms in both the genetic make-up and the potential types of gene-gene and gene-environment interactions. Genetic experimental studies have also long been known to reproduce disease phenotypes (e.g., in mice) that are only a partial approximation of the complex human disease; an example is the Super Oxide Dismutase-1 (SOD1) mutated mouse model for Amyotrophic Lateral Sclerosis (ALS), which has different motor characteristics than the human disease [19].

REFERENCES

1. Ioannidis JP, Trikalinos TA, Khoury MJ: Implications of small effect sizes of individual genetic variants on the design and interpretation of genetic association studies of complex diseases. *Am J Epidemiol* **2006**, 164:609-614.
2. Yang Q, Khoury MJ, Friedman J, Little J, Flanders WD: How many genes underlie the occurrence of common complex diseases in the population? Int J Epidemiol **2005**, 34:1129-1137.
3. Ioannidis JP, Trikalinos TA: Early extreme contradictory estimates may appear in published research: the Proteus phenomenon in molecular genetics research and randomized trials. *J Clin Epidemiol* **2005**, 58:543-549.
4. Lee DH, Jacobs DR Jr, Porta M: Hypothesis: a unifying mechanism for nutrition and chemicals as lifelong modulators of DNA hypomethylation. *Environ Health Perspect* **2009**, 117:1799-1802.
5. Glazier AM, Nadeau JH, Aitman TJ: Finding genes that underlie complex traits. *Science* **2002**, 298:2345-2349.
6. Jirtle RL, Skinner MK: Environmental epigenomics and disease susceptibility. *Nat Rev Genet* **2007**, 8:253-62.
7. Feinberg AP: Phenotypic plasticity and the epigenetics of human disease. *Nature* **2007**, 447:433-440.
8. Edwards TM, Myers JP: Environmental exposures and gene regulation in disease etiology. *Environ Health Perspect* **2007**, 115:1264-1270.
9. Bellivier F, Henry C, Szöke A, Schürhoff F, Nosten-Bertrand M, Feingold J, Launay JM, Leboyer M, Laplanche JL: Serotonin transporter gene polymorphisms in patients with unipolar or bipolar depression. *Neurosci Lett* 1998, 255:143-146.
10. Kim YI: 5,10-Methylenetetrahydrofolate reductase polymorphisms and pharmacogenetics: a new role of single nucleotide polymorphisms in the folate metabolic pathway in human health and disease. *Nutr Rev* **2005**, 63:398-407.
11. Neasham D, Gallo V, Guarrera S, Dunning A, Overvad K, Tjonneland A, Clavel-Chapelon F, Linseisen JP, Malaveille C, Ferrari P, Boeing H, Benetou V, Trichopoulou A, Palli D, Crosignani P, Tumino R, Panico S, Bueno de Mesquita HB, Peeters PH, van Gib CH, Lund E, Gonzalez CA, Martinez C, Dorronsoro M, Barricarte A, Navarro C, Quiros JR, Berglund G, Jarvholm B, Khaw KT, et al.: Double-strand break DNA repair genotype predictive of later mortality and cancer incidence in a cohort of non-smokers. *DNA Repair* **2008**.
12. Marcus PM, Vineis P, Rothman N: NAT2 slow acetylation and bladder cancer risk: a meta-analysis of 22 case-control studies conducted in the general population. *Pharmacogenetics* **2000**, 10:115-122.

13. Vineis P, McMichael A: Interplay between heterocyclic amines in cooked meat and metabolic phenotype in the etiology of colon cancer. *Cancer Causes Control* **1996**, 7:479-486.
14. Vineis P, Pirastu R: Aromatic amines and cancer. *Cancer Causes Control* **1997**, 8:346-355.
15. Ochs-Balcom HM, Wiesner G, Elston RC: A meta-analysis of the association of N-acetyltransferase 2 gene (NAT2) variants with breast cancer. *Am J Epidemiol* **2007**, 166:246-254.
16. Borlak J, Reamon-Buettner SM: N-acetyltransferase 2 (NAT2) gene polymorphisms in colon and lung cancer patients. *BMC Med Genet* **2006**, 7:58.
17. Postmenopausal estrogen use and heart disease *N Engl J Med* 1986, 315:131-136.
18. Vineis P, Anttila S, Benhamou S, Spinola M, Hirvonen A, Kiyohara C, Garte SJ, Puntoni R, Rannug A, Strange RC, Taioli E: Evidence of gene gene interactions in lung carcinogenesis in a large pooled analysis. *Carcinogenesis* **2007**, 28:1902-1905.
19. Nicholson SJ, Witherden AS, Hafezparast M, Martin JE, Fisher EM: Mice, the motor system, and human motor neuron pathology. *Mamm Genome* **2000**, 11:1041-1052.

Geneletti, S., Gallo, V., Porta, M., Khoury, M. J., and Vineis, P. Assessing Causal Relationships in Genomics: From Bradford-Hill Criteria to Complex Gene-Environment Interactions and Directed Acyclic Graphs. Emerging Themes in Epidemiology 2011, 8:5. doi: 10.1186/1742-7622-8-5.

APPENDIX 2

THE CALCULUS OF THE DECISION THEORETIC FRAMEWORK (DTF)

SARA GENELETTI, VALENTINA GALLO, MIQUEL PORTA, MUIN J. KHOURY, and PAOLO VINEIS

Conditional independence [12] is the tool DTF uses to a) express how variables are associated and b) to understand when it is possible to make inferences about causal associations from data that are observational. It is best described as follows: consider 3 variables A, B and C. Say that Pr(A,C|B) = Pr(A|B)Pr(C|B) (where Pr(.) means probability of).

Then we can say that A is independent of C given B; formally: $A \perp\!\!\!\perp C|B$.

This means that if we know what B is, knowing what A is gives us no further information on C; e.g., if we want to know the genetic make up of Alfred (A), we can gain some information by looking at his brother Colin (C). If however, we can see their parents Barry and Barbara (B), then knowing about Colin gives us no further information on Alfred. This shows where the "familial" terminology used in DAGs comes from.

Conditional independence is a non-graphical (and non-causal) equivalent of the d-separation criteria used in the causal DAG approach [1]. It forms the basis for the formal treatment of DTF, and its manipulation allows us to determine under what circumstances we can equate the results of observation to those of experiment [2, 3].

The original role of DAGs in the statistical literature is to encode statistical associations (described, for instance, by Chi-squared tests). Thus, in DTF the lack of directed edges in a DAG is viewed as conditional or marginal independence between variables, not a lack of a causal relationship. There are two problems with interpreting DAGs encoding such associations as causal. The first problem is that often there is more than one DAG representing the same set of conditional independences (see example below). To determine which, if any of them, is causal, we must use knowledge that is not inherent in the data or the DAG (e.g., time ordering). The second problem is that we often do not have data on all the variables that play a role (causal or otherwise) in the problem we are considering. This means that the DAGs only tell us about the relationships between the variables we have observed, making a causal interpretation dangerous.

Consider the following simple example: A and B are proteins produced in the body and C is a cancer thought to be associated to the production of A and B. It is possible to artificially increase the amount of B in the system and we would eventually like to know whether this could prevent the emergence of the cancer C. However, at this point we do not know whether A or B are produced by the presence of C or indeed whether there is any natural ordering to the appearance of the three variables. We obtain the conditional independence $A \perp\!\!\!\perp C|B$ from data on a number of individuals in a case control study investigating possible causes of C. This is encoded by all three DAGs. These three DAGs only tell us one thing, namely that the cancer is not directly associated to protein A (when we only consider these three variables and the individuals in the study). They do not tell us whether treating patients with B will have a positive effect on the incidence of C or indeed how A and B are associated. Thus, trying to determine whether intake of B will act as a preventive agent (i.e., whether B causes C) based only on current knowledge and the DAGs is impossible. When we face a problem that we do not understand fully, interpreting one DAG or even one particular directed edge as causal can be difficult.

One way of determining whether relationships depicted in a DAG describing observational data are causal is to relate it to an equivalent situation under intervention or randomization. It is generally accepted that the ideal for causal inference is the randomized controlled trial because confounding is eliminated or attenuated. It is generally also accepted [4] that when we perform an external intervention, such as randomization on a system in equilibrium, we can view the consequences as causal. Thus, intervention is a formal way of asserting cause-effect relationships.

In DTF we introduce randomization as a variable R. To clarify, consider the following example. Assume that X is a binary variable that can be forced to take on a particular value or "set". It takes on two values: "active" (X = a), or "baseline" (X = b). The randomising variable R has the same settings as X as well as the observational setting R= Φ (the empty set). When R = a then X = a with no uncertainty (imagine forcing X to take on this value, say by administering the treatment to a compliant patient). Similarly, when R= b, X = b with no uncertainty. Finally when R = Φ, X is allowed to arise without intervention and can take on the values a and b as in an observational study. For causal inference in DTF we want to estimate (usually the expected value of) the outcome Y given that an intervention has happened. For example, if we want to know which treatment, active or baseline, is better for Y, we might look at the difference in the expected value of Y given these treatments: E(Y | R = a)- E(Y|R = b). This would then be a measure of the causal effect of a vs b. In observational studies, we do not have E(Y|R = a) the interventional expectation; rather, we have E(Y| X = a, R = Φ) the observational expectation; similarly for b. The question is, therefore, how to make an inference about the former using the latter. One assumption that is often made is that all observed confounders U are observed. However, this is often not possible and other approaches that simulate randomization, such as the instrumental variable approach known as Mendelian randomization [5] can be used. See Dawid [6], Didelez [7], and Geneletti [8] for formal examples.

Introducing randomization can also help us distinguish between intermediate variables and confounders, as when X is randomized the association between X and any confounders U is severed, whilst that with intermediates is not. Statistically, if after randomising X the distribution of U conditional on X remains the same as before randomization, then U is a confounder rather than a mediating variable, as this means that U is independent of X when it is randomized. This corresponds precisely to the situation described by the DAG.

Interventions are represented by decision nodes in augmented DAGs [9], and these can be used to make some causal inferences, as DAGs explicitly represent interventions. By introducing the randomization/intervention variables explicitly into the DAG, we can use conditional independences to determine when it is possible to estimate the causal effect (based R = a,b) from data that are observational (based on R = Φ and X = a,b). Again, as a detailed description of the formal DTF is beyond the scope of this paper, we refer the interested reader to previous work [6–8, 10].

REFERENCES

1. Ioannidis JP, Trikalinos TA, Khoury MJ: Implications of small effect sizes of individual genetic variants on the design and interpretation of genetic association studies of complex diseases. *Am J Epidemiol* **2006**, 164:609-614.
2. Dawid AP: Conditional independence in statistical theory. With discussion. *J Roy Statist Soc B* **1979**, 41:1-31.
3. Lauritzen S: *Graphical models*. Oxford, UK: Oxford University Press, **1996**.
4. Cartwright N: *Nature's capacities and their measurement*. New York: Oxford University Press, **1994**.
5. Davey SG, Ebrahim S: 'Mendelian randomization': can genetic epidemiology contribute to understanding environmental determinants of disease? *Int J Epidemiol* **2003**, 32:1-22.
6. Dawid AP: Causal inference without counterfactuals. *J Am Statist Ass* **2000**, 95:407-448.
7. Didelez V, Sheenan N: Mendelian randomization: why epidemiology needs a formal language for causality. In *Causality and probability in the sciences*. College Publications London, London; **2007**.
8. Geneletti S: Identifying direct and indirect effects in a non-counterfactual framework. *J Roy Stat Soc B* **2007**, 69:199-215.
9. Dawid AP: Influence diagrams for causal modelling and inference. *Intern Statist Rev* **2002**, 70:161-189.
10. Lauritzen S: *Graphical models*. Oxford, UK: Oxford University Press, **1996**.

Geneletti, S., Gallo, V., Porta, M., Khoury, M. J., and Vineis, P. Assessing Causal Relationships in Genomics: From Bradford-Hill Criteria to Complex Gene-Environment Interactions and Directed Acyclic Graphs. Emerging Themes in Epidemiology 2011, 8:5. doi: 10.1186/1742-7622-8-5.

APPENDIX 3

PARKINSON'S DISEASE: ENVIRONMENTAL AND GENETIC RISK FACTORS

SARA GENELETTI, VALENTINA GALLO, MIQUEL PORTA, MUIN J. KHOURY, and PAOLO VINEIS

Large epidemiological studies aimed at identifying risk factors for Parkinson's disease have suggested a role of 1-methyl-4-phenyl-1,2,3,6-tetrahydropyridine (MPTP) (a compound accidentally produced in the manufacture of illegal drugs), of some pesticides, of certain metals and of polychlorinated biphenyls [1]. On the other hand, tea and coffee drinking, use of non-steroidal anti-inflammatory drugs, and high blood levels of uric acid have been suggested to be protective for Parkinson's disease [1].

To date, eleven monogenic forms have been identified (with PARK1 to 11 gene acronyms); they will be selectively discussed below (Table 1) [2]. However, monogenic forms of Parkinson's explain no more than 20% of the early-onset cases of the disease, and less than 3% of the forms with onset in the old ages, a situation that is common to many chronic diseases as breast cancer (e.g., role of BRCA1) or heart disease (e.g., Familial Hypercholesterolemia). Most forms of the disease appear to be caused or at least influenced by complex interactions between several genes, or between genes and environmental factors.

The α-synuclein, encoded by the SNCA gene, is a protein with several functions in signal transduction and vesicle trafficking; it is also a competitive inhibitor of an enzyme involved in the L-Dopa biosynthesis. Three known dominant mutations on the SNCA gene have been identified in families affected by Parkinsonism with dementia characterised pathologically by diffuse Lewy bodies, mainly composed of α-synuclein. The identification of these mutations contributes to the contention as to whether the

TABLE 1. Main identified genes involved in Parkinsonism, with their biological, clinical and pathological main features.

Gene (locus)	Protein	Function	Inheritance	Pathology	Clinical phenotype
1SNCA (PARK1/4)	α-synuclein	Signal transduction, membrane vesicle trafficking, and cytoskeletal dynamics	Dominant	Diffuse Lewy bodies (prominently nigral and hippocampal neuronal loss)	Early onset progressive L-Dopa responsive Parkinsonism, cognitive decline, autonomic dysfunction and dementia
LRRK2 (PARK8)	Dardarin	Cytosolic kinase with several functions (including substrate binding, protein phosphorylation and protein-protein interactions)	Dominant	Predominantly Lewy bodies disease (rare cases with neurofibrillar tangels and/or nigral neuronal loss	Parkinsonism consistent with sporadic Parkinson's Disease. Dystonia, amyotrophy, gaze palsy and dementia occasionally develop
PRKN (PARK2)	Parkin	E3 ligase (conjugating ubiquitine to proteins to target them for degradation by the proteasome)	Recessive (rare "presudo-dominant" cases reported)	Predominantly nigral neuronal loss (compound heterozygotes with Lewy bodies or tau pathology are described)	Early onset Parkinsonism, often presenting with dystopia, with diurnal fluctuations. Typically responsive to very low doses of L-Dopa
PINK1 (PARK6)	-	Mitochondrial kinase	Recessive	Undetermined	Early onset Parkinsonism, slowly progressive and responsive to low doses of L-Dopa
DJ-1 (PARK7)	-	Oxidative stress signalling molecule on mitochondria	Recessive	Undetermined	Slowly progressive early-onset Parkinsonism occasionally with psychiatric disturbances; rare compound heterozygotes with Parkinsonism and dementia or amyotrophy are described

so-called Lewy body disorders (Parkinson's disease, Parkinsonism with dementia, and dementia with Lewy bodies) represent a continuum or have to be considered as distinct diseases [2]. This is thus as well an excellent example of a situation in which researchers try to elucidate the causal relationships between a complex set of genotypes and a rich spectrum of clinical phenotypes.

The LRRK gene encodes for a protein involved in multiple functions; three dominant mutations are known. Pathologically, the disease is characterised by a typical Lewy body pattern consistent with the post mortem diagnosis of Parkinson's disease. However, some cases with tau-positive pathology without Lewy bodies have been observed even within the same family. The pathway leading to one or the other condition is likely to be influenced by genetic and/or environmental factors that remain to be identified [2].

There are more than 50 known variants in the parkin gene and their effect on the disease appears to be recessive. Subjects with homozygous mutations leading to complete loss of parkin expression are found to have a selective loss of dopaminergic neurons in the substantia nigra and in the locus coeruleus without Lewy bodies or neurofibrillar tangles. However, subjects with compound heterozygous mutations (a diploid genotype in which two copies of a gene carry different mutations) may present pathologically with Lewy bodies or neurofibrillar pathology. This behaviour can be due to the fact that the outcome is mutation-specific: some mutations can reduce rather than abolish the protein activity affecting substrate specificity. Otherwise, these two different outcomes can share the primary cause (as for the LRRK case), which is subsequently influenced by gene-gene and/or gene-environment interactions [2].

For the last two recessive mutations, PINK-1 and DJ-1 there is no pathological information available. The protein encoded by PINK-1 gene is a mitochondrial kinase that seems to be involved in protecting the cell from mitochondrial dysfunction and stress-induced apoptosis [2]. The protein encoded by DJ-1 gene also is localised on mitochondria, but it seems to belong to the chaperones family, induced by oxidative stress [3]. This protein has been demonstrated to be involved in cell protection during oxidative stress. Intriguingly, reduced DJ-1 expression in Drosophila melanogaster results in susceptibility to oxidative stress and proteasome inhibition, which leads to a selective sensitivity to the environmental chemical agents paraquat and rotenone.

REFERENCES

1. Kuehn BM: Scientists probe role of genes, environment in Parkinson disease. *JAMA* **2006**, 295:1883-1885.
2. Farrer MJ: Genetics of Parkinson disease: paradigm shifts and future prospects. *Nat Rev Genet* **2006**, 7:306-318.
3. Clements CM, McNally RS, Conti BJ, Mak TW, Ting JP: DJ-1, a cancer- and Parkinson's disease-associated protein, stabilizes the antioxidant transcriptional master regulator Nrf2. *Proc Natl Acad Sci USA* **2006**, 103:15091-15096.

Geneletti, S., Gallo, V., Porta, M., Khoury, M. J., and Vineis, P. Assessing Causal Relationships in Genomics: From Bradford-Hill Criteria to Complex Gene-Environment Interactions and Directed Acyclic Graphs. Emerging Themes in Epidemiology 2011, 8:5. doi: 10.1186/1742-7622-8-5.

AUTHOR ACKNOWLEDGMENTS

Chapter 1
EG-D was supported by a Juan de la Cierva contract from the Spanish Ministry of Science and Innovation and a Marie Curie reintegration grant from the European Commission (ERG-2010-276838). Research in MJ and MAP lab was supported by grants from the Spanish Ministry of Science and Innovation (SAF2011/23638, and CSD2006/49), Generalitat de Catalunya (2009 SGR 1356), and Fundación Salud 2000. AR was financed by the CNRS (France). The funders had no role in study design, data collection and analysis, decision to publish, or preparation of the manuscript.

Chapter 2
This work was funded by research grant UM/MOHE High Impact Research Grant (HIR Grant No. F000009-21001) from the University of Malaya. The authors thank V. Bhavananthan, Nazia Abdul Majid and Sheila Golbabapour for their critical advice on the manuscript.

Chapter 3
The authors thank all the families who participated in the study, and acknowledge the great efforts of Laura Knosp, Susan Berends, Mercedes López, Mirta Gladys Leguizamón, Silvina Argañaraz, Marta Padilla, and Azucena Singh who were critical in contacting and enrolling families and screening samples. They are grateful to the W.M. Keck Foundation Biotechnology Resource Laboratory and the Keck Biostatistics Resource at Yale University for help with conducting the genome-wide methylation experiment and analyzing the raw data. This research was supported by Grants from the National Institutes of Health (R01 HD052953 and HD57192) and the March of Dimes Foundation (FY2006-575 and FY2008-260).

Chapter 4
This work has been made possible by a grant to ECNIS (Environmental Cancer Risk, Nutrition and Individual Susceptibility), a network of excellence operating within the European Union 6th Framework Program, Priority 5: "Food Quality and Safety" (Contract No 513943). Paolo Vineis would like to acknowledge the European Union grant HEALTH-2007-201550 HyperGenes. Sara Geneletti would like to acknowledge the support of the Economic and Social Research

Council (award number RES-576-25-5003). This work does not necessarily represent the official position of the Centers for Disease Control and Prevention.

Chapter 5
This work was supported by grants from the National Institutes of Health (S.M). We wish to indicate that some research work could not be referred to in this article due to space constraints. We are thankful to our colleagues for the critical reading of this manuscript.

Chapter 6
Research support by a grant from Canadian Institutes of Health Research Grant MOP-9186, a Canada Research Chair (to JRD) and a MHRC/CancerCare Manitoba Studentship award (to DK) are gratefully acknowledged.

Chapter 7
This work was supported by NIH grant ES015191 (MST), a DOD Concept Award W81XWH-06-1-0579 (MST), a Medical Research Foundation or Oregon Award (MST) and a CROET summer student fellowship (SB). The funders had no role in study design, data collection and analysis, decision to publish, or preparation of the manuscript.

Chapter 8
This work was supported by the Swiss National Science Foundation, and a grant from the Lotte and Adolf Hotz-Sprenger foundation, Zurich awarded to F.R.A.

Chapter 9
This work was supported by grants from: Associazione Italiana per la Ricerca sul Cancro (AIRC IG 10338) to R.R. and the IG grant AIRC to D.F.; Ministero della Salute Italiano (Ricerca Corrente) to R.R, D.F. and F.L., and partly supported by the AIRC Special Project 5 °— 1000 to F.L.

Chapter 10
This investigation was supported in part by Kaohsiung Medical University Research Foundation (KMUER004) to Steven Shoei-Lung Li.

Chapter 11
Drs Roger Deal, Muthugapatti Kandasamy, Jonathan Arnold, and Mary Anne Della-Fera offered useful editorial comments on the manuscript for which we are grateful. Kip Carter and the UGA's College of Veterinary Medicine's Department of Medical Illustration generously provided the artwork. This study was supported by a grant from the National Institutes of Health (GM36397-25) and the University of Georgia's Research Foundation to RBM and the UGA Graduate Student Association Recruitment Award, the Linton and June Bishop Graduate Fellowship, and the NIH Genetics Training Grant (GM 07103–37) awards to KJM.

Chapter 12
This work was supported by the US National Institutes of Health (NIH) Roadmap Epigenomics Program, NIH grant 5U01ES017154-02 (to M.H and M.A.M.) and Canadian Institutes of Health Research Grant 92093 (to M.H.).

INDEX